Human Interaction with Electromagnetic Fields

Computational Models in Dosimetry

Human Interaction with Electromagnetic Fields

Computational Models in Dosimetry

Dragan Poljak, PhD
Faculty of Electrical Engineering,
Mechanical Engineering and Naval Architecture
Department of Electronics and Computing
University of Split
Split, Croatia

Mario Cvetković, PhD
Faculty of Electrical Engineering,
Mechanical Engineering and Naval Architecture
Department of Electrical Engineering
University of Split
Split, Croatia

ACADEMIC PRESS
An imprint of Elsevier

Elsevier
3251 Riverport Lane
St. Louis, Missouri 63043

Human Interaction with Electromagnetic Fields

ISBN: 978-0-12-816443-3

Notices

Publisher: Mara Conner
Acquisition Editor: Tim Pitts
Editorial Project Manager: Ana Claudia Garcia
Production Project Manager: Kiruthika Govindaraju
Designer: Alan Studholme

Working together
to grow libraries in
developing countries

www.elsevier.com • www.bookaid.org

To our beloved ones, those who are with us, and those who passed away . . .

Contents

About the authors

Dragan Poljak was born on 10 October 1965. He received his BSc in 1990, MSc in 1994, and PhD in electrical engineering in 1996 from the University of Split, Croatia. He is a Full Professor at the Department of Electronics, Faculty of Electrical Engineering, Mechanical Engineering and Naval Architecture at the University of Split, and he is also an Adjunct Professor at Wessex Institute of Technology. His research interests include frequency and time domain computational methods in electromagnetics, particularly in the numerical modeling of wire antenna structures, and numerical modeling applied to environmental aspects of electromagnetic fields. To date prof. Poljak has published nearly 200 journal and conference papers in the area of computational electromagnetics, seven authored books and one edited book, by WIT Press, Southampton–Boston, and one book by Wiley, New Jersey. Professor Poljak is a member of IEEE, a member of the Editorial Board of the journal Engineering Analysis with Boundary Elements, and co-chairman of many WIT International Conferences. He is also an editor of the WIT Press Series Advances in Electrical Engineering and Electromagnetics. He was awarded several prizes for his carrier achievements, such as National Prize for Science (2004), Croatian section of IEEE Annual Award (2016). In 2011 prof. Poljak became a member of WIT Board of Directors. From 2011 to 2015 he was the Vice-dean for research at the Faculty of Electrical Engineering, Mechanical Engineering and Naval Architecture. In June 2013 prof. Poljak became a member of the board of the Croatian Science Foundation. He is currently involved in 3 COST projects, ITER physics EUROfusion collaboration and one national center for excellence in research for technical sciences. He is a co-chair of Working Group 2 of IEEE/International Committee on Electromagnetic Safety (ICES) Technical Committee 95 SC6 EMF Dosimetry Modeling.

Mario Cvetković received his BSc in electrical engineering from the University of Split, Croatia in 2005. In 2009 he obtained MPhil degree from the Wessex Institute of Technology, University of Wales, UK. In December 2013 he received his PhD from University of Split, Croatia. In 2010, he held a seminar to graduate and postgraduate students at the Technical University of Ilmenau, Germany, and in 2014 and 2018 he held seminars on the numerical methods in engineering at the Mälardalen University, Västerås, Sweden. He is a recipient of the "Best Student Paper Award", awarded at the 16th edition of the international conference SoftCOM 2008. At the Scientific Novices Seminar held in 2012, he was recognized for his previous scientific achievements. To date he has published more than 50 journal and conference papers and several book chapters (CRC Press and Springer). He is an assistant professor at the Faculty of Electrical Engineering, Mechanical Engineering and Naval Architecture (FESB), University of Split were he teaches a course on the fundamentals of electrical engineering. He is a member and also secretary of Working Group 2 of IEEE/International Committee on Electromagnetic Safety (ICES) Technical Committee 95 SC6 EMF Dosimetry Modeling.

Preface

There has been a continuous controversy whether the presence of electromagnetic fields pertaining to the non-ionizing part of the spectrum in the environment could be associated with health risk. The biological effects of electromagnetic fields are appreciably dependent on actual intensity and frequency, therefore a rough classification is often related to low frequency (LF) and high frequency (HF) exposures. Consequently, an assessment of distribution of the fields induced in biological bodies is crucial to study the related biological effects.

The present book aims to provide necessary information regarding computational models in electromagnetic and thermal dosimetry.

Chapter 1 provides general considerations of human exposure to electromagnetic fields, while some basics of computational electromagnetics are given in Chapter 2.

Chapter 3 deals with theoretical and experimental procedures on incident field dosimetry covering LF and HF electromagnetic interference (EMI) sources.

Simplified (canonical) models of the human body are presented in Chapter 4.

The central part of the book is given in Chapters 5 and 6, in which realistic models of the human body at LF and HF exposures based on Finite Element Method (FEM) and Boundary Element method (BEM), hybrid FEM/BEM and Method of Moments (MoM) are given. Furthermore, biomedical applications of electromagnetic fields are given in Chapter 7. Therefore, in addition to unwanted human exposure to LF and HF sources, of particular interest are also some biomedical applications of electromagnetic fields.

Finally, some useful mathematical details are available in Appendices A to F.

Rigorous theoretical background accompanied with mathematical details of various formulations and related solution methods being used throughout the book are presented.

The book includes many illustrative computational examples arising from realistic exposure scenarios and a reference list at the end of each chapter.

The intention of the present book is to provide not only useful description of our own expertise concerning bioelectromagnetics, but also to give updated information on some of the latest advances in this area.

We hope that this book will be useful material for undergraduate, graduate and postdoc students, as well as engineers in the industry, to learn about advanced computational models in electromagnetic and thermal dosimetry and to tackle some problems arising from realistic exposure scenarios.

We also think that the book could be used for various university courses involving bioelectromagnetics and computational dosimetry.

The book requires a general background in electrical engineering, involving some topics in basic electromagnetics. Fundamental concepts in bioelectromagnetics as well as numerical modeling principles are given in this book. Thus, the book is convenient for students, specialists, researchers and engineers.

To sum up, we are glad we have managed to compose this material stemming from more than two decades of rather intensive research in bioelectromagnetics. Of course, there are many rather challenging problems we would like to tackle in the future, such as stochastic bioelectromagnetics.

Dragan Poljak and Mario Cvetković
Split
March 2019

CHAPTER 1

On Exposure of Humans to Electromagnetic Fields – General Considerations

1.1 GENERAL CONSIDERATIONS

Technology has become integral part of our lives, permeating all aspects of our everyday existence. From the present day perspective it would be very hard to imagine our lives without technology. At the heart of this ubiquitous technology lies the invisible moving force created as a result of mankind's mastery of the laws of electricity and magnetism. However, this enormous power vested to the humankind does not come without a price.

The use of electricity inevitably results in the generation of electric and magnetic fields. In the 20th century, occurrence of electric, magnetic, and electromagnetic fields in the environment due to the tremendous growth of power grids, radio and television stations, radars, base stations, mobile phones, numerous domestic appliances, and appliances at workplaces has significantly increased. The present century only seems to be showing further explosion of this trend.

However, there is also a continuing public concern regarding the possible adverse health effects due to human exposure to these fields, particularly exposure to high-voltage power lines and radiation from base station towers and mobile phones.

The presence of electromagnetic fields in the environment and their hazard to humans represents a controversial scientific, technical, and, more often than not, public issue. The electromagnetic fields are a product of technology that must be used in everyday life despite the unknown risk.

In everyday life, people are assaulted by the amount of often conflicting information on hazards from power utilities and communication antennas. The widespread use of electrical energy implies that in all residences and workplaces there are levels of electromagnetic fields that would be considered normal.

There is a controversy if field emissions from such fields may cause cancer or other diseases. There are people who are convinced in adverse health effects due to exposure to electromagnetic fields and they are consequently usually cautious about their health and keep protesting. They are often grouped as conservationists.

On the contrary, some people from the industry do not believe in electromagnetic bio-effects.

The public concern is constantly swinging between extremely low frequency (ELF) and high frequency (HF) range, mostly depending on the widely publicized claims of these issues in the media.

In the last few decades public anxiety was directed to the safety of radar equipment at the workplace and microwave ovens in homes. However, it is now accepted that microwave ovens are harmless (at least when used properly), while a number of investigations regarding radar radiation resulted in certain safety precautions that could minimize some established thermal effects.

Today, base station antennas, mobile phones and other wireless communication equipment are the main sources of concern. In particular, the idea of health effects from mobile phones is at the focus of the research in this area.

The starting point of any analysis of possible health risk is the incident field dosimetry, including the evaluation of incident fields generated from various electromagnetic sources, and the internal field dosimetry, comprising various techniques for the determination of an internal electromagnetic field [1,2].

Evaluation of any health risk due to the exposure to electromagnetic fields relies on the results of a well-established body of research based on experimental data from biological systems, epidemiological and human studies, as well as on understanding the various mechanisms of interaction.

Unfortunately, information providing the public with a satisfactory understanding of exposure to electromagnetic fields and the related effect is still rarely available.

The investigation on the effects of electromagnetic fields exposure includes several aspects of electromagnetic fields such the biological, medical, biochemical, epidemiological, environmental, risk assessment, and health policy. The only appropriate approach is to take into account all of the above aspects [1].

Human Interaction with Electromagnetic Fields. https://doi.org/10.1016/B978-0-12-816443-3.00009-1

1.1.1 Environmental Electromagnetic Fields

The electromagnetic spectrum extends from static fields (0 Hz), to time-varying electromagnetic fields with frequencies in the extremely low frequencies (ELF) range, very low frequencies (VLF) to radio frequencies (RF), infrared radiation, visible light, ultraviolet (UV), X-rays, and gamma-ray frequencies exceeding 10^{24} Hz. Depending on the frequency, electromagnetic radiation can be classified as either non-ionizing or ionizing. The separation is generally accepted to be at wavelengths around 10 nm in the far-UV region.

Non-ionizing radiation is a general term for the part of electromagnetic spectrum with weak photon energy, insufficient to cause breaking of atomic bonds in the irradiated tissue or material. Natural sources of non-ionizing radiation include sunlight and lightning discharges. These sources are extremely weak, and with the tremendous proliferation of electricity applications, the density of artificially generated electromagnetic energy in the environment is many times higher than the naturally occurring one.

Compared to non-ionizing radiation, ionizing radiation, with frequencies above 10^{17} Hz, has enough energy to physically change the molecules or atoms it excites, turning them into charged particles and ions that are chemically more active than their electrically neutral counterparts. The resulting chemical changes occurring in biological systems may be cumulative and detrimental or even fatal.

Non-ionizing electromagnetic fields are sometimes split into two main categories: the low frequency (LF) fields, up to 30 kHz, and high frequency (HF) fields, from 30 kHz to 300 GHz. Above this frequency lie the infrared, visible light, ultraviolet, X-ray, and gamma-ray spectra, following an ascending order of frequencies that all belong to the ionizing radiation spectrum. One should note also the static electric and magnetic fields located at the lower end of electromagnetic spectrum.

1.1.1.1 Static Fields

Static electric and magnetic fields are constant fields, which do not change their intensity or direction over time, hence their frequency is 0 Hz.

Due to potential difference of some 300 kV between the ground and the Earth's ionosphere lying some 50 km above, a naturally occurring electric field in the atmosphere is permanently present, varying from around 100 V/m in fair weather to several thousand V/m under thunderclouds. The strength of this field depends on the solar activity, the season, air humidity and various weather conditions.

Other sources of static electric fields of a more modest strength are electrical charges produced via triboelectric effect, i.e., as a result of rubbing or friction.

For example, walking on a non-conducting mat can result in accumulated charges with potential differences of several kilovolts generating fields of up to 500 kV/m. High voltage DC transmission power lines can produce static electric fields of up to 20 kV/m and more. Also, inside DC operated electric trains, static electric fields of up to 300 V/m can be found.

A natural source of magnetic field is the Earth's geomagnetic field, encompassing the planet from pole to pole, while protecting the life on Earth from the ionizing radiation of cosmic rays. This field can be considered static on our time scale, while its magnitude varies with geographic latitude, from 30 μT at the equator to 70 μT near poles.

Much stronger magnetic fields are generated by some types of industrial and medical equipment, such as in Medical Resonance Imaging (MRI) devices, varying from 1.5 to 10 T. Compared to MRI, the household magnets have strengths on the order of several tens of mT.

1.1.1.2 Time-Varying Fields

Aside from a few early experiments on electricity and magnetism, the first artificial electric and magnetic fields originated in the generation, transmission and utilization of electrical energy, followed by the application in telecommunications. Nowadays, technological progress enabled the utilization of electromagnetic fields from the complete frequency spectrum.

At extremely low frequency (up to 3 kHz), the wavelengths are very long (6000 km at 50 Hz and 5000 km at 60 Hz) so that there is no radiated field, and electric and magnetic fields are in no fixed relationship to each other. Since the wavelength of 50/60 Hz fields is much larger than the relevant distances from the field source, the near-field terms, non-radiative in nature, are considerably larger than the radiative terms. ELF fields are generally used for power utilities (transmission, distribution, and applications) and for strategic global communications with submarines submerged in conducting seawater.

Radio frequency (RF) domain covers the range from 3 kHz (100 km wavelength) to 300 GHz (1 mm wavelength). As there are so many applications of HF fields, it would be impossible to list all the sources of these fields. The majority of sources are involved in radio communications applications, including radio and television broadcasting, mobile telephony, local wireless networks, radar, radiofrequency identification (RFID), and various other RF/microwave applications.

1.1.2 Human Exposure to Undesired Radiation

Concern regarding the possible effects due to exposure to electromagnetic fields and related research started after World War II, mainly due to observed heating phenomena during the use of radar. By the mid-1970s the concern was mostly directed toward possible health hazards due to the exposure to radio frequency (RF) fields. In the ensuing years, public concern was directed to extremely low frequency (ELF) fields generated by power lines.

From the beginning of the 1990s, public concern on the presence of electromagnetic fields in the environment has been aroused by a number of articles in the general press. The sources of concern include, but are not limited to, power lines, mobile phones and broadcasting antennas. A controversy was generated with a claim on the possible link between the fields and health risk, particularly with certain forms of cancer in humans. During the last decade, the public concern refocused on the RF exposure from base station towers and mobile phones, primarily due to high penetration rate of this technology.

As a result, continuing research has intensified on the biomedical and health aspects. While safety regulations on the exposure to radiation have been well established in the nuclear power industry, the proposals for limiting the exposure to radio frequency electromagnetic fields have been made in the last two decades.

1.1.3 Biomedical Applications of Electromagnetics Fields

While human exposure to artificial electromagnetic fields has raised many questions regarding potential adverse effects, particularly for the brain exposure to high frequency (HF) radiation, some biomedical applications of electromagnetic fields are of particular importance, as well. The electromagnetic fields are used in medicine for various diagnostic, therapeutic and surgery procedures.

For example, the magneto-therapy is used to relieve joint or muscular pain and diminish stress. It is based on the use of a low-frequency magnetic fields generated by coils applied to parts of the subjects body. The usual fields are of sinusoidal waveform having frequencies between 1 and 300 Hz, with maximum field strength inside the coil between 1 and 10 mT.

Furthermore, there are many established techniques using the electromagnetic fields in medical diagnostic and for therapy purposes, including transcranial magnetic stimulation (TMS), percutaneous electrical nerve stimulation (PENS) or transcutaneous nerve stimulation (TENS). The transcranial electrical stimulation (TES) and direct cortical stimulation (DCS) are examples of intra-operational methods, to name only a few.

For example, transcranial magnetic stimulation (TMS) is a noninvasive and painless technique for excitation or inhibition of brain regions, and in last few decades has become an important tool in preoperative neurosurgical diagnostic/evaluation of patients. TMS is also used in therapeutic purposes for treatment of depression, and is a subject of interest in neurophysiologic research. The technique consists of applying an intense and rapidly varying magnetic field to induce an electric field in the superficial layers of brain tissues, creating stimulation. The field is generated by a coil of specific geometry placed on the patient's scalp, through which a current of several kiloampers flows. Stimulation is created by an isolated pulse or a train of pulses, with frequency lower than 1 Hz or lower than 20 Hz. Each pulse lasts less than 1 ms.

Diathermy is another technique where heat is induced through the use of high frequency electromagnetic currents. It is used as a form of physical therapy and in surgical procedures. Usually, electromagnetic fields of 27.12 and 915 MHz or 2.45 GHz are used. Short wave applicators consist of two electrodes between which an electric field is generated. Radiation can be continuous, with power of around 300 W, or pulsed, at 1.500 W peak. The electric field reaches values of several hundred V/m. The patient is typically exposed during 10–20 minutes. A single applicator with a high frequency antenna is sufficient for introducing microwaves into tissues. The power is usually limited to around 100 W, while radiation parameters depend on the treated body part. As microwaves reach only superficial tissues, precautions must be taken to avoid the indirect effects such as localized burning due to the presence of metal objects, such as piercings and passive or active implanted medical devices.

Electrosurgery procedure involves applying a high frequency electric current to the biological tissues. According to Joule's law, the current flowing between two electrodes produces a temperature elevation proportional to the square of its intensity and to the duration of the application. Temperatures lower than 100°C are required for coagulation, while significantly higher temperatures of around 500°C are required for the incision and cauterization. Therefore, the current parameters are adjusted during the surgical operation. Electrosurgery is utilized in dermatology, gastroenterology, etc. High frequency fields, ranging from 200 kHz to more than 2 MHz, are used to avoid the stimulation of the nervous and muscular tissues and prevent electrolytic processes.

There are many additional examples of biomedical application of electromagnetic fields. Listing all of the techniques is outside the scope of this book. Nonetheless, it is important to emphasize the impact of the electromagnetic fields on the modern medicine and consequently on the quality of our lives. It would be impossible to imagine how the life quality would be degraded if not for so many advances in medicine, in many ways thanks to the use of electromagnetic fields. On the other hand, it is quite easy to imagine, as we further deepen our knowledge, how the consequences of new applications will beneficially transform the quality of our lives.

1.2 COUPLING MECHANISMS AND BIOLOGICAL EFFECTS

A biological effect is an established effect caused by, or in response to, exposure to a biological, chemical or physical agent, including electromagnetic energy. A biological effect occurs when exposure to electromagnetic field cause any noticeable or detectable physiological response in a biological body, such as alterations of the structure, metabolism, or functions of a whole organism, its organs, tissues, and cells [1,3]. These changes are not necessarily harmful to individuals, and may also have beneficial consequences for a persons health or well-being.

The human body has a sophisticated mechanisms to adjust to various influences it encounters in its surroundings. However, it does not possess adequate compensation mechanisms for all biological effects. If some biological effect is outside the range for the human body to compensate, it can result in adverse health effects. For example, if this system is stressed for extended periods of time and the induced changes are irreversible, these condition may be considered as a health hazard.

Therefore, biological effect in itself may or may not result in an adverse health effect, while an adverse health effect results in detectable health impairment of the exposed individual.

These adverse health effects are often the result of accumulated biological effects over time and depend on exposure dose. It is an established fact that electromagnetic fields above certain levels can induce biological effects. Experiments with healthy subjects suggest that short-term exposure at the levels present in the environment do not result in any apparent detrimental effects. On the other hand, the high level exposure that might be harmful is restricted by national and international guidelines.

So far, there is currently no well-established scientific evidence to conclude that low-level long-term exposures to electromagnetic fields at levels found in the environment are adverse to human health, and also there is no confirmed mechanism that could provide a firm basis to predict these adverse effects [4].

A fundamental and detailed knowledge of the biological effects is required to completely understand the potential health risk. The understanding of interaction mechanisms could be used to identify the appropriate dosimetry, to predict dose–response relationships, to design better experiments, and also to determine if detrimental effects are expected at particular exposure levels [5].

It is important to emphasize that the coupling between electromagnetic field and the biological body varies significantly with frequency.

1.2.1 Coupling to Static Fields

1.2.1.1 Coupling to Static Magnetic Fields

People are generally unaffected by static magnetic fields, unless they move around these fields. The magnetic field will thus exert a physical force on electrically charged particles moving through the field. Therefore, movement with respect to magnetic field can induce the electric fields in tissues and these can affect the nervous tissues. The magnitude of the induced electric fields will depend on the total change of the magnetic field.

There are three established physical mechanisms by which static magnetic fields can influence biological systems [6]: magnetic induction, magneto-mechanical interaction, and electron spin interactions.

Magnetic induction arises through the following types of interaction: electrodynamic interactions with moving electrolytes where the static field exerting Lorentz force on a moving ionic charge carriers and cells in the blood will result in the induced electric fields and currents; induced electric fields and currents may also be induced by movement in a static magnetic field, where motion along a field gradient or rotational motion, either in a uniform field or in a field gradient, will result in the change of magnetic flux thereby inducing an electric current.

The magneto-mechanical interactions between static magnetic field and biological bodies are realized by magneto-orientation or magneto-translation, where the former is related to paramagnetic molecules experiencing a torque in a static field orienting them in a way that minimizes their free energy within the field, while the latter is due to a net translational force on both diamagnetic and paramagnetic materials in the presence of field gradients.

The last mechanism is via complex electronic interactions that may affect the rate of specific chemical reactions. These electron spin interactions are related to certain metabolic reactions involving transitional state comprising a radical pair where an applied magnetic field affects the rate and the extent to which the radical pair converts to a state in which recombination is no longer possible.

1.2.1.2 Coupling to Static Electric Fields

The static electric fields do not penetrate the human body because of its high conductivity. The electric field can induce an electric charge on the body surface, which can be sometimes perceived via its interaction with body hair and also as spark discharges. However, apart from this superficial sensory stimulation of hair and skin as the basis for perception of the field, the limited number of animal and human laboratory studies, which have investigated the effects of exposure to static electric fields, have not provided evidence of adverse health effects.

1.2.2 Coupling to Time-Varying Fields

There are three established basic coupling mechanisms through which time-varying electromagnetic fields interact with the biological body [7]:

- Coupling to LF electric fields;
- Coupling to LF magnetic fields;
- Absorption of energy from electromagnetic radiation.

These coupling mechanisms depend on the field characteristics such as frequency, spatial uniformity, propagation and polarization direction, etc., but also on the human body characteristics such as size, morphology, and posture.

Fig. 1.1 illustrate the coupling mechanism of the human body exposed to a low frequency electric and magnetic field, respectively.

1.2.2.1 Coupling to LF Electric Fields

The human body significantly perturbs the spatial distribution of a low frequency electric field [8]. Moreover, the electric field induced inside the body will be considerably smaller compared to the external electric field. As the human body is a good conductor at low frequencies, electric field lines external to the body will be nearly perpendicular to the body surface, as shown by Fig. 1.1A.

The interaction of LF electric fields with humans results in electric current, formation of electrical dipoles, and the reorientation of the already presented electric dipoles in tissue [7]. The intensity of these effects depends on the electrical properties of the body that vary with the type of tissue and also on the frequency of the applied field. External electric fields induce a shift of sur-

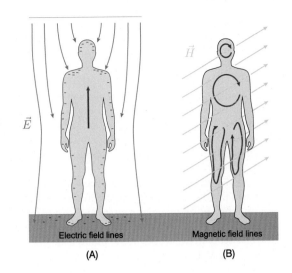

FIG. 1.1 Human body coupling with: (A) low frequency electric field, (B) low frequency magnetic field.

face charges on the body resulting in induced currents in the body, the distribution of which varies with the size and shape of the body.

1.2.2.2 Coupling to LF Magnetic Fields

Contrary to electric field, the human body does not significantly perturb the spatial distribution of a low frequency magnetic field [6]. As the permeability of body tissues is similar to that of air, the internal field is similar to the external field.

The interaction of an LF magnetic field with the human body results in induced electric fields and currents flowing in circular loops inside the body [7], as shown by Fig. 1.1B. The magnitudes of the induced field and the current density are proportional to the loop radius, the tissue conductivity, and the rate of change and magnitude of the magnetic flux density. For a specified magnitude and frequency of magnetic field, the strongest electric fields are induced where the loop of greatest dimensions are formed. The path and the magnitude of the current induced in any part of the body depend on the tissue conductivity.

1.2.2.3 Absorption of Energy From Electromagnetic Radiation

Compared to exposures to LF electric and magnetic fields resulting in negligible energy absorption and thus no measurable temperature rise in the human body, the exposure to high frequency electromagnetic radiation above around 100 kHz can result in a significant absorption of energy. The absorbed energy excites the polarized particles in the tissue sufficiently to transform them into thermal energy resulting in consequent temperature rise. The electromagnetic energy absorbed by the human body is expressed in terms of specific absorption rate (SAR).

Generally, exposure to a plane-wave electromagnetic field can result in a highly nonuniform deposition and distribution of the energy within the body, which has to be assessed by dosimetric calculation and measurement procedures. The energy absorption by the human body can be approximately divided into four frequency ranges [7]:

- 100 kHz–20 MHz, at which absorption in the trunk decreases rapidly with decreasing frequency, and significant absorption may occur in the neck and legs;
- 20 MHz–300 MHz, at which relatively high absorption can occur in the whole body, and to even higher values if partial body (e.g., head) resonances are considered;
- 300 MHz – several gigahertz, at which significant local, nonuniform absorption occurs;
- > 10 GHz, at which energy absorption occurs primarily at the body surface.

The amount of energy absorbed will depend on numerous factors including the dimensions, morphology, and posture of the exposed body. If the human body is not grounded, its resonant absorption frequency is around 70 MHz. For taller individuals, the resonant frequency is somewhat lower, while for shorter adults, children, babies, and seated persons, it is around 100 MHz. For grounded persons, resonant frequencies are lower by a factor of about 2.

The near-field exposures can lead to a high local SAR in the head, wrists, and ankles. The local SAR and whole-body SAR strongly depend on the separation distance between the radiation source and the body.

At frequencies above approximately 10 GHz, the depth of the field penetration into tissues is small, and SAR is not a convenient measure for determining the energy absorption in the body, therefore, a more appropriate dosimetric quantity is the incident power density of the field.

1.2.2.4 Indirect Coupling

There are two established indirect coupling mechanisms [7]. The first is the contact current resulting when the human body comes into contact with an object at a different electric potential. The magnitude and spatial distribution of contact currents will depend on the frequency, size of the object and person, area of contact, and grounding conditions (i.e., when either the body or the object is charged by an EMF). The moment when contact is made between a person and a conducting object or even in case when an individual and a conducting object exposed to a strong field come into close proximity, the transient spark discharge occurs, also known as microshock. The other mechanism is related to the coupling of electromagnetic fields to medical devices worn by or implanted in an individual.

1.2.3 Biological Effects

The biological body response due to electromagnetic field (EMF) exposure depends primarily on the frequency of the applied field. All relevant reported biological effects caused by EMF exposure can be classified as either non-thermal and thermal. The low frequency fields up to 5–10 MHz induce non-thermal effects such as stimulation of muscles, nerves and sensory organs, while high frequency fields in the frequency range from 100 kHz to 300 GHz result in the thermal effects. In the transition region between 100 kHz and 5–10 MHz [3,7], both non-thermal and thermal effects can be produced.

The interaction of electromagnetic fields with living systems can be approximately divided into four frequency ranges, as shown on Fig. 1.2: the static fields, frequencies between 1 Hz–100 kHz (low frequency fields), frequencies between 100 kHz–10 MHz (intermediate frequency fields), and frequencies above 10 MHz (high frequency fields). Although the same biological effect will be induced above a few GHz, compared to lower frequencies, the heating will be restricted to the surface of the body.

The biological response at any given frequency depends on the intensity of the field, where lower level exposure will result mainly in perceptual or sensory effects, while higher level exposure to fields will produce health effects considered to be more serious.

The summary of the sensory and health effects is given in Table 1.1.

1.2.3.1 Biological Effects of Static Magnetic Fields

The people at rest are generally unaffected by static magnetic fields, except at very high intensities when effects on the heart or brain functions may occur. However, movement around static magnetic field may elicit biological effects due to induced electric field affecting the nervous tissues. Some recent studies suggest that these effects may also occur whilst stationary. The magnitude of the induced electric fields depends mainly on the temporal and spatial gradients.

Particularly sensitive are the organs of balance in the ear since walking inside or even quickly moving the head in the static magnetic field might lead to feelings of dizziness or vertigo. The symptoms of nausea and

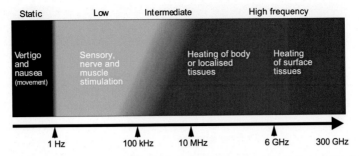

FIG. 1.2 The biological effects from static to 300 GHz. From European Commission, Non-binding guide to good practice for implementing Directive 2013/35/EU Electromagnetic Fields, Volume 1: Practical Guide, doi:10.2767/961464 KE-04-15-140-EN-C © European Union, 2018.

TABLE 1.1
The sensory and health effects related to particular frequency range.

Frequency	Sensory effects	Health effects
Static magnetic field	Vertigo, nausea, metallic taste	Altered blood flow in limbs, altered brain function; altered heart function
1 Hz–10 MHz	Phosphenes (perceived as light flashes); Minor change in brain function (1–400 Hz)	Tingling sensation of pain (nerve stimulation); muscle twitches, disturbed heart rhythm
100 kHz–6 GHz	Microwave hearing effect (200 MHz–6.5 GHz)	Excessive whole-body or localized heating or burns
6–300 GHz		Localized heat damage to eyes and skin

Note that the effects at intermediate frequency fields (100 kHz–10 MHz) are a combination of the effect of LF and HF fields.

other effects such as sensations of taste being produced have also been reported in the vicinity of operating MRI machines. However, all of these effects are temporary, ceasing when movement around strong static magnetic field stops or even slows down. There is no evidence of permanent impairment or severe adverse effect caused by the exposure. Usually, limiting the external magnetic flux density to 2 T or moving slowly inside strong static field might prevent these effects from happening.

1.2.3.2 Biological Effects of LF Fields

Low frequency magnetic field induces an electric field in the human body, which then results in stimulation of sense organs at lower field values, or stimulation of nerves and muscles (particularly in the limbs) in stronger fields. The effects on sensory organs are not harmful but could be annoying or distracting, whereas the effects in stronger fields could be unpleasant or even painful. Different tissues exhibit peak sensitivity at different frequencies, as given in Table 1.2.

The eyes are rather sensitive to the effects of induced electric fields, and the most robustly established effect of exposure is the induction of magnetic phosphenes in the retina, the perception of faint flickering light in the periphery of the visual field. The retina is part of the central nervous system (CNS) and is regarded as an appropriate model for induced electric field effects on CNS neuronal circuitry in general [8].

Exposure to low-frequency electric fields causes well-defined biological responses, ranging from perception to annoyance, through surface electric charge effects. Low frequency electric fields can induce electric fields within the body tissues, that can, in principle, produce similar effects to the fields induced by exposure to low frequency magnetic fields. However, as a consequence of the body shielding effect due to its high conductivity, the induced electric field is usually of a too small magnitude to elicit adverse effects for a typical external electric fields encountered in the environment.

Furthermore, low frequency electric fields produce another effect compared to magnetic fields. A person can experience a prickling or tingling sensation on the skin when standing inside very high intensity electric field such as underneath a high voltage power line on a dry day. This occurs as the low frequency electric field causes the charges to accumulate on the surface of the body, and this electrical charge causes the hairs in the skin to move and vibrate.

TABLE 1.2
Sites of interaction and peak sensitivities for different effects.

Effect	Site of interaction	Peak sensitivity
Metallic taste	Receptors in tongue	< 1 Hz
Vertigo, nausea	Inner ear (vestibular system)	< 0.1–2 Hz
Nerve and muscle stimulation	Blood flow-induced electric fields in tissues	
Phosphenes	Retinal cells in eye	~ 20 Hz
Tactile and pain sensation	Peripheral nerves	~ 50 Hz
Induced muscle contraction	Peripheral nerves and muscles	
Effects on heart	Heart	

1.2.3.3 Biological Effects of HF Radiation

Exposure of humans to electromagnetic fields with frequencies above 100 kHz results in body heating through the absorption of energy. Depending on the exposure scenario, this can result either in heating of the whole body, or localized heating of body part, such as limbs or head.

Healthy individuals usually regulate very efficiently the overall temperature of their bodies, thanks to thermoregulatory capacity of the human body. This protective mechanism is responsible for maintaining a temperature within normal range. However, if the total power absorbed by the body is large enough to cause this protective mechanism for heat control to break down, the uncontrolled rise in the body temperature (hyperthermia) occurs, leading to thermally harmful effects. The most adverse health effects due to HF exposure between 1 MHz and 10 GHz are associated with responses to induced heating, which results in a temperature rise in the tissue higher than 1°C. The prolonged temperature rise of a few degrees or more can be very dangerous.

The human body generates heat from metabolism. The basal metabolic rate (BMR) is defined as the heat production of a human in a thermo-neutral environment at mental and physical rest more than 12 h after the last meal [5]. The standard basal metabolic rate for a 70 kg man is approximately 1.2 W/kg, but it can be altered by changes in active body mass, diets, and endocrine levels.

When humans are exposed to heating from an external thermal source at a much higher rate, thermal damage can occur. However, exposure levels comparable to the BMR might produce thermal effects due to the induction of thermoregulation.

Thermal effects imposed on the body by a given specific absorption rate are strongly affected by the ambient temperature, relative humidity, and airflow. The human body regulates temperature increase due to the thermal effect through perspiration and heat exchange via blood circulation.

Certain areas with limited blood circulatory ability, such as the lens of the eye and the testes, run a particularly high risk of being damaged. The developing fetus is also known to be particularly sensitive to the effects of hyperthermia in the mother.

Other thermal effects may arise around electrically conducting objects, either implanted (nails, screws, artificial hip joints, etc.) or external (watches, bows of spectacles, etc.). For adverse health effects, such as eye cataracts and skin burns, to occur from exposure to high frequency fields, power densities above 1000 W/m^2, existing in close proximity to powerful transmitters such as radars, are needed.

Furthermore, during some exposures a nonuniform distribution of absorbed radio-frequency (RF) power is possible, resulting in nonuniform heating. The points in the body absorbing this power are usually referred to as the *hot spots*. Localized temperatures above 41.6°C may cause protein denaturation and coagulation, increased permeability of cell membranes, or the release of toxins in the immediate vicinity where the hot spots exists [5].

At 6 GHz and above, the electromagnetic fields do not significantly penetrate the body and the resultant heating is largely confined to the surface tissues and skin.

Pulsed radiofrequency fields can give rise to sensory perception in the form of *microwave hearing*. The individuals with normal hearing can perceive pulse-modulated fields with frequencies between 200 MHz and 6.5 GHz. Usually, this effect is described as a buzzing, clicking or popping sound, depending on the modulation characteristics of the field. Typical pulse durations resulting in these effects are on the order of a few tens of microseconds.

To summarize, biological effects of high frequency fields are proved to be hazardous only if the radiation intensity is rather high. In the case of most environmental HF exposures, particularly radio base station antennas and cellular phones, the intensity usually does not exceed the adopted exposure limits.

1.2.3.4 Electromagnetic Fields and Cancer

A considerable number of epidemiological reports, carried out particularly during the 1980s and 1990s, indicated that a long term exposure to 50/60 Hz magnetic fields might be associated with cancer [8]. In general, the initially observed associations between 50/60 Hz magnetic fields and various cancers were not confirmed. However, there is evidence for a relationship between extremely low frequency (ELF) fields and childhood leukemia. Also, the lack of consistent evidences in experimental studies weakens the belief that this association is due to ELF fields. Taking into account available epidemiological evidence concerning ELF exposure, the conclusion is that there is no solid evidence linking electric and magnetic field exposure to cancer.

The World Health Organization's International Agency for Research on Cancer (IARC) evaluation of ELF fields [9], published in June 2002, classified power frequency magnetic fields as possibly carcinogenic to humans (Group 2B). According to the same monograph [9], static electric and magnetic fields and extremely low-frequency electric fields are not classifiable as to their carcinogenicity to humans (Group 3). According to IARC monograph published in May 2011 [10], the radiofrequency fields are classified as possibly carcinogenic to humans (Group 2B). The category 2B is used for agents for which there is limited evidence of carcinogenicity in humans and less than sufficient evidence of carcinogenicity in experimental animals.

Despite many studies, the evidence for any effect remains highly controversial. However, it is clear that if electromagnetic fields do have an effect on cancer, then any increase in risk will be extremely small. The results to date contain many inconsistencies, but no large increases in risk have been found for any cancer in children or adults.

1.3 SAFETY GUIDELINES AND EXPOSURE LIMITS

The protection of humans exposed to electromagnetic fields (EMF) is the ultimate goal of a health-based EMF standards. Safety guidelines for the exposure to electromagnetic fields rely upon a well-established effects based on the experimental data from biological systems, on epidemiological and human studies, as well as on understanding of the various interaction mechanisms.

General steps in the development of exposure standards include the evaluation of a large body of scientific literature, followed by setting the threshold levels, the selection of appropriate safety factors for different categories of populations at risk, and finally derivation of the exposure limits.

The safety or the exposure limit is considered as that threshold below which exposure can be considered safe according to the available scientific knowledge. Nevertheless, the safety limit does not represent an exact boundary between safety and hazard, since the possible risk to human health increases with the increasing exposure levels. For efficient protection against the harmful exposure effects, the regulatory agencies, in addition to setting safety limits, need to incorporate a safety margin to allow for the uncertainty.

However, before going any further, it is necessary first to explain what exactly the standard represents as there is often confusion related to the terminology. The general term "standard" encompasses both guidelines and regulations. It can be considered as a set of specifications and rules promoting the safety of an individual or a group of people. Hence, a standard promoting the safe use of electromagnetic energy and the protection of humans exposed to electromagnetic fields (EMF) can be considered a health-based EMF standard.

1.3.1 EMF Standards

There are various standards related to limiting human exposure to EMF that can be classified as exposure, emission and measurement standards, respectively. These standards can specify either limits of a particular device emitting the EM field or limits of human exposure to the emitted field into human surroundings, such as living or working environment.

The standards related to personal protection generally referring to maximum levels to which complete or partial body exposure is permitted from any EMF emitting devices represent the exposure standards. Such standards are developed by the International Commission on Non-Ionizing Radiation Protection (ICNIRP), the Institute of Electrical and Electronic Engineers International Committee on Electromagnetic Safety (IEEE ICES) and various national authorities. The exposure standards typically include some safety factors and provide a basic guide for limiting personal exposure.

TABLE 1.3
Existence of EMF standards in several countries. SN = Subnational. InP = In preparation. Data from WHO GHO [13].

Frequency		Static		Low-frequency		Radiofrequency	
Country	Year	Public	Workers	Public	Workers	Public	Workers
Australia	2017	InP	InP	Yes	Yes	Yes	Yes
Brazil	2017	No	No	Yes	Yes	Yes	Yes
Canada	2017	No	No	SN	SN	Yes	SN
Croatia	2018		Yes		Yes		Yes
France	2017	No	Yes	Yes	Yes	Yes	Yes
Germany	2017	Yes	Yes	Yes	Yes	Yes	Yes
Japan	2017	No	No	Yes	No	Yes	No
Russian Federation	2017	Yes	Yes	Yes	Yes	Yes	Yes
South Africa	2017	No	No	No	No	Yes	Yes
Switzerland	2017	Yes	Yes	Yes	Yes	Yes	Yes
Turkey	2017	InP	InP	Yes	InP	Yes	InP
UK	2017	Yes	Yes	Yes	Yes	Yes	Yes
USA	2017	SN	Yes	SN	Yes	Yes	Yes

On the other hand, the emission standards developed by the IEEE, the International Electrotechnical Commission (IEC), the European Committee for Electrotechnical Standardization (CENELEC) and many national standardization authorities are not explicitly based on health considerations. Although these standards are intended to ensure compliance with exposure limits, their ultimate goal is to establish specifications taking into account engineering considerations such as to minimize interference of the particular electrical device with other equipment.

Finally, the measurement standards describe how to ensure compliance with the exposure and emission standards by providing instructions on how to carry out the EMF measurements due to some EMF emitting device. The EMF measurement standards are mainly developed by the IEC, the IEEE, the CENELEC, the International Telecommunications Union (ITU) and other standardization bodies.

Currently, as there are no internationally mandated standards for EMF, each country sets its own national standard for electromagnetic fields exposure. In most cases these national standards are based on guidelines set by ICNIRP [11].

The regulations for EMF exposure at the national level could be considered as either voluntary or compulsory. The former category includes various guidelines, recommendations and instructions, such as those developed by the ICNIRP [7,8] or the IEEE [3,4], but having no legal force. These instruments became legally binding only when incorporated into legislation. The latter category represents mandatory or legally binding instruments and includes, e.g., national laws, acts, regulations, and others, and thus requires a legislative framework.

The harmonization of EMF exposure limits is one of major goals of the World Health Organization's (WHO) International EMF Project [12], with currently 54 countries and 8 international organizations involved. The WHO-compiled database including worldwide standards for countries who have legislation on exposure to electromagnetic fields can be found on WHO Global Health Observatory website, a dedicated portal providing access to data and analyses for monitoring the global health situation [13].

An example of existence of standards and legislative status in several world countries can be seen in Tables 1.3 and 1.4, respectively.

Several different governmental exposure limits for low-frequency and radio-frequency fields for general public are listed in Tables 1.5 and 1.6, respectively.

As indicated in Tables 1.5 and 1.6, a number of countries have formulated guidelines establishing limits for occupational and residential EMF exposure. The reason for the large number of guidelines is the way in which they are defined, for example, by frequency, exposure duration, and periodicity of exposure and other factors.

TABLE 1.4
Legislative status of EMF standards in several countries. M = Mandatory. V = Voluntary. R = Recommended. Data from WHO GHO [13].

Frequency	Public			Workers		
	Static	LF	HF	Static	LF	HF
Australia	R	R	M	R	R	M
Brazil		M	M		M	M
Canada		M	M		M	M
Croatia	M	M	M	M	M	M
France			M	M	M	M
Germany	M	M	M	M	M	M
Japan			M			
Russian Federation	M	M	M	M	M	M
South Africa	V	V	R	V	V	R
Switzerland	M	M	M	R	M	M
Turkey	R	R	M	R	R	
UK	R	R	R	M	M	M
USA	V	V	M	V	V	M

TABLE 1.5
Exposure limits for low-frequency fields (public). Data from WHO GHO [13].

	Electric field (kV/m)	Magnetic flux density (µT)	Power frequency (Hz)
Argentina	3[a]	25	50
Australia	[5]/[10][b]	[100]/[1000]	50
Brazil	4.17	83	60
Croatia	2/5[c]	40/100[c]	50
France	5	[1]/100	50
Germany	5[d]	100	50
Japan	3	200	50/60
Russian Federation	0.5	5	50
Switzerland	5	1/100[e]	50
Turkey	[15]	[200]	50
UK	[5]/[9][f]	[100]/[360][f]	50

[a] At the edge of the right-of-way.

[b] 5 kV/m/100 µT: continuous exposure; 5–10 kV/m, 100–1000 µT: exposure limited to a few hours a day; >10 kV/m, > 1000 µT: exposure limited to a few minutes a day, provided the induced current density does not exceed 2 mA/m^2 and precautions are taken to prevent hazardous indirect coupling.

[c] *Public areas* (5 kV/m/100 µT) comprise all locations in urban and rural areas which don't have limited access to general public, and are neither within the increased sensitivity areas nor within the areas of occupational exposure; *Increased sensitivity areas* (2 kV/m/40 µT) comprise residential and commercial purpose buildings, schools, kindergartens, maternity hospitals, hospitals, nursing homes, tourist accommodation objects and children playgrounds.

[d] The prescribed limit applies to places where people are staying permanently with the AC-system operating at maximum capacity and taking into account emissions from other low frequency installations.

[e] 1 µT: Installation limit value as precautionary measure for high-voltage power lines.

[f] 9 kV/m/360 µT applies to power lines and exists in addition to other specific measures that are in place for controlling EMF-related risks from power lines.

TABLE 1.6
Exposure limits for radio-frequency fields (public). Data from WHO GHO [13].

	Electric field (V/m) 900 MHz	Specific absorption rate (SAR) (W/kg) Whole body	Head and trunk	Limbs
Argentina	41.25	0.08	2	4
Australia	41.1[a]	0.08	2	4
Brazil	41.25	0.08	2	4
Canada	32.1[b]	0.08	1.6[c]	4
France	41	0.08	2	4
Germany	41.25	0.08	2	4
Japan	47.55	0.08	2	4
South Africa	[41.0]	[0.08]	[2]	[4]
Switzerland	4/41.25[d]			
Turkey	3/10.23/41.0[e]		2	
UK	[41.25]	[0.08]	[2]	[4]
USA	47.6[f]	0.08	1.6[g]	4[h]

[a] 1302 V/m: instantaneous limit.

[b] The formula for calculating E-field reference levels at frequencies between 300 MHz and 6 GHz is $3.142f$ (exp 0.3417), where f is the frequency in MHz.

[c] Averaged over any 1 g of tissue, 6 minutes.

[d] 4 V/m: Installation limit value per location for new and existing antenna installations at places of sensitive use (buildings in which persons stay for longer periods).

[e] 3 V/m: limit per antenna for schools and hospitals (for hospitals a 3 V/m limit (not frequency dependent) is required inside the building for EMC reasons); 10.23 V/m: limit for a single installation; 41 V/m: limit for cumulative exposure from multiple antenna locations. The total exposure cannot exceed ICNIRP and per antenna it should not exceed 1/4 of ICNIRP field limits (1/16 power density).

[f] Averaging time is 30 minutes.

[g] Local SAR limit for all body parts except extremities (hands, feet, wrists, ankles, and pinnae) (averaged over 1 g, 30 minutes).

[h] Local SAR limit for the extremities applied to hands, feet, wrists, ankles, and pinnae (averaged over 10 g, 30 minutes).

Including every value from every standard at a complete frequency range between 0 Hz and 300 GHz would be very difficult, if at all possible to do with limited space at hand. Therefore, only the two most influential standards around the world today will be briefly reviewed in the following.

1.3.2 ICNIRP Guidelines

The International Commission on Non-Ionizing Radiation Protection (ICNIRP), a non-governmental organization formally recognized by WHO, developed the exposure guidelines for EMF fields, based on the established short-term adverse effects: stimulation of electrically excitable cells in muscle and nervous tissue and heating of the tissue.

In 1998, the ICNIRP published guidelines [7] for limiting exposures in the frequency range from 10 Hz to 300 GHz. The development of the guidelines was based on reviews of a large amount of studies.

In 2010, ICNIRP issued a new exposure guidelines [8] for the low frequency range between 1 Hz and 100 kHz replacing the low-frequency part of the 1998 Guidelines. These guidelines formed the basis of the 2013 European Union's Directive on occupational exposure [14]. A revision of the radio-frequency part of the original 1998 Guidelines corresponding to frequencies from 100 kHz to 300 GHz is currently underway.

Nowadays, most countries that have based their standards on ICNIRP have specifically used the 1998 ICNIRP Guidelines. Currently, the only two countries adopting the 2010 Guidelines are Japan and Germany.

Note that there are also ICNIRP guidelines related to occupational and general public exposure to static magnetic fields published in 2009 [6].

Generally, the guidelines are related to limiting the exposure to electromagnetic radiation in two categories: basic restrictions and reference levels. The basic restrictions represent the limits on the exposure to the time

TABLE 1.7
Basic restrictions for human exposure to low-frequency fields according to the 2010 ICNIRP Guidelines.

Exposure characteristic	Frequency range	Internal electric field [V/m]
Occupational exposure		
CNS tissue of the head	1 Hz–10 Hz	$0.5/f$
	10 Hz–25 Hz	0.05
	25 Hz–400 Hz	$2 \cdot 10^{-3} f$
	400 Hz–3 kHz	0.8
	3 kHz–10 MHz	$2.7 \cdot 10^{-4} f$
All tissues if head and body	1 Hz–3 kHz	0.8
	3 kHz–10 MHz	$2.7 \cdot 10^{-4} f$
General public exposure		
CNS tissue of the head	1 Hz–10 Hz	$0.1/f$
	10 Hz–25 Hz	0.01
	25 Hz–1000 Hz	$4 \cdot 10^{-4} f$
	1000 Hz–3 kHz	0.4
	3 kHz–10 MHz	$1.35 \cdot 10^{-4} f$
All tissues if head and body	1 Hz–3 kHz	0.4
	3 kHz–10 MHz	$1.35 \cdot 10^{-4} f$

Notes: f is the frequency in Hz. All values are rms. In the frequency range above 100 kHz, RF specific basic restrictions need to be considered additionally.

varying electromagnetic fields based directly on the established health effects.

The physical quantities used for basic restrictions reflect the particular concepts of *dose* relevant to the lowest-threshold for a frequency dependent health effect.

The basic restrictions given in the ICNIRP 1998 [7] correspond to the induced current density J [A/m^2] in the low frequency range (1 Hz–10 MHz), while in the high frequency range (100 kHz–10 GHz), the basic restriction is the specific absorption rate or SAR [W/kg]. In the intermediate frequency range (100 kHz–10 MHz) restrictions are on both current density and SAR, while in the very high frequency range (10–300 GHz) the basic restriction is represented by the induced power density S [W/m^2].

In the 2010 Guidelines [8], ICNIRP replaced the physical quantity used to specify the basic restriction at low frequency range (1 Hz–10 MHz). Instead of the induced current density, the internal electric field strength E_i is specified as the basic restrictions on exposure to low frequency EMF, as it is the electric field that affects nerve cells and other electrically sensitive cells.

The basic restrictions for human exposure to low-frequency and high-frequency fields according to [8] and [7], are presented in Tables 1.7 and 1.8, respectively.

As seen from Tables 1.7 and 1.8, the exposure limits are defined as two tiered, to differentiate between occupational or controlled exposure and general population or uncontrolled exposure. The ICNIRP guidelines for the general public differ compared to the occupationally exposed population as additional reduction factor of 5 is introduced to incorporate various uncertainties. For example, the reason for this approach was the possibility that some members of the general public might be exceptionally sensitive to RF radiation.

The fundamental quest in protection of humans exposed to electromagnetic radiation is to satisfy the given basic restrictions. When it is not practical to calculate or measure quantities associated with the basic restrictions which is usually the case in many realistic exposure scenarios comparison with reference levels can be used to estimate if the basic restrictions will be exceeded.

These reference levels given by ICNIRP or the analogous maximum permissible exposure levels (given in the IEEE standards) correspond to basic restrictions under the worst case scenario for one or more of the fol-

TABLE 1.8
Basic restrictions for human exposure to high-frequency EMF according to the 1998 ICNIRP Guidelines.

Exposure characteristic	Frequency range	Whole body average SAR [W/kg]	Localized SAR (head and trunk) [W/kg]	Localized SAR (limbs) [W/kg]	Power density [W/m^2]
Occupational exposure	10 MHz–10 GHz	0.4	10	20	–
	10 GHz–300 GHz	–	–	–	50
General public exposure	10 MHz–10 GHz	0.08	2	4	–
	10 GHz–300 GHz	–	–	–	10

Notes:
1. All SAR values are to be averaged over any 6-min period.
2. Localized SAR averaging mass is any 10 g of contiguous tissue; the maximum SAR so obtained should be the value used for the estimation of exposure.
3. For pulses of duration t_p the equivalent frequency to apply in the basic restrictions should be calculated as $f = 1/(2t_p)$. Additionally, for pulsed exposures in the frequency range 0.3 to 10 GHz and for localized exposure of the head, in order to limit or avoid auditory effects caused by thermoelastic expansion, an additional basic restriction is recommended. This is that the SA should not exceed 10 mJ/kg for workers and 2 mJ/kg for the general public, averaged over 10 g tissue.
4. Power densities are to be averaged over any 20 cm^2 of exposed area and any $68/f^{1.05}$-min period (where f is in GHz) to compensate for progressively shorter penetration depth as the frequency increases.
5. Spatial maximum power densities, averaged over 1 cm^2, should not exceed 20 times the values above.

TABLE 1.9
Reference levels for general public exposure to time varying electric and magnetic fields according to the 2010 ICNIRP Guidelines.

Frequency range	E-field strength [kV/m]	H-field strength [A/m]	B-field [T]
1 Hz–8 Hz	5	$3.2 \cdot 10^4/f^2$	$4 \cdot 10^{-2}/f^2$
8 Hz–25 Hz	5	$4 \cdot 10^3/f$	$5 \cdot 10^{-3}/f$
25 Hz–50 Hz	5	$1.6 \cdot 10^2$	$2 \cdot 10^{-4}$
50 Hz–400 Hz	$2.5 \cdot 10^2/f$	$1.6 \cdot 10^2$	2×10^{-4}
400 Hz–3 kHz	$2.5 \cdot 10^2/f$	$6.4 \cdot 10^4/f$	$8 \times 10^{-2}/f$
3 kHz–10 MHz	$8.3 \cdot 10^{-2}$	21	$2.7 \cdot 10^{-5}$

Notes: f in Hz. See separate sections below for advice on non-sinusoidal and multiple frequency exposure. In the frequency range above 100 kHz, RF specific reference levels need to be considered additionally.

lowing physical quantities: the electric field intensity E [V/m], the magnetic field intensity H [A/m], the magnetic flux density B [T], the power density S [W/m^2], the contact current I_c [mA], and, for pulsed fields, specific energy absorption SA [J/kg].

Basically, if the reference levels are exceeded it does not necessarily mean that the basic restrictions are exceeded, likewise. However, when the reference levels are exceeded, it is necessary to test compliance with the relevant basic restrictions and to determine if the additional protective measures are necessary.

An example of reference levels from [8] for general public exposure is given in Table 1.9. Exposure to a homogeneous field with respect to spatial extension of the human body is assumed.

The contact current reference levels, intended to avoid painful shocks, are presented in Table 1.10.

The reference levels for the general public are lowered by a factor of 2 compared to values for occupational exposure, as the threshold contact currents the result in biological responses in children are approximately half the value of the adult person.

The values of the reference levels, which are necessary to reach the basic restrictions, are oftentimes determined by mathematical modeling and by extrapolation from the results of laboratory investigations at specific frequencies. Table 1.11 summarizes the reference levels for occupational exposure and general public exposure, respectively, in the high-frequency range, given in [7].

TABLE 1.10
Reference levels for contact currents according to the 2010 ICNIRP Guidelines.

Exposure characteristic	Frequency range	Maximum contact current [mA]
Occupational exposure	Up to 2.5 kHz	1.0
	2.5 kHz–100 kHz	0.4 f
	100 kHz–10 MHz	40
General public exposure	Up to 2.5 kHz	0.5
	2.5 kHz–100 kHz	0.2 f
	100 kHz–10 MHz	20

Note: f is the frequency in kHz.

TABLE 1.11
Reference levels for high-frequency fields (given in unperturbed rms values) according to the 1998 ICNIRP Guidelines.

Frequency range	E-field strength [V/m]	H-field strength [A/m]	B-field [µT]	S_{eq} [W/m^2]
Occupational exposure				
10 MHz–400 MHz	61	0.16	0.2	10
400 MHz–2 GHz	3 $f^{1/2}$	0.008 $f^{1/2}$	0.01 $f^{1/2}$	$f/40$
2 GHz–300 GHz	137	0.36	0.45	50
General public exposure				
10 MHz–400 MHz	28	0.073	0.092	2
400 MHz–2 GHz	1.375 $f^{1/2}$	0.0037 $f^{1/2}$	0.0046 $f^{1/2}$	$f/200$
2 GHz–300 GHz	61	0.16	0.20	10

Notes:
1. f as indicated in the frequency range column.
2. Provided that basic restrictions are met and adverse indirect effects can be excluded, field strength values can be exceeded.
3. For frequencies between 100 kHz and 10 GHz, S_{eq}, E^2, H^2, and B^2 are to be averaged over any 6-min period.
4. For frequencies exceeding 10 MHz it is suggested that the peak equivalent plane wave power density, as averaged over the pulse width, does not exceed 1000 times the S_{eq} restrictions, or that the field strength does not exceed 32 times the field strength exposure levels given in the table.
5. For frequencies exceeding 10 GHz, S_{eq}, E^2, H^2, and B^2 are to be averaged over any $68/f^{1.05}$-min period (f in GHz).

The human body is most sensitive in the frequency range between 10 and 400 MHz for whole-body exposure, implying the most strict exposure limits within this frequency range. In addition to the knowledge of the field strength at a specific frequency, important parameters for the estimation of potential adverse effects are the exposure period, the radiation type (continuous or modulated), and the simultaneous exposure to multiple frequency fields.

For pulse-modulated sources, in addition to time averaging, it is necessary to examine the peak value for short pulse durations. For example, according to the guidelines for frequencies exceeding 10 MHz, it is suggested to limit peaks to 1000 times the corresponding limit value of the power density at the specific frequency.

For frequencies in the range from 300 MHz to several GHz, and for localized exposure of the head, the specific absorption from pulses must be limited, to avoid the thermoelastic expansion and the resulting auditory effects. In this frequency range, the threshold specific absorption of 4–16 mJ/kg for producing this effect corresponds, for 30-µs pulses, to 130–520 W/kg peak SAR value in the brain.

1.3.3 IEEE Standards
The International Committee on Electromagnetic Safety (ICES) is a committee under the sponsorship of the Institution of Electrical and Electronic Engineers (IEEE), responsible for developing various standards for different frequency ranges which are published as IEEE standards.

TABLE 1.12
Basic restrictions applying to various regions of the body (rms values) according to [4].

Exposed tissue	f_e [Hz]	General public, E_0 [V/m]	Controlled environment, E_0 [V/m]
Brain	20	$5.89 \cdot 10^{-3}$	$1.77 \cdot 10^{-2}$
Heart	167	0.943	0.943
Hands, wrists, feet and ankles	3350	2.10	2.10
Other tissue	3350	0.701	2.10

Interpretation of table is as follows: $E_i = E_0$ for $f \le f_e$; $E_i = E_0 \, (f/f_e)$ for $f \ge f_e$.
In addition to the listed restrictions, exposure of the head and torso to magnetic fields below 10 Hz shall be restricted to a peak value of 167 mT for the general public, and 500 mT in the controlled environment.

TABLE 1.13
Magnetic MPE levels for head and torso (rms values) according to [4].

Frequency range [Hz]	General public, B [mT]	Controlled environment, B [mT]
< 0.153	118	353
0.153–20	$18.1/f$	$54.3/f$
20–759	0.904	2.71
759–3000	$687/f$	$2060/f$

f is frequency in Hz. MPEs refer to spatial maximum.

The IEEE standards have also dual designations as American National Standards Institute (ANSI) standards, as the original sponsorship of the standards was switched from ANSI to IEEE. As a historical note, ANSI C95.1–1982 standard [15] was the first human exposure standard in the world based on the use of specific absorption rate (SAR). All the major standards in the world today share a similar characteristic frequency-dependent exposure limits to the ANSI 1982 standard.

IEEE standards covering the human exposure to electromagnetic fields are the IEEE C95.6–2002 [4] covering the low frequency range from 0 to 3 kHz, and the IEEE C95.1–2005 [3], covering the radio frequency range from 3 kHz to 300 GHz, respectively.

The IEEE standards [3,4] are based on the same body of evidence as ICNIRP but derive the numerical exposure values from the underlying science in a more detailed way, for example, compared to ICNIRP, they do not round the limits.

The recommendations given in IEEE standards are expressed in terms of basic restrictions (BRs) and maximum permissible exposure (MPE) values.

The IEEE C95.6–2002 [4] standard sets basic restrictions on the in-situ induced electric field, for particular areas of the body, as given in Table 1.12.

Additionally, the basic restriction on the peak magnetic field of 167 mT for the general public and 500 mT in the controlled environment is given for frequencies below 10 Hz. Above 10 Hz, the standard does not specify the basic restriction on the in-situ magnetic field.

The maximum permissible exposure values are given in terms of magnetic flux density B derived from the basic restriction, and the electric field E derived from consideration of indirect effects and not from the basic restrictions.

Table 1.13 lists magnetic flux density MPE limits as an example of the head and torso exposure given in [4]. The magnetic flux density MPEs for extremities, the electric field limits in terms of the undisturbed environmental field E, and the induced and contact current MPEs can be found in [4].

The IEEE C95.6–2002 [4] standard was developed with respect to established mechanisms of biological effects in humans from electric and magnetic field exposures. These fall within the category of short-term effects that are understood in terms of recognized interaction mechanisms. The defined exposure limits are not based on the potential long-term exposure effects.

On the other hand, the IEEE C95.1–2005 [3] standard presents two sets of rules to protect against established adverse health effects associated in the frequency range of 3 kHz to 5 MHz with electrostimulation, and in the frequency range of 100 kHz to 300 GHz with heating. In the transition region between 0.1 and 5 MHz, both sets of rules must be applied.

TABLE 1.14
Basic restrictions for frequencies between 100 kHz and 3 GHz according to [3].

		Action level[a] SAR[b] [W/kg]	Persons in controlled environments SAR[c] [W/kg]
Whole-body exposure	Whole-body average (WBA)	0.08	0.4
Localized exposure	Localized (peak spatial average)	2[c]	10[c]
Localized exposure	Extremities[d] and pinnae	4[c]	20[c]

[a] BR for the general public when an RF safety program is unavailable.
[b] SAR is averaged over the appropriate averaging times as shown in additional tables from [3].
[c] Averaged over any 10 g of tissue (defined as a tissue volume in the shape of a cube).*
[d] The extremities are the arms and legs distal from the elbows and knees, respectively.
* The volume of the cube is approximately 10 cm^3.

The rules and the exposure limits include safety factors accounting for uncertainties and providing a margin of safety for all.

The basic restrictions of IEEE C95.1–2005 [3] represent the limits on internal fields, specific absorption rate (SAR), and current density, while the MPEs, which are derived from the BRs, are limits on external fields and induced and contact current. Two tiers of exposure limits are established as the controlled environment (upper tier) and action level or general public (lower tier). All exposure limits for whole-body and time averaged.

The basic restrictions for whole-body and localized exposure for frequencies between 100 kHz and 3 GHz are given in Table 1.14.

To derive the above exposure limits, a safety factor of 10 has been applied to the established SAR threshold, yielding the SAR value of 0.4 W/kg averaged over the whole body. In case no RF safety program is present, the basic restrictions of the lower tier can also be used for the general public.

During the finalization of this book, the IEEE C95.1–2019 Approved Draft Standard for Safety Levels with Respect to Human Exposure to Electric, Magnetic and Electromagnetic Fields, 0 Hz to 300 GHz [16], superseded the C95.1–2005 Standard [3]. The recommendations in [16] are expressed in terms of exposure reference levels (ERLs) and dosimetric reference levels (DRLs). The DRLs are limits on in situ electric field, specific absorption rate (SAR), and incident power density, while the ERLs are derived from the DRLs, given as limits on external fields and induced and contact current. The new standard is intended for all cases of human exposure except exposure of patients during medical supervision.

1.3.4 Directive 2013/35/EU

The Council of the European Union published a Recommendation 1999/519/EC [17] in 1999 to limit the exposure of the general public to electromagnetic fields. The purpose of this recommendation is to protect the general population from the established adverse effects due to EMF exposure. This recommendation includes the limit values and terminology of the ICNIRP 1998 Guidelines [7]. This document recommends that EU Member States introduce the limits for public exposure and represents the current reference. Although it is not legally binding, several European countries have in fact complied.

In 2013 the EU passed a Directive 2013/35/EU [14] on occupational exposure to electromagnetic fields, closely based on the guidelines published by ICNIRP. Published in June, 2013, the Directive was planned to be transposed into national legislation of different member states of the European Union by July 2016.

The Directive covers the frequency range from static to 300 GHz and applies to all occupational sectors, and introduces the following terminology: exposure limits values (ELVs) equivalent to basic restrictions from the ICNIRP, and action levels (ALs) equivalent to reference level from the ICNIRP. Moreover, it also has two sets of each: the health ELVs corresponding to high action level (high AL), and the sensory ELVs corresponding to low action level (low AL).

The direct effects ALs have been derived from the corresponding ELVs using computer modeling and assuming worst case interactions. This means the compliance with the AL will guarantee compliance with the corresponding ELV. However, in many situations it will be possible to exceed the AL and still comply with the corresponding ELV. The relationship between the AL and ELV is illustrated in Fig. 1.3.

All action levels are specified for fields that are unperturbed by the presence of the worker's body.

Fig. 1.4 illustrates the action levels and exposure limit equivalent electric and magnetic fields at 50 Hz.

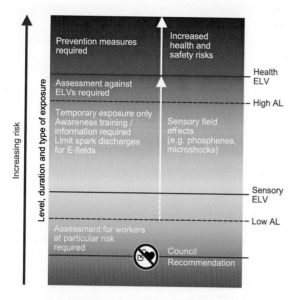

FIG. 1.3 Relationship between exposure limit values and action levels. From European Commission, Non-binding guide to good practice for implementing Directive 2013/35/EU Electromagnetic Fields, Volume 1: Practical Guide, doi:10.2767/961464 KE-04-15-140-EN-C © European Union, 2018.

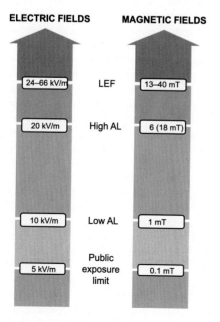

FIG. 1.4 The action levels and exposure limit equivalent electric and magnetic fields at 50 Hz.

1.4 A NOTE ON ELECTROMAGNETIC AND THERMAL DOSIMETRY

The fundamentals of interaction of electromagnetic fields with materials were known by late 19th century in the form of Maxwell equations. However, the implementation of these basic laws of electromagnetics to living systems is an extremely difficult task due to the tremendously high complexity and multiple organizational levels of biological systems. As experiments on humans in the high dose range or for long-term exposures is not possible, irradiation experiments can be performed only on phantoms, tissue probes, and laboratory animals. Theoretical models are then needed to interpret and confirm an experiment and to offer an extrapolation procedure, and thereby establish proper safety guidelines and corresponding exposure limits for humans. The mathematical complexity of the problem has led researchers to investigate simple canonical representations such as plane slabs, cylinders, homogeneous and layered spheres, and prolate spheroids. Spherical models are still being used to study the power deposition characteristics of the heads of humans and animals. On the other hand, sophisticated numerical modeling is required to successfully predict distribution of internal fields. A majority of contemporary realistic, anatomically based, computational models comprising cubical cells mostly use finite-difference time domain (FDTD) method schemes. However, though robust, FDTD suffers from stair-casing approximation error and recently some smooth methods started being used [18]. Thus, in some recent studies, the finite element method (FEM) is used in dosimetry as the method is considered to be more accurate than the FDTD, and a more sophisticated and versatile tool as well, particularly for the treatment of non-homogeneous irregular or curved shape domains. Recent research has also demonstrated that the use of the boundary element method (BEM), method of moments (MoM), fast multipole techniques, and wavelet techniques can be used to reduce the computational task.

1.4.1 Exposure to LF Fields

It is well-known that the induced currents and fields in human organs may give rise to thermal and non-thermal effects. When human is exposed to LF fields, the thermal effects seem to be negligible, and possible non-thermal effects are often related to the cellular level. The knowledge of the current density [7] and the electric field [8] induced inside the body is the key to understanding the interaction of the human being with LF fields. The current density inside the human body can be induced due to an external electric or magnetic field. Internal current densities induced by external electric fields have axial character, whereas induced currents due to external magnetic fields form loops.

1.4.2 Exposure to HF Fields

On the other hand, the key point in HF bioelectromagnetics is the knowledge about the electromagnetic energy absorbed by a biological body due to HF radiation. Furthermore, it is crucial where this energy is deposited, and which biochemical reactions are disturbed and electron transport flow interfered. The basic dosimetric quantity for HF fields is the specific absorption rate (SAR), which denotes the rate of energy absorbed by or dissipated in a unit mass of the body. In a sense of absorbed energy SAR in tissue is proportional to the square of the internal electric field and the main task of dosimetry involves the assessment of the electric field distribution inside the biological body. SAR distribution can be calculated using computational models or estimated from laboratory measurements. Generally, SAR appreciably depends on the incident field parameters and the characteristics of the exposed body. Furthermore, ground effects, and reflector effects and the attenuation of electromagnetic waves in the presence of finitely conducting objects have to be taken into account. When the electric field is oriented parallel to the long axis of the human body, the whole-body SAR reaches maximal values. On the basis of the determined electromagnetic energy absorbed by the human body due to the exposure to HF fields, a corresponding thermal response of the body can be determined by using a thermal model of the body.

1.4.3 Exposure to Transients

Exposure of humans to transient fields may result in higher peak values of internal fields; however, the duration is appreciably shorter. Standard quantity to analyze transient exposures is specific absorption (SA), being temporal integral of SAR. Generally, pulsed exposure have been so studied to a lesser extent than continuous wave (CW) exposures.

1.4.4 Stochastic Modeling

As a matter of fact, all computational models used in electromagnetic-thermal dosimetry suffer from serious drawbacks pertaining to uncertainties in input data propagated to the response of interest such as SAR, or temperature increase due to HF exposures. In the last few years some deterministic models were accompanied with stochastic analysis to partly overcome these difficulties. Furthermore, instead of using a robust Monte Carlo method (MCM), with rather slow convergence rate, the implementation of stochastic collocation method (SCM) has been reported in some publications, e.g., [19], due to its nonintrusive nature and the polynomial representation of the stochastic output, thus significantly reducing the number of simulations.

However, stochastic modeling is out of the scope of this book, but is likely to be systematically addressed by the authors in some forthcoming publications. The interested reader could find an overview and description of the stochastic methods applied to assessment of human exposure to radio-frequency fields in [20].

REFERENCES

1. Riadh W.Y. Habash, Electromagnetic Fields and Radiation: Human Bioeffects and Safety, CRC Press, 2001.
2. Dragan Poljak, Human Exposure to Electromagnetic Fields, WIT Press, Southampton–Boston, 2003.
3. IEEE International Committee on Electromagnetic Safety (SCC39), IEEE standard for safety levels with respect to human exposure to radio frequency electromagnetic fields, 3 kHz to 300 GHz, in: IEEE Std C95.1-2005, 2005, pp. 1–238.
4. IEEE Standards Coordinating Committee 28, IEEE standard for safety levels with respect to human exposure to electromagnetic fields, 0–3 kHz, in: IEEE Std C95.6-2002, 2002, pp. 1–50.
5. Dragan Poljak, Electromagnetic fields: environmental exposure, in: Encyclopedia of Environmental Health, Elsevier, 2011, pp. 259–268.
6. International Commission on Non-Ionizing Radiation Protection (ICNIRP), Guidelines on limits of exposure to static magnetic fields, Health Physics 96 (4) (2009) 504–514.
7. International Commission on Non-Ionizing Radiation Protection (ICNIRP), Guidelines for limiting exposure to time-varying electric, magnetic and electromagnetic fields (up to 300 GHz), Health Physics 74 (4) (1998) 494–522.
8. International Commission on Non-Ionizing Radiation Protection (ICNIRP), Guidelines for limiting exposure to time-varying electric and magnetic fields (1 Hz to 100 kHz), Health Physics 99 (6) (2010) 818–836.
9. International Agency for Research on Cancer, et al., Non-ionizing radiation, part 1: static and extremely low-frequency (ELF) electric and magnetic fields, IARC Monographs on the Evaluation of Carcinogenic Risks to Humans 80 (2002) 1–395.
10. International Agency for Research on Cancer, et al., Non-ionizing radiation, part 2: radiofrequency electromagnetic fields, IARC Monographs on the Evaluation of Carcinogenic Risks to Humans 102 (2013) 1–425.
11. K.R. Foster, Limiting technology: problems in setting exposure guidelines for radiofrequency energy, in: Third Generation Communication Systems, Springer, Berlin, Heidelberg, 2004, pp. 57–78.

12. World Health Organization, The International EMF Project, https://www.who.int/peh-emf/project/en/.

13. World Health Organization, Global Health Observatory (GHO) data, https://www.who.int/gho/phe/emf/en/.

14. Directive 2013/35/EU of the European Parliament and of the council of 26 June 2013 on the minimum health and safety requirements regarding the exposure of workers to the risks arising from physical agents (electromagnetic fields), Official Journal of the European Union L179 (2013) 1–21.

15. American National Standard Institute, Standard safety level of safety levels with respect to human exposure to radio frequency electromagnetic fields, 300 kHz to 100 GHz, in: ANSI Std C95.1-1982, 1982, pp. 1–20.

16. IEEE International Committee on Electromagnetic Safety, IEEE approved draft standard for safety levels with respect to human exposure to electric, magnetic and electromagnetic fields, 0 Hz to 300 GHz, in: IEEE PC95.1/D3.5, October 2018, 2019, pp. 1–312.

17. 1999/519/EC: council recommendation of 12 July 1999 on the limitation of exposure of the general public to electromagnetic fields (0 Hz to 300 GHz), Official Journal of the European Union L199 (1999) 59–70.

18. D. Poljak, M. Cvetković, O. Bottauscio, A. Hirata, I. Laakso, E. Neufeld, S. Reboux, C. Warren, A. Giannopoulos, F. Costen, On the use of conformal models and methods in dosimetry for nonuniform field exposure, IEEE Transactions on Electromagnetic Compatibility 60 (2) (2018) 328–337, https://doi.org/10.1109/TEMC.2017.2723459.

19. Dragan Poljak, Silvestar Šesnić, Mario Cvetković, Anna Šušnjara, Hrvoje Dodig, Sébastien Lalléchère, Khalil El Khamlichi Drissi, Stochastic collocation applications in computational electromagnetics, Mathematical Problems in Engineering 2018 (2018) 1917439, https://doi.org/10.1155/2018/1917439.

20. Joe Wiart, Radio-Frequency Human Exposure Assessment: From Deterministic to Stochastic Methods, John Wiley & Sons, 2016.

CHAPTER 2

Theoretical Background: an Outline of Computational Electromagnetics (CEM)

2.1 FUNDAMENTALS OF COMPUTATIONAL ELECTROMAGNETICS

Rapid progress in the development of digital computers in mid-1960s enabled significant advancement in computational models.

Electromagnetic modeling provides the simulation of an electrical system electromagnetic behavior for a rather wide variety of parameters, for different initial and boundary conditions, excitation types, and different configuration of the system itself. Numerical modeling can be performed within a appreciably shorter time than it would be necessary for building and testing an appropriate prototype via experimental procedures. A basic purpose of a computational model in electromagnetics is to predict an object response to the external excitation generated by a certain EMI source. First, this section outlines some fundamental concepts in electromagnetic theory, then a short introduction to basic numerical methods is presented.

2.1.1 An Outline of Classical Electromagnetic Field Theory

Contributions of Faraday, Maxwell, Heaviside, Hertz and others resulted in the revolutionary concept of a field in classical physics, which was shown to be physically real, not just an abstract mathematical entity. As a consequence, through the notion of a field the *actio in distans* concept was abandoned [1].

In addition to establishing a rigorous electromagnetic field theory, James Clerk Maxwell managed to achieve a grand unification of electricity, magnetism and light. Namely, almost a quarter century before the Hertz experimental verification, Maxwell theoretically anticipated the existence of electromagnetic waves. Light is just an electromagnetic wave, visible to human eye propagating through ether. After Maxwell, H.A. Lorentz extended Maxwell's theory with electrodynamics of a charged particle.

Maxwell's equations were modified a few times [2] in the last 150 years, since they were originally formulated by Maxwell and published for the first time in [3]. The changes pertain to the physical interpretation, mathematical expression and an approach to

the solution methods for different problems. In the mid-1860s, Maxwell originally derived 20 scalar equations, while a set of 4 vector equations was independently derived by Heaviside and Hertz by the end of the 19th century.

Important advancements of Maxwell's theory in the mid-1880s were carried out by Poynting, FitzGerald and Heaviside. Lorentz contribution is related to the development of a microscopic theory by means of Maxwell's equations and inclusion of the force acting on a charged particle arising from the existence of fields.

2.1.2 Maxwell's Equations – Differential and Integral Form

The laws of classical electromagnetism can be expressed in terms of four partial differential equations. There is also an equivalent integral form of these equations.

Originally, the governing equations of electromagnetics were derived in the time domain, but in many practical circumstances, particularly at low frequencies, systems are excited sinusoidally and a time-harmonic variation of electromagnetic fields can be assumed. In such cases it is convenient to represent the variables of interest in a complex phasor form. Thus, an arbitrary time dependent vector field $F(r, t)$ can be expressed as follows [4]:

$$\vec{F}(\vec{r}, t) = \text{Re}\left[\vec{F}_s(\vec{r}) e^{j\omega t}\right] \qquad (2.1)$$

where $F_s(r)$ is the phasor form of $F(r, t)$, and $F_s(r)$ is in general complex with an amplitude and a phase changing with position. Then Re[] implies taking the real part of the quantity in brackets, and ω is the angular frequency of the sinusoidal excitation.

In addition, using the phasor representation, computing the derivatives with respect to time results in

$$\frac{\partial}{\partial t}\left[\vec{F}_s(\vec{r}) e^{j\omega t}\right] = j\omega \vec{F}_s(\vec{r}) e^{j\omega t}. \qquad (2.2)$$

The first Maxwell equation is the differential form of Faraday law (the time-changing magnetic flux density $\vec{B}$

Human Interaction with Electromagnetic Fields. https://doi.org/10.1016/B978-0-12-816443-3.00010-8

causes the curl of electric field $\vec{E}$) given by

$$\nabla \times \vec{E} = -\frac{\partial \vec{B}}{\partial t}. \tag{2.3}$$

Hence, the time varying magnetic fields are vortex sources of electric fields.

The second Maxwell equation is the differential form of generalized Ampere's law stating that either a current density $\vec{J}$ or a time varying electric flux density $\vec{D}$ gives rise to a magnetic field $\vec{H}$:

$$\nabla \times \vec{H} = \vec{J} + \frac{\partial \vec{D}}{\partial t}. \tag{2.4}$$

It is worth noting that the term $\partial \vec{D}/\partial t$ (displacement current density) was originally added by Maxwell to the original expression for Ampere's law, thus making the law consistent with the electric charge conservation.

The third Maxwell equation states that charge densities ρ are the monopole sources of the electric field

$$\nabla \cdot \vec{D} = \rho, \tag{2.5}$$

while the fourth Maxwell equation states that magnetic poles always occur in pairs, and are due to electric currents; no free poles can exist. This is expressed by the divergence Maxwell equation:

$$\nabla \cdot \vec{B} = 0, \tag{2.6}$$

stating that the magnetic field is always solenoidal.

The integral form of the Faraday law states that any change of magnetic flux density $\vec{B}$ through any closed loop induces an electromotive force around the loop. Taking the surface integral of (2.3) and applying the Stokes theorem yields

$$\oint_c \vec{E} \cdot d\vec{s} = -\int_S \frac{\partial \vec{B}}{\partial t} \cdot d\vec{S}, \tag{2.7}$$

where the line integral is taken around the loop and with $d\vec{S} = \hat{n} dS$.

The voltage induced by a varying flux has a polarity such that the induced current in a closed path gives rise to a secondary magnetic flux which opposes the change in the time-varying source magnetic flux.

The integral form of Ampere's law is obtained by integrating (2.4) and applying the Stokes theorem:

$$\oint_c \vec{H} \cdot d\vec{s} = \int_S \vec{J} \cdot d\vec{S} + \int_S \frac{\partial \vec{D}}{\partial t} \cdot d\vec{S}. \tag{2.8}$$

The generalized Ampere's law states that either an electric current or a time-varying electric flux gives rise to a magnetic field. Taking the volume integral of (2.5) and applying the Gauss divergence theorem yields

$$\oint_S \vec{D} \cdot d\vec{S} = \int_V \rho \, dV, \tag{2.9}$$

where right-hand side represents the total charge within the volume V.

Eq. (2.5) is the Gauss flux law for the electric field stating that the flux of $\vec{D}$ vector corresponds to the total electric charge within the domain.

The Gauss flux law for the magnetic field can be derived by taking the volume integral of (2.6) and applying the Gauss divergence theorem, i.e.,

$$\oint_S \vec{B} \cdot d\vec{S} = 0, \tag{2.10}$$

stating that the flux of $\vec{B}$ vector over any closed surface S is identically zero.

In linear, homogeneous and isotropic medium, one deals with the following constitutive equations:

$$\vec{D} = \varepsilon \vec{E}, \tag{2.11}$$
$$\vec{J} = \sigma \vec{E}, \tag{2.12}$$
$$\vec{B} = \mu \vec{H}. \tag{2.13}$$

Furthermore, particle classical electromagnetics requires the Lorentz force equation

$$\vec{F} = q(\vec{v} \times \vec{B}), \tag{2.14}$$

where q denotes the charged particle, $\vec{v}$ is the particle velocity, ε is permittivity, σ is conductivity, and μ is permeability of a medium, respectively.

To solve Maxwell's equations for a given problem, the continuity conditions at the interface of two media with different electrical properties must be specified [5]:

$$\hat{n} \times (\vec{E}_1 - \vec{E}_2) = 0, \tag{2.15}$$
$$\hat{n} \times (\vec{H}_1 - \vec{H}_2) = \vec{J}_s, \tag{2.16}$$
$$\hat{n} \cdot (\vec{D}_1 - \vec{D}_2) = \rho_s, \tag{2.17}$$
$$\hat{n} \cdot (\vec{B}_1 - \vec{B}_2) = 0, \tag{2.18}$$

where $\vec{n}$ is a unit normal vector directed from medium 1 to 2, subscripts 1 and 2 denote fields in regions 1 and 2. Eqs. (2.15) and (2.18) state that the tangential components of $\vec{E}$ and the normal components of $\vec{B}$ are continuous across the boundary. Eq. (2.16) represents that the tangential component of $\vec{H}$ is discontinuous by the surface current density $\vec{J}_s$ on the boundary. Eq. (2.17) means that the discontinuity in the normal component of $\vec{D}$ is the same as the surface charge density ρ_s on the boundary.

In the case of perfect conductor, the electric field $\vec{E}$ and magnetic field $\vec{H}$ vanish within the perfectly conducting medium. These fields are replaced by the surface charge density ρ_s and surface current density $\vec{J}_s$. At higher frequencies there is a well-known effect which confines a current largely to surface regions.

As no time-varying field exists in a perfect conductor, the electric flux density is entirely normal to the conductor and supported by a surface charge density at the interface:

$$D_n = \rho_s. \tag{2.19}$$

The magnetic field is entirely tangential to the perfect conductor and is equilibrated by a surface current density:

$$H_s = J_s. \tag{2.20}$$

Conditions at the extremes of the boundary value problem are obtained by extending the interface conditions.

Assuming the time-harmonic variation of fields, curl Maxwell's equations become:

$$\nabla \times \vec{E} = -j\omega\vec{B}, \tag{2.21}$$
$$\nabla \times \vec{H} = \vec{J} + j\omega\vec{D}. \tag{2.22}$$

It should be observed that the assumption of the time-harmonic variation of fields eliminates the time dependence from Maxwell's equations, thereby reducing the space-time dependence to space dependence only.

2.1.3 The Continuity Equation

The equation of continuity couples the electromagnetic field sources (the charges and current densities) and can be readily derived from Maxwell equation (2.4). Taking the divergence of the Maxwell equation (2.4) yields

$$\nabla \cdot (\nabla \times \vec{H}) = \nabla \cdot \vec{J} + \nabla \cdot (\frac{\partial \vec{D}}{\partial t}). \tag{2.23}$$

As the left-hand side of (2.23) vanishes identically, it follows that

$$\nabla \cdot \vec{J} + \frac{\partial}{\partial t}(\nabla \cdot \vec{D}) = 0. \tag{2.24}$$

Invoking Gauss law (2.5), the equation of continuity is obtained, namely

$$\nabla \cdot \vec{J} = -\frac{\partial \rho}{\partial t}, \tag{2.25}$$

which for time-harmonic dependencies simplifies into

$$\nabla \cdot \vec{J} = -j\omega\rho. \tag{2.26}$$

The rate of charge moving out of a region is equal to the time rate of charge density decrease. The integral form of the continuity equation is obtained by performing the volume integration

$$\int_V \nabla \cdot \vec{J} \, dV = -\frac{\partial}{\partial t} \int_V \rho \, dV \tag{2.27}$$

and applying the Gauss divergence theorem

$$\int_V \nabla \cdot \vec{J} \, dV = \oint_S \vec{J} \cdot d\vec{S}. \tag{2.28}$$

The integral form of the continuity equation is then given by

$$\oint_S \vec{J} \cdot d\vec{S} = -\frac{\partial Q}{\partial t}, \tag{2.29}$$

where the unit normal in $d\vec{S}$ is the outward-directed normal, and Q is the total charge within the volume

$$Q = \int_V \rho \, dV. \tag{2.30}$$

Eq. (2.29) represents the Kirchhoff conservation law widely used in the circuit theory.

2.1.4 Conservation of Electromagnetic Energy – Poynting Theorem

The general conservation law of energy in the macroscopic electromagnetic field can be readily derived from curl Maxwell equations. Starting from divergence of Poynting vector

$$\nabla \cdot (\vec{E} \times \vec{H}) = \vec{H} \cdot \nabla \times \vec{E} - \vec{E} \cdot \nabla \times \vec{H} \tag{2.31}$$

and combining (2.31) with Maxwell equations yields [1,6]

$$\nabla \cdot (\vec{E} \times \vec{H}) = -\frac{\partial}{\partial t}\left(\frac{\vec{E} \cdot \vec{D} + \vec{H} \cdot \vec{B}}{2}\right) - \vec{E} \cdot \vec{J}. \tag{2.32}$$

Taking the volume integral of (2.32) gives

$$\int_V \nabla \cdot (\vec{E} \times \vec{H}) \, dV = -\frac{\partial}{\partial t} \int_V \left(\frac{\vec{E} \cdot \vec{D} + \vec{H} \cdot \vec{B}}{2} \right) dV$$

$$- \int_V \vec{E} \cdot \vec{J} \, dV. \tag{2.33}$$

For a battery with a non-electrostatic field $\vec{E}'$ pumping energy both into heat losses and into a magnetic field, the corresponding current density can be written as

$$\vec{J} = \sigma \left(\vec{E} + \vec{E}' \right). \tag{2.34}$$

Furthermore, applying the Gauss integral theorem to the left-hand side term, the volume integral transforms to the surface integral over the boundary, where $d\vec{S}$ is the outward drawn normal vector surface element, i.e., one obtains

$$\int_V \vec{E}' \cdot \vec{J} \, dV = \frac{\partial}{\partial t} \int_V \frac{1}{2} (\vec{E} \cdot \vec{D} + \vec{H} \cdot \vec{B}) \, dV$$

$$+ \int_V \frac{|\vec{J}|}{\sigma} \cdot \vec{J} \, dV' + \oint_S \left(\vec{E} \times \vec{H} \right) \cdot d\vec{S}. \tag{2.35}$$

The sources within the volume of interest are balanced with the rate of increase of electromagnetic energy in the domain, the rate of flow of energy in through the domain surface and the Joule heat production in the domain.

The time-harmonic complex Poynting vector is given by

$$\vec{S} = \frac{1}{2} \left(\vec{E} \times \vec{H}^* \right). \tag{2.36}$$

Taking the divergence of Poynting vector yields

$$\nabla \cdot \vec{S} = \frac{1}{2} \left(\vec{H}^* \cdot \nabla \times \vec{E} - \vec{E} \cdot \nabla \times \vec{H}^* \right). \tag{2.37}$$

The divergence of complex power density can be expressed in terms of a rate of stored energy, power losses and sources

$$\nabla \cdot \vec{S} = -j\frac{\omega}{2} \left(\mu |\vec{H}|^2 - \varepsilon |\vec{E}|^2 \right) - \frac{|\vec{J}|}{\sigma} + \frac{1}{2} \sigma |\vec{E}'|^2. \tag{2.38}$$

Now the integration over a volume of interest yields

$$\int_V \nabla \cdot \vec{S} \, dV = -j\frac{\omega}{2} \int_V \left(\mu |\vec{H}|^2 - \varepsilon |\vec{E}|^2 \right) dV$$

$$= -\frac{1}{2} \int_V \left| \frac{\vec{J}}{\sigma} \right|^2 dV + \frac{1}{2} \int_V \sigma |\vec{E}'|^2 dV. \tag{2.39}$$

And applying the Gauss theorem, one obtains

$$\frac{1}{2} \int_V \vec{E} \times \vec{H}^* \cdot d\vec{S} = -j\frac{\omega}{2} \int_V \left(\mu |\vec{H}|^2 - \varepsilon |\vec{E}|^2 \right) dV$$

$$= -\frac{1}{2} \int_V \left| \frac{\vec{J}}{\sigma} \right|^2 dV + \frac{1}{2} \int_V \sigma |\vec{E}'|^2 dV. \tag{2.40}$$

Finally, the real and imaginary parts, respectively, of the Poynting flow can be written as

$$\frac{1}{2} \text{Re} \int_V \vec{E} \times \vec{H}^* \cdot d\vec{S} = -\frac{1}{2} \int_V \left| \frac{\vec{J}}{\sigma} \right|^2 dV + \frac{1}{2} \int_V \sigma |\vec{E}'|^2 dV, \tag{2.41}$$

$$\frac{1}{2} \text{Im} \int_V \vec{E} \times \vec{H}^* \cdot d\vec{S} = -\frac{\omega}{2} \int_V \left(\mu |\vec{H}|^2 - \varepsilon |\vec{E}|^2 \right) dV. \tag{2.42}$$

The real part of the integral over Poynting vector represents the total average power while the imaginary part of the integral over Poynting vector is proportional to the difference between average stored magnetic energy in the volume and average stored energy in the electric field.

The $\frac{1}{2}$ factor appears because $\vec{E}$ and $\vec{H}$ fields represent peak values, and it should be omitted for root-mean-square (rms) values. The total average power can, for example, represent the radiated power by an antenna. In addition, the first volume integral on the right-hand side of (2.41) represents a power loss in the conduction currents and is just twice the average power loss.

2.1.5 Electromagnetic Wave Equations

The wave equations are readily derived from the Maxwell curl equations, by differentiation and substitution. Taking curl on both sides of (2.4) yields

$$\nabla \times \nabla \times \vec{H} = \nabla \times \vec{J} + \frac{\partial}{\partial t} \left(\nabla \times \vec{D} \right). \tag{2.43}$$

Using constitutive equations (2.11) and (2.12) and assuming uniform scalar material properties yields

$$\nabla \times \nabla \times \vec{H} = \sigma \nabla \times \vec{E} + \varepsilon \frac{\partial}{\partial t}\left(\nabla \times \vec{E}\right). \qquad (2.44)$$

According to the Maxwell equation (2.3), curl of $\vec{E}$ is replaced by the rate of change of magnetic flux density:

$$\nabla \times \nabla \times \vec{H} = -\mu\sigma \frac{\partial \vec{H}}{\partial t} - \mu\varepsilon \frac{\partial^2 \vec{H}}{\partial t^2}. \qquad (2.45)$$

Performing some mathematical manipulations, similar equations can be derived for the electric field. Using the standard vector identity, valid for any vector $\vec{E}$, namely

$$\nabla \times \nabla \times \vec{H} = \nabla \cdot \left(\nabla \cdot \vec{H}\right) - \nabla^2 \vec{H}, \qquad (2.46)$$

and taking into account solenoidal nature of the magnetic field (2.6), the wave equation is obtained as

$$\nabla^2 \vec{H} - \mu\sigma \frac{\partial \vec{H}}{\partial t} - \mu\varepsilon \frac{\partial^2 \vec{H}}{\partial t^2} = 0. \qquad (2.47)$$

If a linear, isotropic, homogeneous, source-free medium is considered, then the set of equations (2.47) simplifies into

$$\nabla^2 \vec{H} - \frac{1}{v^2} \frac{\partial^2 \vec{H}}{\partial t^2} = 0, \qquad (2.48)$$

where v denotes the wave propagation velocity in lossless homogeneous medium,

$$v = \frac{1}{\sqrt{\mu\varepsilon}}. \qquad (2.49)$$

The velocity of wave propagation in free space is the velocity of light,

$$c = \frac{1}{\sqrt{\mu_0 \varepsilon_0}}, \qquad (2.50)$$

where $c = 3 \times 10^8$ m/s, approximately.

The complex phasor representation of the wave equation (2.47) results in the following equation of the Helmholtz type:

$$\nabla^2 \vec{H} - \gamma^2 \vec{H} = 0, \qquad (2.51)$$

where γ is the complex propagation constant given by

$$\gamma = \sqrt{j\omega\mu\sigma - \omega^2\mu\varepsilon}. \qquad (2.52)$$

For a linear, isotropic, homogeneous, source-free medium, the Helmholtz equation (2.51) simplifies into

$$\nabla^2 \vec{H} + k^2 \vec{H} = 0, \qquad (2.53)$$

where k is a wave number of a lossless medium,

$$k = \omega\sqrt{\mu\varepsilon}. \qquad (2.54)$$

The complex form of potential wave equations could be derived similarly.

2.1.6 Electromagnetic Potentials

Instead of using fields, the analysis of classical electromagnetic phenomena can be simplified by using auxiliary potential functions, such as the electric scalar potential φ, or the magnetic vector potential $\vec{A}$, which are readily derived from the Maxwell equations.

Thus, the divergence Maxwell equation (2.6) is satisfied if the flux density $\vec{B}$ is expressed in terms of an auxiliary vector function $\vec{A}$, i.e.,

$$\vec{B} = \nabla \times \vec{A}. \qquad (2.55)$$

Maxwell curl equation (2.5) then becomes

$$\nabla \times \vec{E} = -\frac{\partial}{\partial t}(\nabla \times \vec{A}), \qquad (2.56)$$

and, by rearranging (2.56), it follows that

$$\nabla \times (\vec{E} + \frac{\partial \vec{A}}{\partial t}) = 0. \qquad (2.57)$$

Now, the bracket quantity in (2.57) can be written as the gradient of the scalar potential function φ:

$$\vec{E} + \frac{\partial \vec{A}}{\partial t} = -\nabla\varphi, \qquad (2.58)$$

or

$$\vec{E} = -\frac{\partial \vec{A}}{\partial t} - \nabla\varphi. \qquad (2.59)$$

By using further mathematical manipulations, potential wave functions can be obtained. Thus, taking the divergence of (2.59), one obtains

$$\nabla \cdot \vec{E} = -\frac{\partial \left(\nabla \cdot \vec{A}\right)}{\partial t} - \nabla \cdot (\nabla\varphi). \qquad (2.60)$$

Utilizing Gauss law yields

$$\nabla^2\varphi + \frac{\partial \left(\nabla \cdot \vec{A}\right)}{\partial t} = -\frac{\rho}{\varepsilon}. \qquad (2.61)$$

Furthermore, from the second curl Maxwell equation (2.4) with (2.55), it follows that

$$\nabla \times \nabla \times \vec{A} = \mu \vec{J} + \mu \varepsilon \left[-\nabla \left(\frac{\partial \varphi}{\partial t} \right) - \frac{\partial \vec{A}}{\partial t} \right] = 0. \quad (2.62)$$

Now, using the vector identity

$$\nabla \times \nabla \times \vec{A} = \nabla \left(\nabla \cdot \vec{A} \right) - \nabla^2 \vec{A}, \quad (2.63)$$

expression (2.62) can be written as

$$\nabla \left(\nabla \cdot \vec{A} \right) - \nabla^2 \vec{A} = \mu \vec{J} - \mu \varepsilon \nabla \left(\frac{\partial \varphi}{\partial t} \right) - \mu \varepsilon \frac{\partial^2 \vec{A}}{\partial t^2}. \quad (2.64)$$

Now choosing the divergence of $\vec{A}$ as

$$\nabla \cdot \vec{A} + \mu \varepsilon \frac{\partial \varphi}{\partial t} = 0, \quad (2.65)$$

expressions (2.61) and (2.64) become

$$\nabla^2 \varphi - \mu \varepsilon \frac{\partial^2 \varphi}{\partial t^2} = -\frac{\rho}{\varepsilon}, \quad (2.66)$$

$$\nabla^2 \vec{A} - \mu \varepsilon \frac{\partial^2 \vec{A}}{\partial t^2} = -\mu \vec{J}. \quad (2.67)$$

The set of Eqs. (2.66)–(2.67) can be regarded as a set of inhomogeneous potential wave equations for lossless media.

Thus, knowing the potential functions $\vec{A}$ and φ, the magnetic and electric fields can be determined from Eqs. (2.55) and (2.59).

The potential wave equations can be solved completely in a relatively small number of special cases. Integral solutions to potential wave equations (2.66) and (2.67) are given in the form of so-called retarded potentials [5,6]:

$$\varphi(\vec{r}, t) = \frac{1}{4\pi\varepsilon} \int_{V'} \frac{\rho(\vec{r}', t - R/c)}{R} dV', \quad (2.68)$$

$$\vec{A}(r, t) = \frac{\mu}{4\pi} \int_{V'} \frac{\vec{J}(\vec{r}', t - R/c)}{R} dV', \quad (2.69)$$

where R is a distance from the source point to the observation point.

The solution of Eq. (2.69) is carried out by separating this vector equation into its Cartesian components. The result is a set of equations identical in form to the scalar potential equation. Recombining these components results in the solution (2.69).

If all electromagnetic quantities of interest are varying harmonically in time, the particular integrals for the retarded potentials are given by [5,6]:

$$\varphi(\vec{r}) = \frac{1}{4\pi\varepsilon} \int_{V'} \frac{\rho(\vec{r}') e^{-jkR}}{R} dV', \quad (2.70)$$

$$\vec{A}(r) = \frac{\mu}{4\pi} \int_{V'} \frac{\vec{J}(\vec{r}') e^{-jkR}}{R} dV', \quad (2.71)$$

while the complex notation $e^{j\omega t}$ is understood and omitted.

2.1.7 Plane Wave Propagation

Plane waves are a satisfactory approximation for many realistic scenarios, e.g., radio waves at large distances from the transmitter, or from scattering obstacles having negligible curvature, and are well represented. Moreover, complicated wave patterns can be considered as a superposition of plane waves. Finally, the basic ideas of propagation, reflection and refraction in the plane wave approach are useful in understanding of more complex wave problems.

Considering the case of source-free homogeneous medium and the x-component of the vector field intensity and assuming there is no variation of the fields in both x and y directions, the corresponding Helmholtz equation for the electric field is given by

$$\frac{\partial^2 E_y}{\partial x^2} = -k^2 E_y. \quad (2.72)$$

The related solution can be written as

$$E_y = A e^{-jkx} + B e^{+jkx}, \quad (2.73)$$

where k is the wavenumber of the corresponding lossless medium defined by relation (2.54), A and B are the magnitudes of forward and backward wave, respectively.

There is no variation of the field quantities in the plane perpendicular to the propagation direction, therefore the wave is simply called a plane wave. As a medium is unbounded, only the forward wave exists:

$$E_y = E_0 e^{-jkx}, \quad (2.74)$$

where E_0 denotes the magnitude of the electric field.

The corresponding plane wave is shown in Fig. 2.1.

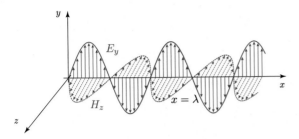

FIG. 2.1 The electric and magnetic field of an x-directed plane wave.

The corresponding magnetic field can be obtained from the time-harmonic curl Maxwell equation

$$\nabla \times \vec{E} = -j\omega\mu\vec{H}. \qquad (2.75)$$

Furthermore, it follows that

$$\vec{H} = -\frac{1}{j\omega\mu}\begin{vmatrix} \hat{e}_x & \hat{e}_y & \hat{e}_z \\ \frac{\partial}{\partial x} & 0 & 0 \\ 0 & E_y & 0 \end{vmatrix} = -\hat{e}_z\frac{1}{j\omega\mu}\frac{\partial E_y}{\partial x}, \qquad (2.76)$$

and the corresponding magnetic field is

$$H_z = \frac{E_0}{Z_0}e^{-jkx} = H_0\,e^{-jkx}, \qquad (2.77)$$

where

$$Z_0 = \frac{E_0}{H_0} = \sqrt{\frac{\mu}{\varepsilon}} \qquad (2.78)$$

is the wave impedance of the medium. For free space it is approximately $Z_0 = 120\pi$.

Now the space-time notation can be used:

$$E_y(x,t) = \mathrm{Re}\left[E_0\,e^{j(\omega t - kx)}\right], \qquad (2.79)$$

and the electric and magnetic fields can be written as follows:

$$E_y(x,t) = E_0\cos(\omega t - kx), \qquad (2.80)$$

$$H_z(x,t) = \frac{E_0}{Z_0}\cos(\omega t - kx). \qquad (2.81)$$

Note that in circuit theory, contrary to electromagnetism, a sinusoidal dependence, rather than cosinusoidal, is chosen.

2.1.8 Radiation and Hertz Dipole

Electromagnetic radiation is a phenomenon caused due to the acceleration of charged particles. Fields generated by accelerated charges can be analyzed by using microscopic approach, dealing with individual particles, or by using the macroscopic approach within which average fields over the charge distributions are considered [7,8].

The choice of the particular approach depends on whether the observation times and distances are smaller or higher than the characteristic times and distances associated with the sources [7,8].

A macroscopic viewpoint of radiation is provided by Maxwell equations from which the radiated fields due to their sources in terms of charge and current densities are presented. In particular, radiation from thin wires can be rigorously analyzed by a corresponding type of either space-frequency or space-time integral equation, respectively.

Thus, radiation of electromagnetic energy is an undesired leakage phenomenon or a desired process for exciting waves in space. In the case of desired radiation, the goal is to excite waves from the given source in the required direction, as efficiently as possible. The matching unit between the source and waves in space is known as the radiator, antenna or areal. The results developed for radiating or transmitting antennas can be applied to the same antenna when used for receiving applications if antenna does not contain active components (the principle of reciprocity). The relation of the radiating case to the receiving situation can be made rigorously by means of the so-called principle of reciprocity [9].

The simplest radiating system is that of an ideal short linear element (Hertz dipole) with current considered uniform over its entire length. More complex antenna structures can be considered to be composed from infinite number of such elementary antennas with the corresponding magnitudes and phases of their currents.

The current element, usually called Hertzian dipole, oriented in the z-direction, with its location at the origin of a set of spherical coordinates, is shown on Fig. 2.2.

The length h of this electrically short antenna is very small compared to wavelength. As the wire radius a is small compared to the wavelength, the particular integral for retarded potential can be written in the form

$$\vec{A}(r) = \frac{\mu}{4\pi}\int_{s'}\frac{\vec{J}(\vec{r}')\,\vec{S}e^{-jkR}}{R}\,d\vec{s}', \qquad (2.82)$$

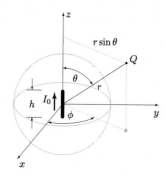

FIG. 2.2 Hertzian dipole.

which results in

$$\vec{A}(z) = \frac{\mu}{4\pi} \int_L \frac{I(z')\,e^{-jkR}}{R} \hat{e}\,dz', \qquad (2.83)$$

and finally, if the current along the short antenna is assumed to be constant and expressed in the following phasor form:

$$I(z') = I_0, \qquad (2.84)$$

one obtains the final expression for the related retarded potential:

$$A_z(z) = \frac{\mu}{4\pi} \int_L \frac{I(z')\,e^{-jkR}}{R}\,dz' = \frac{\mu I_0\,e^{-jkr}}{4\pi r} \int_{-h}^{h} dz'$$

$$= \frac{\mu h I_0}{4\pi r} e^{-jkr}. \qquad (2.85)$$

If the system of spherical coordinates is considered, then

$$A_r = A_z \cos\theta = \mu \frac{h I_0}{4\pi r} e^{-jkr} \cos\theta, \qquad (2.86)$$

$$A_\theta = -A_z \sin\theta = -\mu \frac{h I_0}{4\pi r} e^{-jkr} \sin\theta. \qquad (2.87)$$

Combining the complex phasor form of Eqs. (2.59) and (2.65) gives

$$\vec{E} = -j\omega\vec{A} - \nabla\varphi, \qquad (2.88)$$

$$\nabla\vec{A} = -j\omega\mu\varepsilon\varphi, \qquad (2.89)$$

and then the electric field can be expressed as follows:

$$\vec{E} = -j\omega\vec{A} + \frac{1}{j\omega\mu\varepsilon}\nabla(\nabla\cdot\vec{A}). \qquad (2.90)$$

Finally, the field components become:

$$H_\phi = \frac{h I_0}{4\pi} e^{-jkr}\left(\frac{jk}{r} + \frac{1}{r^2}\right)\sin\theta, \qquad (2.91)$$

$$E_r = \frac{h I_0}{4\pi} e^{-jkr}\left(\frac{2Z_0}{r^2} + \frac{2}{j\omega\varepsilon r^3}\right)\cos\theta, \qquad (2.92)$$

$$E_\theta = \frac{h I_0}{4\pi} e^{-jkr}\left(\frac{j\omega\mu}{r} + \frac{1}{j\omega\varepsilon r^3} + \frac{Z_0}{r^2}\right)\sin\theta. \qquad (2.93)$$

At large distances from the source ($r \gg \lambda$), the only significant terms for E and H are those varying as $1/r$. This is the region where the far field is significant and the corresponding components are:

$$H_\phi = \frac{jkh I_0}{4\pi r} e^{-jkr}\sin\theta, \qquad (2.94)$$

$$E_\theta = \frac{j\omega\mu h I_0}{4\pi r} e^{-jkr}\sin\theta = Z_0 H_\phi. \qquad (2.95)$$

The total power collected in the far field can be obtained by using expression (2.41). Namely, the total power is the integral of the time average Poynting vector over any surrounding surface. For simplicity this surface is a sphere of radius r:

$$P = \oint_S \frac{1}{2}\mathrm{Re}(\vec{E} \times \vec{H}^*)\,d\vec{S}$$

$$= \int_0^{2\pi}\int_0^{\pi} \frac{1}{2}\mathrm{Re}(E_\theta H_\varphi^*)\,r^2 \sin\theta\,d\theta\,d\varphi$$

$$= \frac{Z_0 k^2 I_0^2 h^2}{16\pi} \int_0^\pi \sin^3\theta\,d\theta = \frac{Z_0\pi I_0^2}{3}\left(\frac{h}{\lambda}\right)^2. \qquad (2.96)$$

As the power radiated by the electrically short antenna is proportional to the squared value of the ratio h/λ, the Hertzian dipole with the entire length h is rather small compared to wavelength λ and represents a radiator with a very small efficiency.

2.1.9 Fundamental Antenna Parameters

To describe antenna properties, it is necessary to define various parameters. Some of the parameter definitions specified in [5] are given in this chapter.

2.1.9.1 Radiation Power Density

The power density of an antenna corresponds to the real part of the time average Poynting vector, and for the case of a free space it follows that

$$\overline{P}_d = \frac{1}{2}\mathrm{Re}(E_\theta H_\phi) = \frac{E_\theta^2}{2Z_0}, \qquad (2.97)$$

where E_θ and H_ϕ are the maximal (peak) values of the electric and magnetic fields, respectively. If the rms values are considered, then (2.97) becomes

$$\overline{P}_d = \frac{E_\theta^2}{Z_0}. \qquad (2.98)$$

In the far field region, E_θ and H_ϕ vary as $1/r$ and thus $\overline{P}_d$ varies as $1/r^2$.

2.1.9.2 Radiation Intensity

Radiation intensity, or the antenna power pattern, in a given direction is defined as the power radiated from an antenna per unit solid angle. The radiation intensity is a far field parameter which can be obtained by simply multiplying the radiation power density by the square distance, i.e.,

$$U = r^2 \frac{1}{2} \mathrm{Re}(E_\theta H_\phi) = r^2 \frac{1}{2} \frac{E_\theta}{Z_0}. \qquad (2.99)$$

The relationship between the total power and radiation intensity is given by

$$P_{rad} = \oint_\Omega U \, d\Omega = \int_0^{2\pi} \int_0^\pi U \sin\theta \, d\theta \, d\phi, \qquad (2.100)$$

where $d\Omega = \sin\theta \, d\theta \, d\phi$.

In particular, for an isotropic source, Eq. (2.100) becomes

$$P_{rad} = \oint_\Omega U_0 \, d\Omega = 4\pi U_0, \qquad (2.101)$$

or equivalently, the radiation intensity of an isotropic source is given by

$$U_0 = \frac{P_{rad}}{4\pi}. \qquad (2.102)$$

As is obvious, the radiation intensity U is independent of the angles ϕ and θ.

2.1.9.3 Directivity

Directivity of an antenna is defined as the ratio of the radiation intensity in a given direction from the antenna to the radiation intensity averaged over all directions. The average radiation intensity is equal to the total power radiated by the antenna divided by 4π. If the direction is not strictly specified, the direction of maximum radiation intensity is implied.

Namely, the directivity D of a non-isotropic source is equal to the ratio of its radiation intensity in a given direction over that of an isotropic source:

$$D = \frac{U}{U_0} = \frac{4\pi U}{P_{rad}}. \qquad (2.103)$$

If the direction is not specified, it implies the direction of maximum radiation intensity (maximum directivity) determined as

$$D_{max} = \frac{U_{max}}{U_0} = \frac{4\pi U_{max}}{P_{rad}}, \qquad (2.104)$$

where D_{max} is the maximum directivity, U_0 is radiation intensity of isotropic source and U_{max} is the maximum radiation intensity. To summarize, directivity is a measure that describes only the directional properties of the antenna, and it is therefore controlled only by the pattern.

2.1.9.4 Input Impedance, Radiation and Loss Resistance

Input impedance is defined as the ratio of the voltage and current at the pair of the input antenna terminals:

$$Z_a = R_a + jX_a, \qquad (2.105)$$

where R_a is the resistance at antenna terminals and X_a is the reactance at antenna terminals.

In addition, the resistive part of (2.105) consists of two components:

$$R_a = R_r + R_L, \qquad (2.106)$$

where R_r is the radiation resistance, and R_L is the loss resistance of the antenna.

The radiation resistance R_r is defined as the ratio of the total power radiated by the antenna P_0 and the square of the rms antenna input current I:

$$R_r = \frac{P_0}{I^2}. \qquad (2.107)$$

If the peak value of the current is considered, then

$$R_r = \frac{P_0}{2I_m^2}. \qquad (2.108)$$

This is an equivalent resistance that would dissipate a power equal to the total radiated power when the current through it were equal to the antenna input current.

2.1.9.5 Gain and Radiation Efficiency

Antenna gain is closely related to the directivity, but it is also a measure that takes into account the efficiency of the antenna. The gain of an antenna in a given direction is defined as the ratio of the intensity, in a given direction, and the radiation intensity that would be obtained if the power accepted by the antenna were radiated isotropically. This is called the absolute gain. The radiation intensity corresponding to the isotropically radiated power is equal to the power input by the antenna divided by 4π. Thus,

$$G(\phi, \theta) = 4\pi \frac{U(\phi, \theta)}{P_{in}}, \tag{2.109}$$

where P_{in} is the total input power.

However, in most cases, one deals with a relative gain which is often defined as the ratio of the power gain in a given direction and the power gain of a reference in its referenced direction. The power input has to be the same for both antennas. The reference antenna is usually a dipole, horn, or any other antenna whose gain can be determined or is known.

In most cases, however, the reference antenna is a lossless isotropic source.

When the direction is not stated, the power gain is usually taken in the direction of maximum radiation.

In addition, the total radiated power P_{rad} is related to the total input power P_{in}, hence, it follows that

$$P_{rad} = e P_{in}, \tag{2.110}$$

where e is the antenna radiation efficiency defined as the ratio of the power delivered to the radiation resistance R_r and radiation losses R_L:

$$e = \frac{R_r}{R_r + R_L}. \tag{2.111}$$

Radiation efficiency can be considered as a measure of the dissipative or heat losses in an antenna. It is worth noting that impedance mismatch losses at the input port of the antenna are not included in the radiation efficiency.

Taking into account (2.110), Eq. (2.111) becomes

$$G(\phi, \theta) = e\left[4\pi \frac{U(\phi, \theta)}{P_{rad}}\right], \tag{2.112}$$

which is related to the directivity by the expression

$$G(\phi, \theta) = e D(\phi, \theta). \tag{2.113}$$

In many applications, partial gains G_ϕ and G_θ are used. These partial gains are defined as:

$$G_\phi = 4\pi \frac{U_\phi}{P_{in}}, \tag{2.114}$$

$$G_\theta = 4\pi \frac{U_\theta}{P_{in}}, \tag{2.115}$$

where U_ϕ is the radiation intensity in a given direction contained in the ϕ-field component, U_θ is the radiation intensity in a given direction contained in the θ-field component, and P_{in} is total input power.

Usually, the gain is given in terms of decibels instead of the dimensionless quantity.

2.1.10 Dipole Antennas

The finite length wire antennas (linear antennas) can be modeled as a superposition of infinitesimal radiation sources. Antenna problem can be considered in two different modes: radiation (transmitting) mode and scattering (receiving) mode.

The key to understanding the behavior of radiated or scattered fields is the knowledge of the current distribution induced along the antenna. For the case of electrically short wires where the length is small compared to the wavelength, the current can be approximated as a constant (as in the case of Hertz dipole) or linear function. On the other hand, if the wire length or frequency increases, the current distribution changes significantly. There are several approaches to determine the current distribution of a linear antenna. The simplest way is to assume the current waveform. Even though such an approach could provide satisfactory results in many applications, if thin wires are considered, there are situations where an accurate current distribution calculation is necessary. A more rigorous approach to this problem is to obtain the current distribution by solving the corresponding integral equation, which is discussed in details in Sect. 2.1.11. This section deals with the approximate current distributions.

One of the simplest antennas most commonly used in practice and also as an EMC model is the center-fed dipole antenna shown in Fig. 2.3. The entire length of the wire is L.

At high frequencies the reasonable and traditionally widely used approximation for the antenna current is the sinusoidal distribution defined as

$$I(z) = I_0 \sin\left[k\left(\frac{L}{2} - |z|\right)\right]. \tag{2.116}$$

At sufficiently low frequencies where $\lambda > L/2$, i.e., $k(L/2 - |z|) \ll 1$, Eq. (2.116) becomes

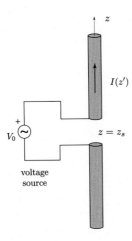

FIG. 2.3 Center-fed dipole antenna.

$$I(z) = I_0 \left(1 - \frac{2|z|}{L}\right). \tag{2.117}$$

Approximating the distance from the source to the observation point gives

$$R = r - z'\cos\theta. \tag{2.118}$$

The radiated far field E_θ, where the following condition is satisfied:

$$kr = 2\pi\frac{r}{\lambda} \gg 1, \tag{2.119}$$

can be determined from the magnetic vector potential

$$E_\theta = -j\omega A_\theta = j\omega A_z \sin\theta, \tag{2.120}$$

where the magnetic vector potential for the case of sinusoidal current distribution is given by

$$A_z(z) = \frac{\mu}{4\pi} \int_L \frac{I(z')e^{-jkR}}{R} dz'$$

$$= \frac{\mu I_0}{4\pi r} e^{-jkr} \int_{-L/2}^{L/2} \sin\left[k\left(\frac{L}{2} - |z'|\right)\right] e^{jkz'\cos\theta} dz'. \tag{2.121}$$

Combining Eqs. (2.120) and (2.121) yields

$$E_\theta = jk\frac{Z_0}{4\pi r} e^{-jkr} \sin\theta$$

$$\times \int_{-L/2}^{L/2} \sin\left[k\left(\frac{L}{2} - |z'|\right)\right] e^{jkz'\cos\theta} dz'$$

$$= jZ_0\frac{I_0}{2\pi r} e^{-jkr} \frac{\cos\left(k\frac{L}{2}\cos\theta\right) - \cos\left(k\frac{L}{2}\right)}{\sin\theta}. \tag{2.122}$$

In the far-field region, the magnetic and electric field are related, and one obtains

$$H_\phi = \frac{E}{Z_0} = j\frac{I_0}{2\pi r} e^{-jkr} \frac{\cos\left(k\frac{L}{2}\cos\theta\right) - \cos\left(k\frac{L}{2}\right)}{\sin\theta}. \tag{2.123}$$

Finally, the total power radiated by the dipole antenna can be obtained from integral (2.41), i.e.,

$$P_{rad} = \int_0^{2\pi}\int_0^\pi \frac{|E_\theta|^2}{2Z_0} r^2 \sin\theta\, d\theta\, d\phi$$

$$= Z_0\frac{I_0^2}{4\pi} \int_0^\pi \frac{\left[\cos\left(k\frac{L}{2}\cos\theta\right) - \cos\left(k\frac{L}{2}\right)\right]}{\sin\theta} d\theta. \tag{2.124}$$

The radiation resistance is defined by relation (2.108). For the case of half-length dipole, the total power is

$$P_{rad} = Z_0\frac{I_0^2}{4\pi} \int_0^\pi \frac{\cos^2\left(k\frac{L}{2}\cos\theta\right)}{\sin\theta} d\theta. \tag{2.125}$$

The radiation resistance of half-length dipole is approximately equal 73Ω.

2.1.11 Pocklington Integro-Differential Equation for a Straight Thin Wire

Henry Cabourn Pocklington was the first who formulated the frequency domain integro-differential equation for a total current flowing along a straight thin wire antenna in 1897 [10]. He also presented the first approximate solution of this equation. During the last 120 years, many outstanding researchers have investigated both the formulation and the numerical solution of the Pocklington equation. Maybe the most important advance in the formulation was carried out by Erik Hallén in the late 1930s. Having started from Pocklington integro-differential equation in the frequency domain, Hallén managed to derive a new type of integral equation for thin wire configurations. Since than, many numerical techniques for solving the Pocklington and Hallén equation, respectively, were reported by different authors [6].

Numerical modeling of the wire antennas and scatterers started in 1965 with the classical paper by K.K. Mei [11]. Mei derived certain types of Pocklington and Hallén equation, having also reported a related numerical technique for solving these equations. Today his technique could be referred to as the point-matching technique. A number of important contributions to the numerical solution of Pocklington and Hallén integral equations were given in the 1970s: Silvester and Chan (1972, 1973) [12,13], proposed the use of the strong finite element formulation to the Hallén and Pocklington equations, while Butler and Wilton (1975, 1976) [14,15], proposed several moment method techniques for solving these equations.

In addition to the improvement in development of numerical solution methods, there have been important achievements in the formulation of the problem. Thus, the most important outcome is the extension of the original Pocklington formulation to the wires radiating in the presence of an imperfectly conducting half-space. This was worked out by E.K. Miller et al. (1972) [16], T. Sarkar (1977) [17] and by Parhami and Mittra [18]. The numerical solution of Pocklington equation via Galerkin–Bubnov Indirect Boundary Element Method (GB-IBEM) was reported elsewhere, e.g., in [6].

Consider a dipole antenna insulated in free-space having length L and radius a, as shown in Fig. 2.4.

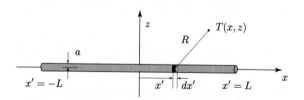

FIG. 2.4 Single wire antenna insulated in free-space.

The dipole antenna parameters of interest describing its radiating behavior can be determined, provided its axial current distribution is known. This current flowing along the thin wire antenna is governed by the frequency domain Pocklington integro-differential equation. The Pocklington equation can be derived in a few ways starting from Maxwell equation for time harmonic fields.

The thin wire approximation requires wire dimensions to satisfy the conditions:

$$a \ll \lambda_0 \quad \text{and} \quad a \ll L, \qquad (2.126)$$

where λ_0 is the wavelength of a plane wave in free-space.

Consequently, only axial component of $\vec{A}$ exists along the wire, and (2.92) can be written as

$$E_x = \frac{1}{j\omega\mu\varepsilon_0}\left[\frac{\partial^2 A_x}{\partial x^2} + k^2 A_x\right]. \qquad (2.127)$$

The magnetic vector potential A_x is expressed by a particular integral over the unknown axial current $I(x)$ along the wire:

$$A_x = \frac{\mu}{4\pi}\int_{-L}^{L} I(x')\frac{e^{-jkR}}{R}\,dx', \qquad (2.128)$$

where R is a distance from the source point to the observation point and k is the wave number of free-space.

Assuming the wire to be perfectly conducting (PEC), and according to the continuity conditions for the tangential electric field components, the total tangential electric field vanishes on the PEC wire surface:

$$E_x^{inc} + E_x^{sct} = 0, \qquad (2.129)$$

where E_x^{inc} is the incident field and E_x^{sct} is the scattered field due to the presence of the PEC surface. Inserting (2.128) into (2.127), the electric field scattered from the antenna surface becomes

$$E_x^{sca} = \frac{1}{j4\pi\omega\varepsilon_0}\int_{-L}^{L}\left(\frac{\partial^2}{\partial x^2} + k^2\right)g_0(x,x')\,I(x')\,dx'. \qquad (2.130)$$

Combining (2.129) and (2.130) leads to the Pocklington integro-differential equation for the unknown current distribution along the single straight wire antenna insulated in free-space:

$$E_x^{inc} = -\frac{1}{j4\pi\omega\varepsilon_0}\int_{-L}^{L}\left(\frac{\partial^2}{\partial x^2} + k^2\right)g_0(x,x')\,I(x')\,dx', \qquad (2.131)$$

where $g_0(x,x')$ is the free-space Green function,

$$g_0(x,x') = \frac{e^{-jkR}}{R}, \qquad (2.132)$$

while the distance R from the source to the observation point is

$$R = \sqrt{(x - x')^2 + a^2}. \qquad (2.133)$$

Once the axial current on the antenna is determined, other important antenna parameters, such as radiated field, radiation pattern, or input impedance, can be calculated. The details can be found elsewhere, e.g., in [6].

Very similar derivation could be undertaken for imperfectly conducting wires. Studies for thin wires in the presence of inhomogeneous media could be found elsewhere, e.g., in [1] and [6]. Numerical solution of integro-differential equation (2.131) is discussed in Sect. 2.2.

2.2 INTRODUCTION TO NUMERICAL METHODS IN ELECTROMAGNETICS

Problems arising in electromagnetics can be formulated in terms of differential, integral or variational equations. Generally, there are two basic approaches to solve problems in electromagnetics; the differential, or the field approach, and integral, or the source approach.

The field approach deals with a solution of a corresponding differential equation with associated boundary conditions, specified at a boundary of a computational domain. Such an approach is rather useful for handling the interior field problems; see Fig. 2.5.

A standard boundary-value problem can be formulated in terms of the operator equation

$$\mathcal{L}(u) = p \qquad (2.134)$$

on the domain Ω with conditions $F(u) = q|_\Gamma$ prescribed on the boundary Γ (see Fig. 2.5), where $\mathcal{L}$ is a linear differential operator, u solution of the problem, and p is the excitation function representing the known sources inside the domain. Methods for the solution of the interior field problem are generally referred to as differential methods, or field methods.

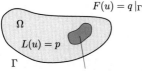

$$F(u) = q|_\Gamma$$

Ω

$L(u) = p$

Γ

p − known sources inside the domain

FIG. 2.5 Differential approach.

The source, or the integral approach, is based on the solution of a corresponding integral equation. The source approach is convenient for the treatment of the exterior field problems, see Fig. 2.6.

The integral formulation can be generally written as follows:

$$g(u) = h, \qquad (2.135)$$

where g represents an integral operator, unknowns u are related to field sources, i.e., charge densities or current densities distributed along the boundary Γ', while h is an excitation function (see Fig. 2.6), e.g., the electric field illuminating the metallic object, thus inducing the charge density. Once the sources are determined, the field at an arbitrary point, inside or outside the domain, can be obtained by integrating the sources.

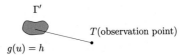

Γ'

T(observation point)

$g(u) = h$

FIG. 2.6 Integral approach.

The methods of solutions for the exterior field problems are referred to as the integral methods, or methods of sources.

The boundary conditions used in electromagnetic field problems are usually of the Dirichlet (forced) and Neumann (natural) type, or their combination (mixed boundary conditions) [6,19]. These boundary conditions can be either homogeneous or inhomogeneous.

Operator equations can be handled analytically and/or numerically. Analytical solution methods yield exact solutions, but are limited to a narrow range of applications, mostly related to canonical problems. There are few practical engineering problems that can be solved in closed form. Numerical methods are applicable to almost all scientific engineering problems, with the main drawbacks pertaining to the approximation limit in the model itself, space and time discretization. Moreover, the criteria for accuracy, stability and convergence are not always straightforward and clear to the researcher. The most commonly used methods in computational electromagnetics (CEM), among others, are the Finite Difference Method (FDM), Finite Element Method (FEM), Boundary Element Method (BEM), and Method of Moments (MoM).

The problems being analyzed can be regarded as steady state or transient, and the solution methods are usually classified as frequency or time domain. The frequency and time domain techniques for solving transient electromagnetic phenomena have been fully documented in [6].

A frequency domain solution is commonly applied for many sources and a single frequency, whereas with the time domain, it is for a single source and many frequencies [6].

The time domain solution obtained is specific to the temporal variation of the excitation source. The transient response of a structure when subjected to different excitations, for example, a step-function or a Gaussian voltage source, will require the computation process to be repeated for the respective solutions, whereas in the frequency domain approach, the solution from one set of computations can be applied to obtain transient results from different sources if the geometry of the structure is unchanged. This difference is a significant factor when considering the relative merits of the two approaches.

Generally, a deeper physical insight is obtained when using the time domain approach. However, an understanding of the resonant characteristic can only be obtained from the frequency spectrum. Furthermore, nonlinearities are more conveniently handled in the time domain. Frequency domain formulation is definitely simpler and easier to use, thus allowing more complex structures to be analyzed more conveniently as for such geometries larger computing effort is required in the time domain approach.

According to the differential and integral formulation, numerical methods per se can be classified as domain, boundary or source simulation methods.

It is important to clarify some principles and ideas of how to describe field problems via partial differential or integral equations. Namely, there are some basic differences between domain methods (e.g., finite element method), boundary methods (e.g., boundary element method) and source simulation methods (charge or current simulation method). This section outlines some basics of Finite Element Method (FEM), as well as direct and indirect Boundary Element Method (BEM). Some basics regarding modeling via the Finite Difference Method (FDM) could be found elsewhere, e.g., in [6].

Finite element method (FEM) modeling of partial differential equations is undertaken by discretizing the entire calculation domain Ω, and integrating any known sources Ω_s (if any) within the domain, see Fig. 2.7A. Dirichlet and Neumann boundary conditions are specified along the boundary Γ for both FDM and FEM. Finally, an algebraic equation system provides a sparse matrix, usually banded and in many cases symmetric.

Modeling via the Boundary Integral Equation Method involves direct and indirect approach. Integral formulations of partial differential equations along the boundary are carried out using the Green integral theorem. The resulting equations are modeled discretizing only the boundary and by integrating any known sources Ω_s within a given subdomain. Typically, modeling of partial differential equations in terms of re-

lated boundary integral formulations results in less unknowns but dense matrices [6].

The boundary element approach method using field and potential quantities rather than field sources is usually referred to as the direct BEM formulation. However, there are many applications in computational electromagnetics of boundary integral formulation for which it is more convenient to deal with sources. Such an approach is known as direct BEM, and related integral equations are posed in terms of sources distributed over the boundary. This method can be considered as a special case of the direct BEM approach involving integration over unknown charge density or current density. Integral equations over unknown sources Ω_s' (see Fig. 2.7B) can also be derived from the Green integral theorem, and the solution method can be referred to as a special variant of the boundary element method – indirect boundary element method.

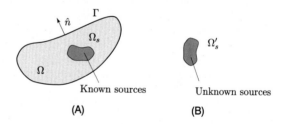

FIG. 2.7 Method of fields versus method of sources.

However, as classic BEM uses potential or field on the domain boundary and this integral equation approach deals with unknown sources, some authors use the term finite elements for integral operators [19]. On the other hand, in order to stress the integration over sources, the term source element method (SEM) or source integration method (SIM) is suggested [6].

2.2.1 Weighted Residual Approach

Partial differential equations (PDEs), integral equations (IEs) and integro-differential equations (IDEs) are modeled using the weighted residual approach, also referred to as a projection approach [19].

Assume one seeks a solution of an operator differential equation

$$\mathcal{L}(q, t) = 0, \qquad (2.136)$$

where $\mathcal{L}$ is a corresponding differential operator.

Now, within the variational formalism, differential (operator) equation can be regarded as a Lagrange equation, thus it can be written as

$$\delta F = \int_{t_1}^{t_2} \mathcal{L}(q,t)\delta q \, dt. \qquad (2.137)$$

On the other hand, in the framework of Galerkin procedure, construction of a corresponding functional is not required, therefore, it is of interest to write the variation of the functional in a way that the differential equation is multiplied by an arbitrary function and integrated.

Provided that the variation of the functional is zero, it can be written as

$$\int_{t_1}^{t_2} \mathcal{L}(q,t)\,\psi(t)\,dt = 0, \qquad (2.138)$$

where the expression

$$\int_{t_1}^{t_2} \mathcal{L}(q,t)\,\psi(t)\,dt \qquad (2.139)$$

is regarded as a scalar product of functions $\mathcal{L}$ and ψ.

It is worth noting that the requirement for the integral to vanish is equivalent to the orthogonality of these functions. The first step is to derive the formulation for an approximation of functions.

2.2.1.1 Fundamental Lemma of Variational Calculus

The fundamental lemma of variational calculus can be outlined as follows: scalar product of functions u and W in an n-dimensional Hilbert space is defined by the integral

$$\langle u, W \rangle = \int_{\Omega} u(x) \cdot W^*(x)\,d\Omega, \quad x \in \Omega, \qquad (2.140)$$

where $W^*(x)$ stands for the complex conjugate of $W(x)$. If for a given function u and an arbitrary function W, the scalar product vanishes

$$\int_{\Omega} u(x) \cdot W^*(x)\,d\Omega = 0, \quad x \in \Omega, \qquad (2.141)$$

then it follows that

$$u(x) \equiv 0, \quad x \in \Omega. \qquad (2.142)$$

Defining $u(x)$ as a residual (difference between an exact function f and approximate function $\tilde{f}$),

$$u(x) = f - \tilde{f}, \qquad (2.143)$$

according to the fundamental lemma of variational calculus, the residual integral vanishes

$$\int_{\Omega} (f - \tilde{f})\,W\,d\Omega = 0 \qquad (2.144)$$

for an arbitrary function W only if $f = \tilde{f}$.

An approximate function can be expressed in terms of a linear combination

$$\tilde{f} = \sum_{i=1}^{n} \alpha_i N_i(x). \qquad (2.145)$$

Note that the series converges only in the case

$$\lim_{n \to \infty} \left(\sum_{i=1}^{n} \alpha_i N_i(x) \right) = f. \qquad (2.146)$$

The residual integral can be written in the following form:

$$\int_{\Omega} \left[f - \sum_{i=1}^{n} \alpha_i N_i(x) \right] W_j \, d\Omega = 0, \quad j = 1, 2, \dots, n, \qquad (2.147)$$

where W_j stands for a set of test (weighting) functions.

The weighted residual integral is transformed into a set of linear equations:

$$\sum_{i=1}^{n} a_{ji}\alpha_i = b_j, \quad j = 1, 2, \dots, n. \qquad (2.148)$$

In matrix notation one has

$$\begin{bmatrix} a_{11} & a_{12} & \cdots & a_{in} \\ a_{21} & a_{22} & \cdots & a_{2n} \\ \vdots & \vdots & \ddots & \vdots \\ a_{n1} & a_{n2} & \cdots & a_{nn} \end{bmatrix} \begin{Bmatrix} \alpha_1 \\ \alpha_2 \\ \vdots \\ \alpha_n \end{Bmatrix} = \begin{Bmatrix} b_1 \\ b_2 \\ \vdots \\ b_n \end{Bmatrix}. \qquad (2.149)$$

The general term of the system matrix and the right-hand side vector are given by:

$$a_{ji} = \int_{\Omega} N_i(x)\,W_j(x)\,d\Omega, \qquad (2.150)$$

$$b_j = \int_\Omega f \cdot W_j(x) \, d\Omega. \qquad (2.151)$$

Choosing different base and test functions, respectively, one deals with different approximations of the original functions.

The presented formalism of approximation of functions can be readily applied to the approximate solution of differential equations.

Any differential equation can be considered in the general implicit form of

$$A(u) = 0, \qquad (2.152)$$

where

$$A(u) \equiv \mathcal{L}(u) - p. \qquad (2.153)$$

Therefore, an operator differential equation can be written as

$$\mathcal{L}(u) - p = 0, \qquad (2.154)$$

while the boundary conditions are

$$\mathcal{M}(u) - r = 0. \qquad (2.155)$$

Therefore, the corresponding residuals are given by:

$$\xi_\Omega = \mathcal{L}(\tilde{u}) - p, \qquad (2.156)$$

$$\xi_\Gamma = \mathcal{M}(\tilde{u}) - r. \qquad (2.157)$$

And the error due to approximate solution is taken into account by satisfying the fundamental lemma of the variational calculus.

Thus, *weighted residual integrals* over the domain Ω and boundary Γ vanish, i.e.,

$$\int_\Omega \left[\mathcal{L}(\tilde{u}) - p \right] W_j \, d\Omega + \int_\Gamma \left[\mathcal{M}(\tilde{u}) - r \right] W_j \, d\Gamma = 0.$$

$$(2.158)$$

In principle, without loss of generality, it is possible to choose base functions to automatically satisfy

$$\mathcal{M}(u) - r = 0. \qquad (2.159)$$

Thus, one has

$$\int_\Omega \left[\mathcal{L}(\tilde{u}) - p \right] W_j \, d\Omega = 0, \qquad (2.160)$$

and writing an approximate solution in the form

$$\tilde{u}(x) = \sum_{i=1}^n \alpha_i \, N_i(x), \qquad (2.161)$$

it follows that

$$\int_\Omega \left[\mathcal{L} \left(\sum_{i=1}^n \alpha_i \, N_i(x) \right) - p \right] W_j \, d\Omega = 0. \qquad (2.162)$$

Finally, an original operator (differential) equation is transformed into a linear equation system:

$$\sum_{i=1}^n \alpha_i \int_\Omega \mathcal{L}(N_i(x)) \, W_j \, d\Omega = \int_\Omega p \, W_j \, d\Omega, \quad j = 1, 2, \ldots, n,$$

$$(2.163)$$

which can be written as

$$\sum_{i=1}^n a_{ji} \, \alpha_i = b_j, \quad j = 1, 2, \ldots, n, \qquad (2.164)$$

or in matrix notation,

$$[a] \{\alpha\} = \{b\}, \qquad (2.165)$$

where the general matrix term and the right-hand side vector are:

$$a_{ji} = \int_\Omega \mathcal{L}(N_i) \, W_j \, d\Omega, \qquad (2.166)$$

$$b_j = \int_\Omega p \, W_j \, d\Omega. \qquad (2.167)$$

Choosing certain type of base and test functions, one deals with different numerical techniques for the solution of partial differential equations (PDEs).

2.2.2 The Finite Element Method (FEM)

The Finite Element Method (FEM) is one of the most commonly used numerical methods in science and engineering. The method is highly automatized and rather suitable for computer implementation based on step by step algorithms. The special features of FEM are related to efficient modeling of complex shape geometries and inhomogeneous domains, and also to the relatively simple mathematical formulation providing a highly banded and symmetric matrix, same accuracy refinement as by a higher order approximation. The method also provides automatic inclusion of natural (Neumann) boundary conditions. This method generally gives better results than highly robust Finite Difference Method (FDM) approaches [5] in modeling complicated boundaries and is particularly suited for problems with closed boundaries.

2.2.2.1 Basic Concepts of FEM

The calculation domain is discretized into sufficiently small segments – finite elements. The unknown solution over a finite element is expressed in terms of a linear combination of local interpolation functions (shape functions), for 1D problem, as it is shown in Fig. 2.8.

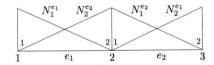

FIG. 2.8 Linear shape functions over 2 elements.

The global base functions assigned to nodes are assembled from local shape functions, assigned to elements, as it is shown in Fig. 2.9 for the case of linear approximation.

The approximate solution of a problem of interest can be written as

$$\tilde{f} = \sum_{i=1}^{n} \alpha_i N_i, \qquad (2.168)$$

where coefficients α_i represent the solution at the global nodes, while n denotes the total number of nodes.

The approximate solution along two finite elements is shown in Fig. 2.10.

Elements of local and global nodes and linear approximation of the unknown solution over a domain of interest are shown in Fig. 2.11.

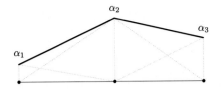

FIG. 2.10 Approximate solution above two finite elements.

The accuracy can be improved by a finer discretization, or by implementation of higher order approximation.

2.2.2.2 One-Dimensional FEM

Many problems in science and engineering can be formulated in terms of second-order differential equations of the following form:

$$\frac{d}{dx}\left[\lambda\frac{du}{dx}\right] + k^2 u - p = 0, \qquad (2.169)$$

where λ and k depend on the properties of a medium and p represents the excitation function, i.e., the sources within the domain of interest. For the one-dimensional case, the domain is related to interval $[a, b]$.

Substituting an approximate solution $\tilde{u}$ into the differential equation (2.169) and integrating over the calculation domain according to the weighted residual approach [6], we get

$$\int_a^b \left[\frac{d}{dx}\left(\lambda\frac{d\tilde{u}}{dx}\right) + k^2\tilde{u} - p\right] W_j \, dx = 0, \quad j = 1, 2, \ldots, n, \qquad (2.170)$$

which can also be written as

$$\int_a^b \frac{d}{dx}\left(\lambda\frac{d\tilde{u}}{dx}\right) W_j \, dx + \int_a^b k^2\tilde{u}\, W_j \, dx = \int_a^b p\, W_j \, dx. \qquad (2.171)$$

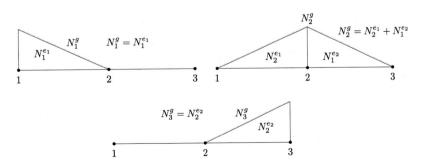

FIG. 2.9 Assembling of global functions from local shape functions.

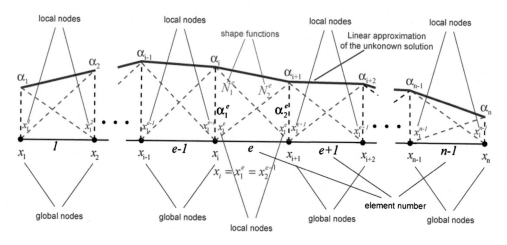

FIG. 2.11 Local and global quantities.

Eq. (2.171) represents the strong formulation of the problem. Within strong formulation, the base functions must be in the domain of the differential operator, and automatically satisfy the prescribed boundary conditions. The strong requirements can be avoided moving to the weak formulation of the problem.

The order of differentiation can be decreased by carefully performing integration by parts. Differentiation of a product of two functions can be written as

$$\frac{d}{dx}\left(\lambda\frac{d\tilde{u}}{dx}W_j\right) = \frac{d}{dx}\left(\lambda\frac{d\tilde{u}}{dx}\right)W_j + \lambda\frac{d\tilde{u}}{dx}\frac{dW_j}{dx}, \quad \lambda = \lambda(x). \tag{2.172}$$

Somewhat rearranging Eq. (2.172) and integrating along the interval yields

$$\frac{d}{dx}\left(\lambda\frac{d\tilde{u}}{dx}\right)W_j = \frac{d}{dx}\left(\lambda\frac{d\tilde{u}}{dx}W_j\right)$$

$$-\lambda\frac{d\tilde{u}}{dx}\frac{dW_j}{dx} \quad \Big/ \cdot\int_a^b dx, \tag{2.173}$$

and so one obtains

$$\int_a^b \frac{d}{dx}\left(\lambda\frac{d\tilde{u}}{dx}\right)W_j\,dx = \lambda\frac{d\tilde{u}}{dx}W_j\Big|_a^b - \int_a^b \lambda\frac{d\tilde{u}}{dx}\frac{dW_j}{dx}\,dx. \tag{2.174}$$

Substituting expression (2.174) into (2.171), the weak formulation is obtained as

$$\int_a^b k^2\tilde{u}\,W_j\,dx + \lambda\frac{d\tilde{u}}{dx}W_j\Big|_a^b - \int_a^b \lambda\frac{d\tilde{u}}{dx}\frac{dW_j}{dx}\,dx$$

$$= \int_a^b p\,W_j\,dx. \tag{2.175}$$

The second term on the right-hand side of Eq. (2.175) is the natural boundary (Neumann) condition, thus being directly included into the weak formulation and representing the flux density at the ends of the interval.

Applying the finite element algorithm, the unknown solution $\tilde{u}$ is expanded into a linear combination of basis functions. Implementing the Galerkin–Bubnov procedure (the same choice of base and test functions) yields the following matrix equation:

$$[a]\{\alpha\} = \{b\} + \{Q\}, \tag{2.176}$$

where $[a]$ is the global matrix of the system, $\{\alpha\}$ is the solution vector, $\{b\}$ represents the excitation vector, and $\{Q\}$ denotes the flux density.

A local approximation for the unknown function over a finite element is given by

$$\tilde{u}^e = \alpha_1^e N_1^e(x) + \alpha_2^e N_2^e(x), \tag{2.177}$$

where $N_1^e(x)$ and $N_2^e(x)$ are the linear shape functions.

Finite element matrix and vector are given by the following integrals:

$$a_{ji}^e = \int_{x_1^e}^{x_2^e} k^2 N_i^e N_j^e\,dx - \int_{x_1^e}^{x_2^e} \lambda\frac{dN_i^e}{dx}\frac{dN_j^e}{dx}\,dx, \tag{2.178}$$

$$b_j^e = \int_{x_1^e}^{x_2^e} p\, N_j^e\, dx, \qquad (2.179)$$

and the form of the local matrix equation is then

$$\begin{bmatrix} a_{11}^e & a_{12}^e \\ a_{21}^e & a_{22}^e \end{bmatrix} \cdot \begin{bmatrix} \alpha_1^e \\ \alpha_2^e \end{bmatrix} = \begin{bmatrix} b_1^e \\ b_2^e \end{bmatrix} + \begin{bmatrix} -\lambda \dfrac{du}{dx}\Big|_{x=x_1^e} \\[2mm] \lambda \dfrac{du}{dx}\Big|_{x=x_2^e} \end{bmatrix}.$$
$$(2.180)$$

The resulting global matrix system is assembled from the local ones and it is given by

$$\begin{bmatrix} a_{11}^{e_1} & a_{12}^{e_1} & 0 & \cdots & 0 \\ a_{21}^{e_1} & a_{22}^{e_1}+a_{11}^{e_2} & a_{12}^{e_2} & & 0 \\ 0 & a_{21}^{e_2} & a_{22}^{e_2}+a_{11}^{e_3} & & \vdots \\ \vdots & & & \ddots & a_{12}^{e_n} \\ 0 & 0 & \cdots & a_{21}^{e_n} & a_{22}^{e_n} \end{bmatrix} \cdot \begin{bmatrix} \alpha_1 \\ \alpha_2 \\ \alpha_3 \\ \vdots \\ \alpha_n \end{bmatrix}$$

$$= \begin{bmatrix} b_1^{e_1} \\ b_2^{e_1}+b_1^{e_2} \\ b_2^{e_2}+b_1^{e_3} \\ \vdots \\ b_2^{e_n} \end{bmatrix} + \begin{bmatrix} -\lambda \dfrac{du}{dx}\Big|_{x=a} \\ 0 \\ 0 \\ \vdots \\ \lambda \dfrac{du}{dx}\Big|_{x=b} \end{bmatrix}. \qquad (2.181)$$

Note that the flux densities vanish at internal nodes, and only the values related to the domain boundary (in the 1D case, interval ends) are not equal zero.

2.2.2.3 Incorporation of Boundary Conditions

Contrary to the Neumann boundary conditions which are automatically included into the weak formulation of the finite element method, the Dirichlet (forced) boundary conditions are incorporated into the matrix system subsequently, i.e.,

$$\alpha_1 = u(a), \quad \alpha_n = u(b), \qquad (2.182)$$

thus decreasing the number of unknowns, i.e., the first and last rows of the matrix equations are omitted, and the global matrix equation results in a matrix system of $n-2$ unknowns:

$$\begin{bmatrix} a_{22}^{e_1}+a_{11}^{e_2} & a_{12}^{e_2} & 0 & \cdots & 0 \\ a_{21}^{e_2} & a_{22}^{e_2}+a_{11}^{e_3} & a_{12}^{e_3} & & 0 \\ 0 & a_{21}^{e_3} & a_{22}^{e_3}+a_{11}^{e_4} & & \vdots \\ \vdots & & & \ddots & a_{12}^{e_{n-1}} \\ 0 & 0 & \cdots & a_{21}^{e_{n-1}} & a_{22}^{e_{n-1}}+a_{11}^{e_n} \end{bmatrix}$$

$$\cdot \begin{bmatrix} \alpha_2 \\ \alpha_3 \\ \alpha_4 \\ \vdots \\ \alpha_{n-1} \end{bmatrix} = \begin{bmatrix} b_2^{e_1}+b_1^{e_2} \\ b_2^{e_2}+b_1^{e_3} \\ b_2^{e_3}+b_1^{e_4} \\ \vdots \\ b_2^{e_{n-1}}+b_1^{e_n} \end{bmatrix} + \begin{bmatrix} 0 \\ 0 \\ 0 \\ \vdots \\ 0 \end{bmatrix}$$

$$- \begin{bmatrix} \alpha_1 a_{21}^{e_1} \\ 0 \\ 0 \\ \vdots \\ 0 \end{bmatrix} - \begin{bmatrix} 0 \\ 0 \\ 0 \\ \vdots \\ \alpha_n a_{12}^{e_n} \end{bmatrix}. \qquad (2.183)$$

Once the unknown coefficients, α_2 to α_{n-1}, are determined, it is possible from the first and last rows (equations) to obtain the flux densities at the ends of the interval:

$$\lambda \frac{du}{dx}\Big|_{x=a} = b_1^{e_1} - \alpha_1 a_{11}^{e_1} - \alpha_2 a_{12}^{e_1},$$

$$\lambda \frac{du}{dx}\Big|_{x=b} = -b_2^{e_n} + \alpha_{n-1} a_{21}^{e_n} + \alpha_n a_{22}^{e_n}. \qquad (2.184)$$

If the Dirichlet condition is prescribed at one end of the interval, and the Neumann condition at the other, then the global system consists of $n-1$ unknowns and is given by a combination of system (2.181) and (2.183).

2.2.2.4 Computational Example: 1D Problem

Determine the distribution of time harmonic magnetic field within a slab of length h (h is negligible compared to other dimensions, Fig. 2.12) by discretizing the domain into 2 linear finite elements. The known values are: $k=1$, $h=2$ m, $H_0 = 120$ A/m.

For the 1D wave problem, the time harmonic magnetic field is governed by the following Helmholtz equation:

$$\frac{\partial^2 H_z}{\partial x^2} + k^2 H_z = 0.$$

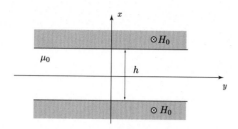

FIG. 2.12 Geometry of the problem.

Applying the weighted residual approach gives

$$\int_{-h/2}^{h/2} \left[\frac{\partial^2 H_z}{\partial x^2} + k^2 H_z \right] W_j \, dx = 0,$$

and, having performed integration by parts, one obtains the weak formulation

$$\int_{-h/2}^{h/2} \frac{\partial H_z}{\partial x} \frac{\partial W_j}{\partial x} \, dx - \int_{-h/2}^{h/2} k^2 H_z W_j \, dx = \frac{\partial H_z}{\partial x} W_j \bigg|_{-h/2}^{h/2}.$$

Now, discretizing the domain to finite elements with Galerkin–Bubnov procedure, $W_j = N_j$, the general term of FEM matrix is obtained:

$$a_{ji} = \int_{x_1}^{x_2} \frac{\partial N_j}{\partial x} \frac{\partial N_i}{\partial x} \, dx - \int_{x_1}^{x_2} k^2 N_i N_j \, dx.$$

Provided the linear shape functions are chosen as

$$N_1(x) = \frac{x_2 - x}{\Delta x}, \quad N_2(x) = \frac{x - x_1}{\Delta x},$$

the terms of FEM matrix become:

$$a_{11} = \int_{x_1}^{x_2} \left[\frac{\partial N_1}{\partial x} \frac{\partial N_1}{\partial x} - N_1 N_1 \right] dx = \frac{1}{\Delta x} - \frac{\Delta x}{3},$$

$$a_{12} = a_{21} = \int_{x_1}^{x_2} \left[\frac{\partial N_1}{\partial x} \frac{\partial N_2}{\partial x} - N_1 N_2 \right] dx$$

$$= -\frac{1}{\Delta x} - \frac{\Delta x}{6},$$

$$a_{22} = \int_{x_1}^{x_2} \left[\frac{\partial N_2}{\partial x} \frac{\partial N_2}{\partial x} - N_2 N_2 \right] dx = \frac{1}{\Delta x} - \frac{\Delta x}{3}.$$

The global matrix, assembled from local matrices, is of the form

$$[a] = \begin{bmatrix} \dfrac{1}{\Delta x} - \dfrac{\Delta x}{3} & -\dfrac{1}{\Delta x} - \dfrac{\Delta x}{6} & 0 \\ -\dfrac{1}{\Delta x} - \dfrac{\Delta x}{6} & \dfrac{2}{\Delta x} - \dfrac{2\Delta x}{3} & -\dfrac{1}{\Delta x} - \dfrac{\Delta x}{6} \\ 0 & -\dfrac{1}{\Delta x} - \dfrac{\Delta x}{6} & \dfrac{1}{\Delta x} - \dfrac{\Delta x}{3} \end{bmatrix},$$

and, for the given problem geometry, the global matrix system is given by

$$\begin{bmatrix} \dfrac{2}{3} & -\dfrac{7}{6} & 0 \\ -\dfrac{7}{6} & \dfrac{4}{3} & -\dfrac{7}{6} \\ 0 & -\dfrac{7}{6} & \dfrac{2}{3} \end{bmatrix} \begin{bmatrix} \alpha_1 \\ \alpha_2 \\ \alpha_3 \end{bmatrix} = \begin{bmatrix} 0 \\ 0 \\ 0 \end{bmatrix} + \begin{bmatrix} q_1 \\ 0 \\ q_3 \end{bmatrix},$$

where the corresponding flux densities are:

$$q_1 = -\frac{\partial H_z}{\partial x}\bigg|_{x=-\frac{h}{2}}, \quad q_3 = -\frac{\partial H_z}{\partial x}\bigg|_{x=\frac{h}{2}}.$$

Finally, inserting the actual boundary conditions gives

$$\alpha_1 = \alpha_3 = H_0,$$

and, solving the matrix system

$$-\frac{7}{6} H_0 + \frac{4}{3}\alpha_2 - \frac{7}{6} H_0 = 0,$$

the following value of magnetic field strength is obtained:

$$\alpha_2 = \frac{7}{4} H_0 = 210 \, \text{A/m}.$$

Now, the flux densities q_1 and q_3 can be determined from the first and third rows of the matrix equation.

2.2.2.5 Two-Dimensional FEM

The simplest discretization of a 2D domain can be performed using the so-called triangular elements; see Fig. 2.13. The shape functions are given by equations of planes in 3D space. An approximate solution is shown in Fig. 2.14, while the corresponding shape functions over a triangle are shown in Fig. 2.15. Fig. 2.16 shows the global functions assigned to the ith node, assembled from neighboring shape functions. Note that in the 1D case, global bases always consist of two neighboring shape functions only.

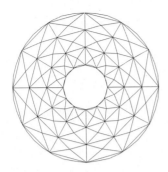

FIG. 2.13 Discretization of a 2D domain via triangular elements.

FIG. 2.16 Global base function assigned to the ith node.

$$N_2(x, y) = \frac{1}{2A} (2A_2 + b_2 x + a_2 y), \qquad (2.186\text{b})$$

$$N_3(x, y) = \frac{1}{2A} (2A_3 + b_3 x + a_3 y). \qquad (2.186\text{c})$$

Thus, the ith shape function can be written as

$$N_i(x, y) = \frac{1}{2A} (2A_i + b_i x + a_i y), \quad i = 1, 2, 3, \qquad (2.187)$$

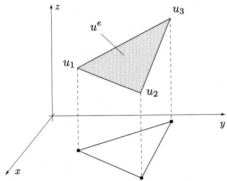

FIG. 2.14 Approximate solution for 2D problem.

where A denotes the area of a triangle,

$$2A = 2(A_1 + A_2 + A_3), \qquad (2.188)$$

According to Fig. 2.15, the solution on a triangular element is given by

$$u^e = \alpha_1^e N_1^e(x, y) + \alpha_2^e N_2^e(x, y) + \alpha_3^e N_3^e(x, y), \quad (2.185)$$

where N_1, N_2, N_3 are 2D shape functions determined by:

$$N_1(x, y) = \frac{1}{2A} (2A_1 + b_1 x + a_1 y), \qquad (2.186\text{a})$$

and A_1–A_3, a_1–a_3 and b_1–b_3 are auxiliary variables:

$$2A_1 = x_2 y_3 - x_3 y_2, \quad a_1 = x_3 - x_2, \quad b_1 = y_2 - y_3,$$
$$(2.189\text{a})$$

$$2A_2 = x_3 y_1 - x_1 y_3, \quad a_2 = x_1 - x_3, \quad b_2 = y_3 - y_1,$$
$$(2.189\text{b})$$

$$2A_3 = x_1 y_2 - x_2 y_1, \quad a_3 = x_2 - x_1, \quad b_3 = y_1 - y_2,$$
$$(2.189\text{c})$$

or it can be simply written as

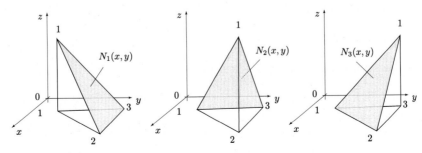

FIG. 2.15 Shape functions over a triangle.

$$2A_i = x_j y_k - x_k y_j, \quad a_i = x_k - x_j,$$
$$b_i = y_j - y_k, \quad i = 1, 2, 3; \; j = 2, 3, 1; \; k = 3, 1, 2. \tag{2.190}$$

Combining relations (2.185)–(2.186c), the solution on a triangle is

$$u^e = \frac{1}{2A} \sum_{i=1}^{3} (2A_i + b_i x + a_i y) \alpha_i, \tag{2.191}$$

where $n_e = 3$.

2.2.2.6 The Weak Formulation for Generalized Helmholtz Equation

Many problems in science and engineering can be formulated via the generalized Helmholtz equation. As in the 1D case, FEM is implemented through the weak formulation of the problem.

The generalized inhomogeneous Helmholtz equation can be written in the following form:

$$\nabla(k\nabla u) + ru = p, \tag{2.192}$$

where u is the unknown solution, k and r depend on the material properties, while p represents the sources inside the domain of interest.

Applying the weighted residual approach yields

$$\int_{\Omega} [\nabla(k\nabla u) + ru - p] \, W_j \, d\Omega = 0, \tag{2.193}$$

i.e., one obtains

$$\int_{\Omega} \nabla(k\nabla u) \, W_j \, d\Omega + \int_{\Omega} ru W_j \, d\Omega - \int_{\Omega} p \, W_j \, d\Omega = 0. \tag{2.194}$$

Applying the simple differentiation rule,

$$\nabla(k\nabla u) W_j = \nabla \left[(k\nabla u) W_j \right] - (k\nabla u) \nabla W_j, \tag{2.195}$$

and generalized Gauss integral theorem,

$$\int_{\Omega} \nabla \vec{A} \, d\Omega = \oint_{\Gamma} \vec{A} \cdot \hat{n} \, d\Gamma, \tag{2.196}$$

the weak formulation of the Helmholtz equation (2.192) is obtained as

$$\int_{\Omega} \left[(k\nabla u)\nabla W_j - ru W_j \right] d\Omega$$
$$= \int_{\Gamma} k \frac{\partial u}{\partial n} W_j \, d\Gamma - \int_{\Omega} p \, W_j \, d\Omega. \tag{2.197}$$

Expression (2.197) is usually referred to as the *variational equation* [6].

The term on the left-hand side gives rise to the finite element matrix while the first term on the right-hand side is the flux through the part of the domain boundary in which the Neumann boundary condition (flux density) is prescribed. The second term on the right-hand side contains the known sources in the domain if such sources exist.

Applying the finite element algorithm, the unknown solution over an element is expressed in terms of a linear combination of shape functions.

In the matrix form, the approximate solution can be written as follows:

$$u^e = \{N\}^T \{\alpha\} = \begin{bmatrix} N_1 & N_2 & N_3 \end{bmatrix} \begin{bmatrix} \alpha_1 \\ \alpha_2 \\ \alpha_3 \end{bmatrix}, \tag{2.198}$$

where $\{\alpha\}$ denotes the unknown solution coefficients.

The gradient of scalar function u in 2D is simply determined by the relation

$$\nabla u = \frac{\partial u}{\partial x} \hat{e}_x + \frac{\partial u}{\partial y} \hat{e}_y. \tag{2.199}$$

Inserting (2.198) into (2.199) yields

$$\nabla u = \begin{bmatrix} \dfrac{\partial u}{\partial x} \\ \dfrac{\partial u}{\partial y} \end{bmatrix} = \begin{bmatrix} \dfrac{\partial N_1}{\partial x} & \dfrac{\partial N_2}{\partial x} & \dfrac{\partial N_3}{\partial x} \\ \dfrac{\partial N_1}{\partial y} & \dfrac{\partial N_2}{\partial y} & \dfrac{\partial N_3}{\partial y} \end{bmatrix} \begin{bmatrix} \alpha_1 \\ \alpha_2 \\ \alpha_3 \end{bmatrix}. \tag{2.200}$$

Implementation of the Galerkin–Bubnov procedure $(W_j = N_j)$ leads to the following finite element matrix:

$$[a]^e = k_e \int_{\Omega} \begin{bmatrix} \dfrac{\partial N_1}{\partial x} & \dfrac{\partial N_1}{\partial y} \\ \dfrac{\partial N_2}{\partial x} & \dfrac{\partial N_2}{\partial y} \\ \dfrac{\partial N_3}{\partial x} & \dfrac{\partial N_3}{\partial y} \end{bmatrix} \begin{bmatrix} \dfrac{\partial N_1}{\partial x} & \dfrac{\partial N_2}{\partial x} & \dfrac{\partial N_3}{\partial x} \\ \dfrac{\partial N_1}{\partial y} & \dfrac{\partial N_2}{\partial y} & \dfrac{\partial N_3}{\partial y} \end{bmatrix} d\Omega$$
$$- r_e \int_{\Omega} \begin{bmatrix} N_1 \\ N_2 \\ N_3 \end{bmatrix} \begin{bmatrix} N_3 & N_2 & N_3 \end{bmatrix} d\Omega. \tag{2.201}$$

Note that it is necessary to discretize the domain into sufficiently small elements thus ensuring the constant quantities over an element. Otherwise, parameters k and r become spatially dependent, which increases the complexity of integration.

The derivatives of shape functions are simply given by:

$$\frac{\partial N_i}{\partial x} = \frac{b_i}{2A}, \quad \frac{\partial N_i}{\partial y} = \frac{a_i}{2A}. \tag{2.202}$$

Performing certain mathematical manipulations yields

$$[a]^e = \frac{k_e}{4A} \begin{bmatrix} a_1^2 + b_1^2 & a_1 a_2 + b_1 b_2 & a_1 a_3 + b_1 b_3 \\ a_2 a_1 + b_2 b_1 & a_2^2 + b_2^2 & a_2 a_3 + b_2 b_3 \\ a_3 a_1 + b_3 b_1 & a_3 a_2 + b_3 b_2 & a_3^2 + b_3^2 \end{bmatrix}$$
$$- r_e \frac{A}{12} \begin{bmatrix} 2 & 1 & 1 \\ 1 & 2 & 1 \\ 1 & 1 & 2 \end{bmatrix}. \tag{2.203}$$

The related global matrix is obtained by assembling the contributions from the local ones.

2.2.2.7 Computation of Fluxes on the Domain Boundary

When determining the entire solution for a scalar potential (all coefficients α_i are known), it is possible to compute Q_i, which stem from the potentials. The total flux Q on the part of the domain boundary is defined by the integral

$$Q = \int_\Gamma k \frac{\partial u}{\partial n} N_j \, d\Gamma. \tag{2.204}$$

Quantity $q = k \partial u / \partial n$ represents the flux density, i.e., the Neumann or natural boundary condition.

On the finite element located on a part of the boundary, the flux can be expressed by the integral

$$Q_e = \int_{\Gamma_e} q_e N_j \, d\Gamma. \tag{2.205}$$

The flux density q generally varies, but can be assumed constant over an element, provided the element is sufficiently small. Usually, the value of the flux density on the center of the element is taken as the average flux value for the whole element. Note that within the finite element algorithm, the flux Q represents the concentrated value of the flux assigned to the node.

The flux on a finite element located on a part of the boundary can be written as

$$\{Q\}^e = \int_{\Delta\Gamma} q_e \begin{bmatrix} N_1 \\ N_2 \end{bmatrix} d\Gamma. \tag{2.206}$$

On the other hand, the concentrated flux on the ith node consists of contributions from neighboring (adjacent) elements, as indicated in Fig. 2.17, is given by the expression

$$Q_i = \int_{\Delta\Gamma_1} q_1 N_2^{\Gamma_1} \, d\Gamma + \int_{\Delta\Gamma_2} q_2 N_1^{\Gamma_2} \, d\Gamma. \tag{2.207}$$

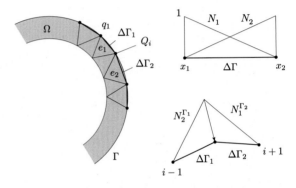

FIG. 2.17 Elements on the boundary along which the flux density q is distributed, while the flux Q is concentrated in nodes.

Assuming constant densities q_1 and q_2, one obtains

$$Q_i = q_1 \frac{\Delta\Gamma_1}{2} + q_2 \frac{\Delta\Gamma_2}{2}. \tag{2.208}$$

For the case when $\Delta\Gamma_1 = \Delta\Gamma_2 = \Delta\Gamma$, Eq. (2.208) simplifies into

$$Q_i = q_1 \frac{\Delta\Gamma}{2} + q_2 \frac{\Delta\Gamma}{2} = \frac{\Delta\Gamma}{2}(q_1 + q_2). \tag{2.209}$$

For example, if the boundary consists of 3 finite elements with constant flux density, i.e., $q_1 = q_2 = q_3 = q_0$, the contributions in nodes are as follows:

$$Q_1 = q_0 \frac{\Delta\Gamma}{2}, \quad Q_2 = q_0 \Delta\Gamma, \quad Q_3 = q_0 \Delta\Gamma,$$
$$Q_4 = q_0 \frac{\Delta\Gamma}{2}. \tag{2.210}$$

Therefore, only the first and last contributions represent half the value of the other nodes.

2.2.2.8 Computation of Sources on a Finite Element

Contribution of the source on a 2D finite element requires the evaluation of the integral

$$\{p\}^e = \int\limits_{\Omega_e} p_e \{N\} \, d\Omega = \int\limits_{\Omega_e} p_e \begin{Bmatrix} N_1 \\ N_2 \\ N_3 \end{Bmatrix} d\Omega, \qquad (2.211)$$

where p denotes the source density inside the domain Ω.

Again, performing sufficiently fine domain discretization, the constant value of the source density over a triangle can be assumed. According to the Galerkin procedure where $W_j = N_j$, the right-hand side is then

$$\{p\}^e = p_e \int\limits_{\Omega_e} \{N\} \, d\Omega. \qquad (2.212)$$

The integral of the shape functions over a triangular element is then

$$\int\limits_{\Omega_e} N_i \, d\Omega \quad \Rightarrow \quad \int\limits_{\Omega_e} N_i(x, y) \, dx \, dy, \qquad (2.213)$$

corresponding to the volume of the pyramid with the height equal to 1 and the area of the base (triangle) being A, as depicted in Fig. 2.18.

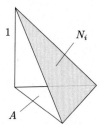

FIG. 2.18 Integral of shape function over a finite element.

The right-hand side is now given by the local vector

$$\{p\}^e = \frac{1}{3} p_e \begin{Bmatrix} 1 \\ 1 \\ 1 \end{Bmatrix} A. \qquad (2.214)$$

The assembling of the global system is carried out by using standard FEM algorithm.

2.2.2.9 Three-Dimensional Elements

Similarly as a 2D finite element is constructed from a 1D element, a 3D finite element can be obtained from a 2D element. Thus, by triangle expansion, a three-sided tetrahedral element is obtained, as the simplest 3D element, shown in Fig. 2.19 [20,21].

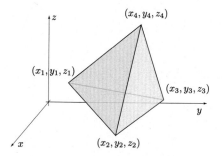

FIG. 2.19 Four-node tetrahedral element.

Shape functions can be derived through the same procedure as in the case of 2D triangle elements. The solution on the tetrahedral element is approximated linearly as

$$\tilde{u}^e = a_0' + a_1'x + a_2'y + a_3'z, \quad (x, y, z) \in e. \qquad (2.215)$$

Elements a_0', a_1', a_2' and a_3' are determined from the criterion of function collocation on the vertices of tetrahedron, resulting in the following linear equation system:

$$\begin{aligned} \tilde{u}_1 &= a_0' + a_1'x_1 + a_2'y_1 + a_3'z_1, \\ \tilde{u}_2 &= a_0' + a_1'x_2 + a_2'y_2 + a_3'z_2, \\ \tilde{u}_3 &= a_0' + a_1'x_3 + a_2'y_3 + a_3'z_3, \\ \tilde{u}_4 &= a_0' + a_1'x_4 + a_2'y_4 + a_3'z_4. \end{aligned} \qquad (2.216)$$

Cramer's rule yields

$$\tilde{u}^e = \frac{1}{D} \left[D_0 + D_1 x + D_2 y + D_3 z \right], \qquad (2.217)$$

where

$$D = \begin{vmatrix} 1 & x_1 & y_1 & z_1 \\ 1 & x_2 & y_2 & z_2 \\ 1 & x_3 & y_3 & z_3 \\ 1 & x_4 & y_4 & z_4 \end{vmatrix}, \qquad (2.218)$$

$$D_0 = \tilde{u}_1 \underbrace{\begin{vmatrix} x_2 & x_3 & x_4 \\ y_2 & y_3 & y_4 \\ z_2 & z_3 & z_4 \end{vmatrix}}_{V_1} - \tilde{u}_2 \underbrace{\begin{vmatrix} x_1 & x_3 & x_4 \\ y_1 & y_3 & y_4 \\ z_1 & z_3 & z_4 \end{vmatrix}}_{-V_2}$$

$$+ \tilde{u}_3 \underbrace{\begin{vmatrix} x_1 & x_2 & x_4 \\ y_1 & y_2 & y_4 \\ z_1 & z_2 & z_4 \end{vmatrix}}_{V_3} - \tilde{u}_4 \underbrace{\begin{vmatrix} x_1 & x_2 & x_3 \\ y_1 & y_2 & y_3 \\ z_1 & z_2 & z_3 \end{vmatrix}}_{-V_4}, \quad (2.219a)$$

$$D_1 = - \tilde{u}_1 \underbrace{\begin{vmatrix} 1 & y_2 & z_2 \\ 1 & y_3 & z_3 \\ 1 & y_4 & z_4 \end{vmatrix}}_{-a_1} + \tilde{u}_2 \underbrace{\begin{vmatrix} 1 & y_1 & z_1 \\ 1 & y_3 & z_3 \\ 1 & y_4 & z_4 \end{vmatrix}}_{a_2}$$

$$- \tilde{u}_3 \underbrace{\begin{vmatrix} 1 & y_1 & z_1 \\ 1 & y_2 & z_2 \\ 1 & y_4 & z_4 \end{vmatrix}}_{-a_3} + \tilde{u}_4 \underbrace{\begin{vmatrix} 1 & y_1 & z_1 \\ 1 & y_2 & z_2 \\ 1 & y_3 & z_3 \end{vmatrix}}_{a_4}, \quad (2.219b)$$

$$D_2 = \tilde{u}_1 \underbrace{\begin{vmatrix} 1 & x_2 & z_2 \\ 1 & x_3 & z_3 \\ 1 & x_4 & z_4 \end{vmatrix}}_{b_1} - \tilde{u}_2 \underbrace{\begin{vmatrix} 1 & x_1 & z_1 \\ 1 & x_3 & z_3 \\ 1 & x_4 & z_4 \end{vmatrix}}_{-b_2}$$

$$+ \tilde{u}_3 \underbrace{\begin{vmatrix} 1 & x_1 & z_1 \\ 1 & x_2 & z_2 \\ 1 & x_4 & z_4 \end{vmatrix}}_{b_3} - \tilde{u}_4 \underbrace{\begin{vmatrix} 1 & x_1 & z_1 \\ 1 & x_2 & z_2 \\ 1 & x_3 & z_3 \end{vmatrix}}_{-b_4}, \quad (2.219c)$$

$$D_3 = - \tilde{u}_1 \underbrace{\begin{vmatrix} 1 & x_2 & y_2 \\ 1 & x_3 & y_3 \\ 1 & x_4 & y_4 \end{vmatrix}}_{-c_1} + \tilde{u}_2 \underbrace{\begin{vmatrix} 1 & x_1 & y_1 \\ 1 & x_3 & y_3 \\ 1 & x_4 & y_4 \end{vmatrix}}_{c_2}$$

$$- \tilde{u}_3 \underbrace{\begin{vmatrix} 1 & x_1 & y_1 \\ 1 & x_2 & y_2 \\ 1 & x_4 & y_4 \end{vmatrix}}_{-c_3} + \tilde{u}_4 \underbrace{\begin{vmatrix} 1 & x_1 & y_1 \\ 1 & x_2 & y_2 \\ 1 & x_3 & y_3 \end{vmatrix}}_{c_4}. \quad (2.219d)$$

Interpolating expressions (2.218)–(2.219d) into (2.217) yields the solution over the considered element:

$$\tilde{u}^e = \frac{1}{D} \left[\sum_{i=1}^{4} (V_i + a_i x + b_i y + c_i z) \tilde{u}_i \right], \quad (2.220)$$

where the shape functions are

$$N_i(x, y, z) = \frac{1}{D} (V_i + a_i x + b_i y + c_i z), \quad i = 1, 2, 3, 4. \quad (2.221)$$

Finally, the solution on the element is expressed as follows:

$$\tilde{u}^e = \sum_{i=1}^{4} \tilde{u}_i N_i. \quad (2.222)$$

The derivatives of the shape functions are simply given by:

$$\frac{\partial N_i}{\partial x} = \frac{a_i}{D}, \quad \frac{\partial N_i}{\partial y} = \frac{b_i}{D}, \quad \frac{\partial N_i}{\partial z} = \frac{c_i}{D}, \quad i = 1, 2, 3, 4. \quad (2.223)$$

Note that a consistent order of numbering should be used, often in anti-clockwise direction, as it is indicated in Fig. 2.19.

2.2.3 The Boundary Element Method (BEM)

Since the middle of the 1980s, the boundary element method (BEM) has been used to model a variety of problems in electromagnetics. The basic idea of BEM is to discretize the integral equation using boundary elements [6]. BEM can be regarded as a combination of classical boundary integral equation method and the discretization concepts originated from FEM.

2.2.3.1 Integral Equation Formulation

The first step in solving a problem via BEM is deriving the integral equation formulation of the differential equation governing the problem.

The governing differential equation for the static field problem, either electrostatic or magnetostatic, for source-free domains is defined by the Laplace equation

$$\nabla^2 u = 0, \quad (2.224)$$

or the Poisson equation if sources p are present within the domain

$$\nabla^2 u = -p. \quad (2.225)$$

A typical calculation domain Ω with the related boundary Γ is shown in Fig. 2.20A, where $\hat{n}$ is the external normal vector to the boundary, and R denotes the distance from the source to the observation point.

It is worth mentioning that the observation point P can be also located on the boundary itself, as shown in Fig. 2.20B.

The boundary conditions associated with these problems can be divided into essential condition (Dirichlet), where $u = u|_\Gamma$, defined on Γ_1, and natural condition

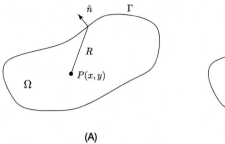

(A)   (B)

FIG. 2.20 The geometry of the problem.

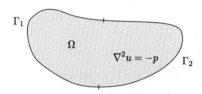

FIG. 2.21 Calculation domain with boundary conditions.

(Neumann), $\frac{\partial u}{\partial n}|_\Gamma = q$, defined on Γ_2, as shown in Fig. 2.21. The total boundary is then given by $\Gamma_1 \cup \Gamma_2$.

For simplicity, the case of Laplace equation (2.224) is considered first, and afterwards the procedure can be extended to the solution of Poisson equation (2.225).

Applying the weighted residual approach, Eq. (2.224) can be integrated over the calculation domain Ω:

$$\int_\Omega \nabla^2 u \cdot W \, d\Omega = 0, \qquad (2.226)$$

where W is the weighting function.

Performing some mathematical manipulations and applying the generalized Gauss theorem, Eq. (2.226) becomes

$$\int_\Omega W \cdot \nabla^2 u \, d\Omega = \int_\Gamma W \frac{\partial u}{\partial n} \, d\Gamma - \int_\Gamma u \frac{\partial W}{\partial n} \, d\Gamma$$
$$+ \int_\Omega u \cdot \nabla^2 W \, d\Omega, \qquad (2.227)$$

which can be also rewritten as follows:

$$\int_\Gamma W \frac{\partial u}{\partial n} \, d\Gamma - \int_\Gamma u \frac{\partial W}{\partial n} \, d\Gamma + \int_\Omega \nabla^2 W \cdot u \, d\Omega = 0.$$
$$(2.228)$$

The weighting function W can be chosen to be the solution of the differential equation, i.e.,

$$\nabla^2 W - \delta(\vec{r} - \vec{r}\,') = 0, \qquad (2.229)$$

where δ is the Dirac delta function, $\vec{r}$ denotes the observation points, and $\vec{r}\,'$ denotes the source points.

The solution of (2.229) represents the fundamental solution, or Green function.

Thus, the domain integral in (2.228) becomes

$$\int_\Omega u \cdot \nabla^2 W \, d\Omega = - \int_\Omega u \, \delta(\vec{r} - \vec{r}\,') \, d\Omega = -u_i, \quad (2.230)$$

and, combining Eqs. (2.228)–(2.230), the following integral relation is obtained:

$$u_i = \int_\Gamma \psi \frac{\partial u}{\partial n} \, d\Gamma - \int_\Gamma u \frac{\partial \psi}{\partial n} \, d\Gamma. \qquad (2.231)$$

The integral expression (2.231) is the Green representation of function u, where W can be replaced by function ψ.

Function ψ is the fundamental solution of (2.229). For two-dimensional problems, it is given by

$$\psi = -\frac{1}{2\pi} \ln R, \qquad (2.232)$$

while for the three-dimensional case it is

$$\psi = \frac{1}{4\pi R}, \qquad (2.233)$$

where $R = |\vec{r} - \vec{r}\,'|$ denotes the distance from the source point (boundary point) to the observation point.

The corresponding Green integral representation of Poisson equation (2.225) can be obtained starting from the weighted residual integral

$$\int_{\Omega} \left[\nabla^2 u + p \right] \cdot W \, d\Omega = 0 \qquad (2.234)$$

and performing similar mathematical manipulations, i.e.,

$$u_i = \int_{\Gamma} \psi \frac{\partial u}{\partial n} \, d\Gamma - \int_{\Gamma} u \frac{\partial \psi}{\partial n} \, d\Gamma - \int_{\Omega} p \psi \, d\Omega. \qquad (2.235)$$

When the observation point P is located on the boundary Γ, the boundary integral becomes singular as R approaches zero. Performing certain procedures to extract the singularity, relation (2.235) can be written as

$$c_i u_i = \int_{\Gamma} \psi \frac{\partial u}{\partial n} d\Gamma - \int_{\Gamma} u \frac{\partial \psi}{\partial n} d\Gamma + \int_{\Omega} p \psi d\Omega, \qquad (2.236)$$

where

$$c_i = \begin{cases} 1, & i \in \Omega, \\ 1 - \dfrac{\theta_2 - \theta_1}{2\pi}, & i \in \Gamma, \\ 0, & i \notin \Omega. \end{cases} \qquad (2.237)$$

For a well-posed static field problem, either u or $\frac{\partial u}{\partial n}$ on the boundary Γ must be known, which is described by the forced (Dirichlet), natural (Neumann) or mixed (Cauchy) boundary condition.

Knowing all values of potential u and its normal derivative $\frac{\partial u}{\partial n}$ on the boundary, the potential at an arbitrary point of the domain can be calculated.

2.2.3.2 Boundary Element Discretization
The boundary can be discretized into a series of constant, linear or quadratic elements; see Fig. 2.22.

2.2.3.3 Constant Boundary Elements
The simplest solution can be obtained by using constant boundary elements. The geometry of the constant boundary element for two-dimensional problems is shown in Fig. 2.23.

The next step in the boundary element solution procedure is the transformation of the global coordinates into the local ones; see Fig. 2.24.

This transformation of coordinates is given by the following set of $x = x(\xi)$ and $y = y(\xi)$:

$$x(\xi) = \frac{x_1 - x_2}{2} \xi + \frac{x_1 + x_2}{2}, \qquad (2.238a)$$

$$y(\xi) = \frac{y_1 - y_2}{2} \xi + \frac{y_1 + y_2}{2}, \qquad (2.238b)$$

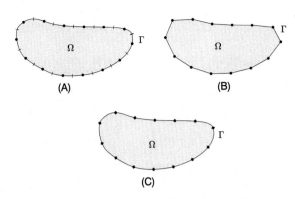

FIG. 2.22 (A) Constant element approximation, (B) Linear element approximation, (C) Quadratic element approximation.

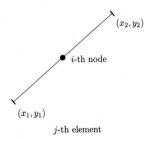

FIG. 2.23 Constant boundary element approximation.

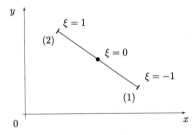

FIG. 2.24 Global and local coordinates.

where (x_1, y_1) and (x_2, y_2) are the global coordinates of the element.

In addition, it follows that

$$d\Gamma = \sqrt{dx^2 + dy^2} = \sqrt{\left(\frac{dx}{d\xi}\right)^2 + \left(\frac{dy}{d\xi}\right)^2} \, d\xi = \frac{\Delta\Gamma}{2} d\xi, \qquad (2.239)$$

where $\Delta\Gamma$ is the segment length defined by

$$\Delta\Gamma = \sqrt{(x_2 - x_1)^2 + (y_2 - y_1)^2}. \qquad (2.240)$$

Using the constant boundary element approximation, the integral equation formulation becomes

$$c_i u_i = \sum_{j=1}^{M} \left[Q_j \int_{\Gamma_j} \psi \, d\Gamma - U_j \int_{\Gamma_j} \frac{\partial \psi}{\partial n} d\Gamma \right] + \int_{\Omega_S} p \psi \, d\Omega, \tag{2.241}$$

where i denotes the ith boundary node and j stands for the jth, and p is the constant value of the source on the segment of the domain containing sources.

The resulting algebraic equation system is

$$P_i + \sum_{j=1}^{M} H_{ij} U_j = \sum_{j=1}^{M} Q_j G_{ij}, \tag{2.242}$$

or in the matrix form,

$$\{P\} + [H]\{U\} = [G]\{Q\}, \tag{2.243}$$

where

$$P_i = \int_{\Omega_S} p \psi \, d\Omega, \tag{2.244}$$

$$\frac{\partial u}{\partial n}\Big|_j = Q_j, \tag{2.245}$$

$$u = U_j, \tag{2.246}$$

$$G_{ij} = \int_{\Gamma_j} \psi_{ij} \, d\Gamma$$

$$= \begin{cases} \dfrac{1}{2\pi} \displaystyle\int_{\Gamma_j} \ln\left(\dfrac{1}{R_{ij}}\right) d\Gamma, & \text{for 2D problems,} \\[18pt] \displaystyle\int_{\Gamma_j} \dfrac{1}{4\pi R_{ij}} \, d\Gamma, & \text{for 3D problems,} \end{cases} \tag{2.247}$$

$$H_{ij} = \int_{\Gamma_j} \frac{\partial \psi_{ij}}{\partial n} \, d\Gamma$$

$$= \begin{cases} \dfrac{1}{2\pi} \displaystyle\int_{\Gamma_j} \dfrac{\partial}{\partial n} \ln\left(\dfrac{1}{R_{ij}}\right) d\Gamma, & \text{for 2D problems,} \\[18pt] \dfrac{1}{4\pi} \displaystyle\int_{\Gamma_j} \dfrac{\partial}{\partial n}\left(\dfrac{1}{R_{ij}}\right) d\Gamma, & \text{for 3D problems.} \end{cases} \tag{2.248}$$

The matrix system (2.243) can be solved once the set of boundary conditions is prescribed. If the domain of interest contains unknown sources, then a coupling of

BEM with some domain discretization method, such as FEM, is required, which leads to hybrid methods.

2.2.3.4 Linear and Quadratic Elements

A higher accuracy and faster convergence can be achieved by applying linear or quadratic elements. Note that the geometry of the elements is also modeled by means or quadratic functions. Such elements are then referred to as isoparametric elements [6]. When using isoparametric elements, the global coordinate x is a function of the local parametric coordinate ξ on the element.

Function $x(\xi)$ can be written as

$$x = \sum_{i=1}^{N} x_i f_i(\xi), \tag{2.249}$$

where approximating functions $f_i(\xi)$ are usually polynomials.

Furthermore, the unknowns along elements are interpolated as follows:

$$u = \sum_{j=1}^{n_e} f_j(\xi) U_j, \tag{2.250}$$

$$\frac{\partial u}{\partial n} = \sum_{j=1}^{n_e} f_j(\xi) \frac{\partial u}{\partial n}\Big|_j = \sum_{j=1}^{n_e} f_j(\xi) Q_j, \tag{2.251}$$

where U_j denotes the unknown coefficients of the potential distribution, and Q_j is the value of the normal derivative at the jth node.

Hence, for a linear approximation, it follows that

$$u = f_1(\xi) U_1 + f_2(\xi) U_2, \tag{2.252}$$

$$\frac{\partial u}{\partial n} = f_1(\xi) Q_1 + f_2(\xi) Q_2, \tag{2.253}$$

where U_1, U_2, Q_1, Q_2 are the values of the vector potential and its normal derivative on the nodes $j = 1$ and $j = 2$, respectively.

The linear shape functions are given by:

$$f_1(\xi) = \frac{1}{2}(1 - \xi), \tag{2.254}$$

$$f_1(\xi) = \frac{1}{2}(1 + \xi). \tag{2.255}$$

For linear elements (see Fig. 2.25), the geometry is a linear function of the coordinates, i.e.,

$$x = f_1(\xi) x_1 + f_2(\xi) x_2, \tag{2.256}$$
$$y = f_1(\xi) y_1 + f_2(\xi) y_2. \tag{2.257}$$

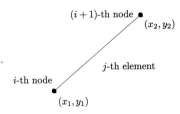

FIG. 2.25 Linear boundary element approximation.

For a quadratic interpolation, it follows that

$$u = f_1(\xi)U_1 + f_2(\xi)U_2 + f_3(\xi)U_3, \qquad (2.258)$$

$$\frac{\partial u}{\partial n} = f_1(\xi)Q_1 + f_1(\xi)Q_2 + f_3(\xi)Q_3, \qquad (2.259)$$

where the shape functions are defined as:

$$f_1(\xi) = \frac{1}{2}\xi(\xi - 1), \qquad (2.260)$$

$$f_2(\xi) = \frac{1}{2}(1 + \xi)(1 - \xi), \qquad (2.261)$$

$$f_3(\xi) = \frac{1}{2}\xi(\xi + 1), \qquad (2.262)$$

and U_j and Q_j are the values of the vector potential and its normal derivative at the given node j, respectively.

For the case of quadratic elements (see Fig. 2.26), the geometry is represented by following functions:

$$x = f_1(\xi)x_1 + f_2(\xi)x_2 + f_3(\xi)x_3, \qquad (2.263)$$

$$y = f_1(\xi)y_1 + f_2(\xi)y_2 + f_3(\xi)y_3. \qquad (2.264)$$

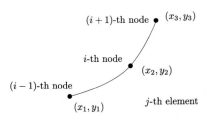

FIG. 2.26 Quadratic boundary element approximation.

The resulting matrix equation is

$$\{P\} + [H]\{U\} = [G]\{Q\}, \qquad (2.265)$$

while the corresponding coefficients are now given by

$$G_{ij} = \int_{\Gamma_j} f_j \psi_{ij} \, d\Gamma = \int_{\Gamma_j} f_j \psi_{ij} \frac{d\Gamma}{d\xi} \, d\xi, \qquad (2.266)$$

$$H_{ij} = \int_{\Gamma_j} f_j \frac{\partial \psi_{ij}}{\partial n} \, d\Gamma = \int_{\Gamma_j} f_j \frac{\partial \psi_{ij}}{\partial n} \frac{d\Gamma}{d\xi} \, d\xi, \qquad (2.267)$$

where $\{f\}$ denotes the corresponding vector of linear or quadratic shape functions.

The BEM procedures presented so far pertain to the solution of two-dimensional potential problems. If three-dimensional problems are analyzed, then triangular or quadrilateral surface elements have to be applied [19].

2.2.4 Numerical Solution of Integral Equations Over Unknown Sources

The solution of integral equations over unknown sources (indirect boundary element approach) is illustrated on the treatment of Pocklington integro-differential equation (2.131). More mathematical details on the subject can be found elsewhere, e.g., in [6,19]. For simplicity, the integral equation (2.131) can be written in an operator form

$$\mathcal{K}(I) = E, \qquad (2.268)$$

where $\mathcal{K}$ is a linear operator, I is the unknown current to be found for a given excitation E.

The Galerkin–Bubnov Indirect Boundary Element Method (GB-IBEM) of solution starts by expanding the unknown current $I(x)$ into a finite sum of linearly independent basis functions $\{f_i\}$ with unknown complex coefficients α_i, i.e.,

$$I_n(x') = \sum_{n=1}^{N_g} I_n f_n(x'), \qquad (2.269)$$

where $f_n(x)$ is the linear elements shape function, and I_n stands for the unknown coefficients of the solution, N_g denotes the total number of basis functions. Substituting (2.269) into (2.268) yields

$$\mathcal{K}(I) = \sum_{n=1}^{N_g} I_n \mathcal{K}(f_n). \qquad (2.270)$$

The residual $\mathcal{R}$ is given by

$$\mathcal{R} = \sum_{n=1}^{N_g} I_n \mathcal{K}(f_n) - E. \qquad (2.271)$$

According to the definition of the scalar product of functions in a Hilbert function space, the error $\mathcal{R}$ is weighted to zero with respect to certain weighting functions $\{W_j\}$, i.e.,

$$\int_L \mathcal{R} W_m^* \, dx = 0, \quad m = 1, 2, \ldots, N_g, \qquad (2.272)$$

where (*) means complex conjugate.

As the operator $\mathcal{K}$ is linear, performing some mathematical manipulation and by choosing $W_m = f_m$ (the Galerkin–Bubnov procedure), the system of algebraic equations is obtained:

$$\sum_{n=1}^{N_g} \int_L \mathcal{K}(f_n) \, f_m \, dx = \int_L E \, f_m \, dx, \quad m = 1, 2, \ldots, N_g. \qquad (2.273)$$

Featuring the weak formulation by carefully performing integration by parts yields

$$\int_{-L}^{L} \int_{-L}^{L} \frac{df_m(x)}{dx} \frac{df_n(x')}{dx'} g_0(x, x') \, dx' dx$$

$$+ k^2 \int_{-L}^{L} \int_{-L}^{L} f_m(x) \, f_n(x') \, g_0(x, x') \, dx' dx$$

$$= -j4\pi\omega\varepsilon \int_{-L}^{L} E_x^{inc}(x) \, f_m(x) \, dx, \quad m = 1, 2, \ldots, N_g. \qquad (2.274)$$

Eq. (2.274) represents the weak Galerkin–Bubnov formulation of the integral equation (2.131). The second order differential operator is replaced by trivial differentiation over basis and weight functions which must be chosen from the class of once differentiable functions.

This formulation is convenient for implementation of GB-IBEM, and boundary conditions are subsequently incorporated into the global matrix of the linear equation system.

Applying the GB-IBEM algorithm and discretizing the wire to segments yields the global system of equations assembled from local ones in a manner similar to finite element step-by-step procedures:

$$\sum_{i=1}^{M} [Z]_{ji} \, \{I\}_i = \{V\}_j, \quad j = 1, 2, \ldots, M, \qquad (2.275)$$

where M is the total number of segments, $[Z]_{ji}$ is the mutual impedance matrix representing the interaction

of the ith source to the j the observation segment, and $\{V\}_j$ is the right-side voltage vector for the jth observation segment.

As functions $f(x)$ are required to be once differentiable, a reasonable choice for the shape functions over the segments is the family of Lagrange's polynomials

$$L_i(x) = \prod_{j=1}^{m} \frac{x - x_j}{x_i - x_j}, \quad j \neq i, \qquad (2.276)$$

and one has

$$f_1(x) = \frac{x_2 - x}{\Delta x}, \qquad (2.277a)$$

$$f_2(x) = \frac{x - x_1}{\Delta x}, \qquad (2.277b)$$

where x_1 and x_2 are the coordinates of the segment nodes and $\Delta x = x_2 - x_1$ is the segment length.

Now $[Z]_{ji}$ and $\{V\}_j$ are given by

$$[Z]_{ji} = \int_{\Delta l_j} \int_{\Delta l_i} \begin{bmatrix} \dfrac{df_1(x)}{dx} \dfrac{df_1(x')}{dx'} & \dfrac{df_1(x)}{dx} \dfrac{df_2(x')}{dx'} \\[2mm] \dfrac{df_2(x)}{dx} \dfrac{df_1(x')}{dx'} & \dfrac{df_2(x)}{dx} \dfrac{df_2(x')}{dx'} \end{bmatrix}$$

$$\times g_0(x, x') \, dx' dx$$

$$+ k^2 \int_{\Delta l_j} \int_{\Delta l_i} \begin{bmatrix} f_1(x) f_1(x') & f_1(x) f_2(x') \\ f_2(x) f_1(x') & f_2(x) f_2(x') \end{bmatrix}$$

$$\times g_0(x, x') \, dx' dx \qquad (2.278)$$

$$= \frac{1}{\Delta x^2} \frac{df_1(x')}{dx'} \int_{x_1}^{x_2} \int_{x_1}^{x_2} \begin{bmatrix} 1 & -1 \\ -1 & 1 \end{bmatrix} g_0(x, x') \, dx' dx$$

$$+ \frac{k^2}{\Delta x^2} \int_{x_1}^{x_2} \int_{x_1}^{x_2} \begin{bmatrix} (x_2 - x)(x_2 - x') & (x_2 - x)(x' - x_1) \\ (x - x_1)(x_2 - x') & (x - x_1)(x' - x_1) \end{bmatrix}$$

$$\times g_0(x, x') \, dx' dx,$$

where $\{V\}_j$ represents the right-hand side voltage vector for the jth observation segment,

$$\{V\}_j = -j4\pi\omega\varepsilon \int_{\Delta l_j} E_x^{inc}(x) \begin{bmatrix} f_1(x) \\ f_2(x) \end{bmatrix} dx$$

$$= -\frac{j4\pi\omega\varepsilon}{\Delta x} \int_{x_1}^{x_2} E_x^{inc}(x) \begin{bmatrix} (x_2 - x) \\ (x - x_1) \end{bmatrix} dx, \qquad (2.279)$$

where Δl_i, Δl_j assign the widths of the ith and jth segments, respectively.

The evaluation of the right-hand side vector can be undertaken in closed form if the delta-function voltage generator is used (antenna mode), or the plane wave excitation (scatterer mode). In the radiation mode, the right-hand side vector is different from zero only in the feed gap area. The x-component of the impressed (incident) electric field is given by

$$E_x^{inc}(x) = \frac{V_g}{\Delta l_g}, \qquad (2.280)$$

where V_g is the feed voltage and $\Delta l_g = \Delta x$ (for convenience) is the feed-gap width.

Using the linear shape functions, we obtain

$$\{V\}_j = -\frac{j4\pi\omega\varepsilon}{\Delta l_g} \int_{x_1=-\frac{\Delta l_g}{2}}^{x_2=\frac{\Delta l_g}{2}} \frac{V_g}{\Delta l_g} \begin{bmatrix} (x_2 - x) \\ (x - x_1) \end{bmatrix} dx$$

$$= -j2\pi\omega\varepsilon V_g \begin{Bmatrix} 1 \\ 1 \end{Bmatrix}. \qquad (2.281)$$

If the scattering mode for the simple case of normal incidence is considered, the wire is illuminated by the plane wave, i.e.,

$$E_x^{inc}(x) = E_0, \qquad (2.282)$$

and then the right-hand side vector differs from zero on an each segment and the local voltage vector is

$$\{V\}_j = -\frac{j4\pi\omega\varepsilon}{\Delta x} \int_{x_1}^{x_2} E_0 \begin{bmatrix} (x_2 - x) \\ (x - x_1) \end{bmatrix} dx$$

$$= -j2\pi\omega\varepsilon E_0\Delta x \begin{Bmatrix} 1 \\ 1 \end{Bmatrix}. \qquad (2.283)$$

More mathematical details on the method could be found elsewhere, e.g., in [6].

A computational example deals with a classical radiation problem of a dipole antenna radiating in freespace [22] with several applications in computational electromagnetics (CEM).

The wire radius is $a = 2$ mm, operating frequency is $f = 300$ MHz and length is $L = 0.5, 1$, and 1.25 m which actually corresponds to $L = 0.5\lambda$, $L = \lambda$ and $L = 1.25\lambda$, where λ is the actual wavelength. The dipole is excited by the unit voltage at its center.

Fig. 2.27 shows the corresponding current distribution along the wire.

Once the current distribution along the dipole is obtained, all other parameters of interest could be calculated.

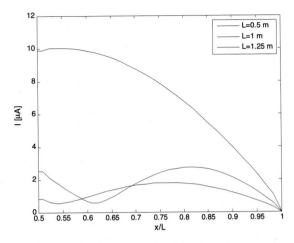

FIG. 2.27 Current distribution along the dipole antenna at $f = 300$ MHz.

REFERENCES

1. Dragan Poljak, Khalil E. Drissi, Computational Methods in Electromagnetic Compatibility: Antenna Theory Approach Versus Transmission Line Models, John Wiley & Sons, New York, 2018.
2. John W. Arthur, The evolution of Maxwell's equations from 1862 to the present day, IEEE Antennas and Propagation Magazine 55 (3) (2013) 61–81.
3. James Clerk Maxwell, VIII. A dynamical theory of the electromagnetic field, Philosophical Transactions of the Royal Society of London 155 (1865) 459–512.
4. William Band, Introduction to Mathematical Physics, Van Nostrand Company, New York, 1959.
5. Matthew N.O. Sadiku, Numerical Techniques in Electromagnetics, CRC Press, Boca Raton, 2000.
6. Dragan Poljak, Advanced Modeling in Computational Electromagnetic Compatibility, John Wiley & Sons, New York, 2007.
7. Edmund K. Miller, Jeremy A. Landt, Direct time-domain techniques for transient radiation and scattering from wires, Proceedings of the IEEE 68 (11) (1980) 1396–1423.
8. D. Poljak, On radiation mechanism and modeling of dipole antenna in classical electromagnetics: a tribute to first 120 years of the Pocklington integral equation, in: 2017 25th International Conference on Software, Telecommunications and Computer Networks, SoftCOM, 2017, pp. 1–5.
9. Simon Ramo, John R. Whinnery, Theodore Van Duzer, Fields and Waves in Communication Electronics, John Wiley & Sons, New York, 1994.
10. L. Rosenhead, Henry Cabourn Pocklington. 1870-1952, Obituary Notices of Fellows of the Royal Society 8 (22) (1953) 555–565.
11. K. Mei, On the integral equations of thin wire antennas, IEEE Transactions on Antennas and Propagation 13 (3) (1965) 374–378.

12. P. Silvester, K.K. Chan IET, Bubnov–Galerkin solutions to wire-antenna problems, Proceedings of the Institution of Electrical Engineers 119 (8) (1972) 1095–1099.

13. P. Silvester, K.K. Chan, IET, Analysis of antenna structures assembled from arbitrarily located straight wires, Proceedings of the Institution of Electrical Engineers 120 (1) (1973) 21–26.

14. Chalmers Butler, D. Wilton, Analysis of various numerical techniques applied to thin-wire scatterers, IEEE Transactions on Antennas and Propagation 23 (4) (1975) 534–540.

15. D. Wilton, C. Butler, Efficient numerical techniques for solving Pocklington's equation and their relationships to other methods, IEEE Transactions on Antennas and Propagation 24 (1) (1976) 83–86.

16. E.K. Miller, A.J. Poggio, G.J. Burke, E.S. Selden, Analysis of wire antennas in the presence of a conducting half-space. Part II. The horizontal antenna in free space, Canadian Journal of Physics 50 (21) (1972) 2614–2627.

17. T.K. Sarkar, Analysis of arbitrarily oriented thin wire antennas over a plane imperfect ground, Archiv Elektronik und Übertragungstechnik 31 (1977) 449–457.

18. Parviz Parhami, Raj Mittra, Wire antennas over a lossy half-space, IEEE Transactions on Antennas and Propagation 28 (3) (1980) 397–403.

19. P.P. Silvester, R.L. Ferrari, Finite Elements for Electrical Engineers, third edition, Cambridge University Press, 1996, 516 pp.

20. John Leonidas Volakis, John L. Volakis, Arindam Chatterjee, Leo C. Kempel, Finite Element Method for Electromagnetics, Wiley-IEEE Press, 1998.

21. O.C. Zienkiewicz, R.L. Taylor, The Finite Element Method, Volume 1: The Basis, fifth edition, Butterworth-Heinemann, Oxford, 2000, 689 pp.

22. Dragan Poljak, Milica Rančić, On the frequency domain analysis of straight thin wire radiating above a lossy half-space: Pocklington equation versus Hallén equation revisited: 80th anniversary of the Hallén integral equation, in: 2018 26th International Conference on Software, Telecommunications and Computer Networks, SoftCOM, IEEE, 2018, pp. 1–6.

CHAPTER 3

Incident Electromagnetic Field Dosimetry

Incident field dosimetry involves procedures for the assessment of LF and HF electric and magnetic fields. Some techniques pertaining to theoretical and experimental dosimetry are presented.

3.1 ASSESSMENT OF EXTERNAL ELECTRIC AND MAGNETIC FIELDS AT LOW FREQUENCIES

Human exposure to extremely low frequency (ELF) electric and magnetic fields generated by power lines and substations, respectively, initiated an increasing public concern regarding possible adverse health effects. A controversy has been caused due to the possible link between the low frequency fields and leukemia, or certain forms of tumor (e.g., nervous tissue tumor) in humans [1,2]. As the displacement currents at extremely low frequencies are negligible, the electric and magnetic fields can be analyzed separately. While the electric field can be easily shielded and its magnitude is reduced in the surrounding environment by the presence of trees, houses, etc., the related effects on the conductive parts of the human body thus may be appreciably attenuated; the magnetic field, on the other hand, penetrates through almost all known materials.

In the case of the magnetic field exposure, the internal currents form close loops, while in the electric field exposure the currents induced in the body have the axial character.

Note that ELF fields range from 3 Hz to 3 kHz, thus comprising fields emitted by power lines at frequencies 50 and 60 Hz, where 50 Hz power utilities are used in Europe and 60 Hz in the United States.

The electric and magnetic fields in the vicinity of ELF sources can be determined analytically, numerically or by measurement depending on the complexity of the particular configuration.

Generally, simpler configurations of power lines, in which conductor sag is not taken into account, can be analyzed by using analytical formulas [3,4], while the analysis of the fields generated by power substations often requires the use of numerical techniques [5,6].

3.1.1 Fields Generated by Power Lines

A general configuration of a 3-phase power line composed of straight conductors above a lossy ground is given in Fig. 3.1.

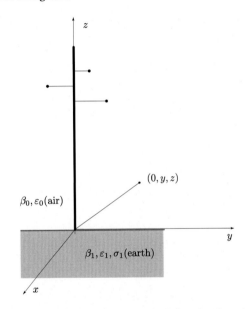

FIG. 3.1 Coordinate system for calculating electric and magnetic fields.

As the conductors are considered to be rather long along the x-axis, only the points in y–z plane are considered. The currents flowing along the conductors are given by

$$|I_1(x)| = |I_2(x)| = |I_3(x)|, \qquad (3.1)$$

where

$$I_2(x) = I_1(x) e^{j2\pi/3}, \quad I_3(x) = I_1(x) e^{-j2\pi/3}. \quad (3.2)$$

The total current is then

$$I(x) = I_1(x) + I_2(x) + I_3(x) = 0. \qquad (3.3)$$

It is worth noting that the quasi-static analysis of power lines can be also carried out by knowing the

Human Interaction with Electromagnetic Fields. https://doi.org/10.1016/B978-0-12-816443-3.00011-X

current density induced along the conductors. This is discussed in Sect. 3.1.2.

3.1.1.1 The Electric Field

Being one of typical configurations of the 110 kV three-phase, a three-wire power line consists of three conductors in a configuration presented in Fig. 3.2.

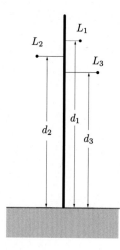

FIG. 3.2 Configuration of high voltage, three-phase, three-wire power line.

The related electric field of a configuration shown in Fig. 3.1 is given by the following formulas [1,3]:

$$E_{0x}(x, y, z) = \frac{\omega\mu_0 I}{4\pi\beta_0} \left\{ e^{j\beta_0 r_1} \left[\frac{j\beta_0}{r_1} - \frac{1}{r_1^2} - \frac{j}{\beta_0 r_1^3} \right] \right.$$

$$- \frac{x^2}{r_1^2} \left(\frac{j\beta_0}{r_1} - \frac{3}{r_1^2} - \frac{3j}{\beta_0 r_1^3} \right) \right]$$

$$- e^{j\beta_0 r_2} \left[\frac{j\beta_0}{r_2} - \frac{1}{r_2^2} - \frac{j}{\beta_0 r_2^3} \right.$$

$$- \frac{x^2}{r_2^2} \left(\frac{j\beta_0}{r_2} - \frac{3}{r_2^2} - \frac{3j}{\beta_0 r_2^3} \right) \right]$$

$$+ 2e^{j\beta_0 r_2} \left[\frac{\beta_0}{\beta_1} \left(\frac{z+d}{r_2} \right) \left(\frac{j\beta_0}{r_2} - \frac{1}{r_2^2} \right) \right.$$

$$- \frac{\beta_0^2}{\beta_1^2} \left[\frac{j\beta_0}{r_2} - \frac{1}{r_2^2} - \frac{j}{\beta_0 r_2^3} \right.$$

$$\left. \left. - \frac{y^2}{r_2^2} \left(\frac{j\beta_0}{r_2} - \frac{3}{r_2^2} - \frac{3j}{\beta_0 r_2^3} \right) \right] \right\}, \quad (3.4)$$

$$E_{0y}(x, y, z) = -\frac{\omega\mu_0 I}{4\pi\beta_0}$$

$$\times \left[e^{j\beta_0 r_1} \frac{xy}{r_1^2} \left(\frac{j\beta_0}{r_1} - \frac{3}{r_1^2} - \frac{3j}{\beta_0 r_1^3} \right) \right.$$

$$- e^{j\beta_0 r_2} \frac{xy}{r_2^2} \left(\frac{j\beta_0}{r_2} - \frac{3}{r_2^2} - \frac{3j}{\beta_0 r_2^3} \right)$$

$$\left. \times \left(1 - \frac{2\beta_0^2}{\beta_1^2} \right) \right], \quad (3.5)$$

$$E_{0z}(x, y, z) = -\frac{\omega\mu_0 I}{4\pi\beta_0} \left[e^{j\beta_0 r_1} \left(\frac{x}{r_1} \right) \left(\frac{z-d}{r_1} \right) \right.$$

$$\times \left(\frac{j\beta_0}{r_1} - \frac{3}{r_1^2} - \frac{3j}{\beta_0 r_1^3} \right)$$

$$- e^{j\beta_0 r_2} \left(\frac{x}{r_2} \right) \left(\frac{z+d}{r_2} \right)$$

$$\times \left(\frac{j\beta_0}{r_2} - \frac{3}{r_2^2} - \frac{3j}{\beta_0 r_2^3} \right)$$

$$\left. + 2\frac{\beta_0}{\beta_1} e^{j\beta_0 r_2} \left(\frac{x}{r_2} \right) \left(\frac{j\beta_0}{r_2} - \frac{1}{r_2^2} \right) \right],$$

$$(3.6)$$

where $\beta_0 = \omega/c$ and $j\omega\mu_0\sigma_1 = (1+j)(\frac{\omega\mu_0\sigma_1}{2})^{1/2}$.

Note that r_1 is the distance from the element of a current at $(x, 0, d)$ to the observation point, and r_2 is the distance from the image on the earth to the point of observation.

The field in the plane $x = 0$ due to all three conductors is

$$\left[E_{0j}(0, y, z) \right]_t = \int_{-\infty}^{+\infty} \left[E_{0j}(0, y - L_1, z) \right.$$

$$+ e^{j2\pi/3} E_{0j}(0, y + L_2, z)$$

$$\left. + e^{-j2\pi/3} E_{0j}(0, y - L_3, z) \right] e^{j\beta_0 x'} dx',$$

$$(3.7)$$

where $j = x, y, z$.

As $\beta_0 r_1 = 1$ and $\beta_0 r_2 = 1$, the $1/r_1^3$ and $1/r_2^3$ terms are dominant at distances r_1 and r_2, negligible when compared to $L_2 + L_3$. Having performed analytical integration, the following set of the field formulas is obtained:

$$\left[E_{0x}(0, y, z) \right]_t$$

$$\sim \frac{j\omega\mu_0 I}{\pi\beta_1^2} (\beta_0 d) \left\{ \frac{(z+d_1)^2 - (y-L_1)^2}{\left[(z+d_1)^2 + (y-L_1)^2 \right]^2} \right.$$

$$+ e^{j2\pi/3} \left[\frac{(z+d_2)^2 - (y+L_2)^2}{\left[(z+d_2)^2 + (y+L_2)^2 \right]^2} \right]$$

$$+ e^{-j2\pi/3} \left[\frac{(z+d_3)^2 - (y-L_3)^2}{\left[(z+d_3)^2 + (y-L_3)^2 \right]^2} \right] \right\}, \quad (3.8)$$

$$\left[E_{0y}(0, y, z) \right]_t$$

$$\sim \frac{\omega \mu_0 I}{2\pi \beta_0} \left\{ (y - L_1) \left[\frac{1}{(z-d_1)^2 + (y-L_1)^2} \right. \right.$$

$$\left. - \frac{1 - 2\beta_0^2/\beta_1^2}{(z+d_1)^2 + (y-L_1)^2} \right]$$

$$+ e^{j2\pi/3}(y + L_2) \left[\frac{1}{(z-d_2)^2 + (y+L_2)^2} \right.$$

$$\left. - \frac{1 - 2\beta_0^2/\beta_1^2}{(z+d_2)^2 + (y+L_2)^2} \right]$$

$$+ e^{-j2\pi/3}(y - L_3) \left[\frac{1}{(z-d_3)^2 + (y-L_3)^2} \right.$$

$$\left. \left. - \frac{1 - 2\beta_0^2/\beta_1^2}{(z+d_3)^2 + (y-L_3)^2} \right] \right\}, \quad (3.9)$$

$$\left[E_{0z}(0, y, z) \right]_t \sim \frac{\omega \mu_0 I}{2\pi \beta_0} \left\{ \frac{z - d_1}{(z-d_1)^2 + (y-L_1)^2} \right.$$

$$- \frac{z + d_1}{(z+d_1)^2 + (y-L_1)^2}$$

$$+ e^{j2\pi/3} \left[\frac{z - d_2}{(z-d_2)^2 + (y+L_2)^2} \right.$$

$$\left. - \frac{z + d_2}{(z+d_2)^2 + (y+L_2)^2} \right]$$

$$+ e^{-j2\pi/3} \left[\frac{z - d_3}{(z-d_3)^2 + (y-L_3)^2} \right.$$

$$\left. \left. - \frac{z + d_3}{(z+d_3)^2 + (y-L_3)^2} \right] \right\}. \quad (3.10)$$

An illustrative computational example is related to the 3-phase power line with $L_1 = 2.7$ m, $L_2 = 3.2$ m, $L_3 = 3.7$ m, $d_1 = 25.2$ m, $d_2 = 22.7$ m, $d_3 = 20.2$ m, $I = 90$ A, and $U = 110$ kV.

The following figures show the spatial distribution of the electric field components in the $x = 0$ plane, where y coordinate takes values from -20 to 20 m (note that $y = 0$ is the a position of the post carrying the line) while z coordinate takes values from 0 to 40 m (note that $z = 0$ represent ground plane level). Fig. 3.3

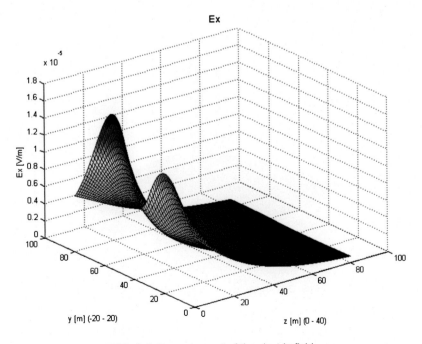

FIG. 3.3 E_x component of the electric field.

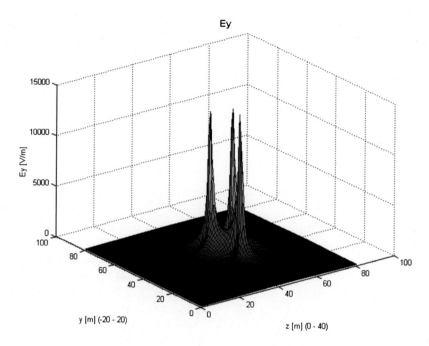

FIG. 3.4 E_y component of the electric field.

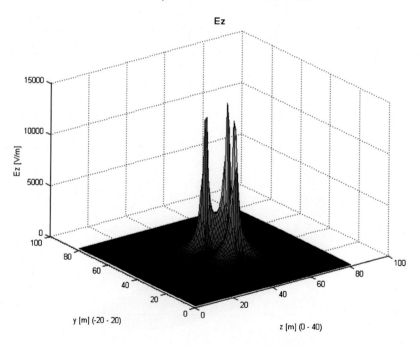

FIG. 3.5 E_z component of the electric field.

shows E_x component, while E_y and E_z are depicted in Figs. 3.4 and 3.5.

It is obvious that the magnitude of E_x component is rather small, therefore it is possible to neglect it.

On the other hand, E_y component peaks are rather high near the lines itself, when its magnitude rapidly attenuates as the distance from the lines increases.

Finally, E_z component shows a similar behavior as E_y component. This is the component collinear to the longest dimension of the body and therefore it is expected to have a significant influence on the induced current.

Furthermore, Figs. 3.6 and 3.7 show the behavior of E_z component on different heights, respectively.

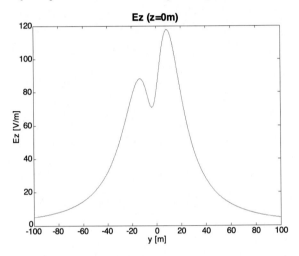

FIG. 3.6 E_z component of the electric field on the ground level.

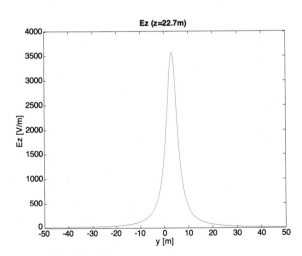

FIG. 3.7 E_z component of the electric field on the level of the middle conductor.

It can be observed that the field on the ground level is highly dependent on actual configuration of power lines, i.e., the fields from different lines create constructive and destructive interference, thus generating some local maxima and minima, respectively. Such interference does not exist on a higher level, where the influence of closest conductor is dominant.

The electric field rapidly decreases with distance from the power line and on the ground level reaches values that are below exposure limits proposed by safety guidelines.

3.1.1.2 The Magnetic Field

The components of the magnetic field generated by a 3-phase power line can be readily obtained from the curl Maxwell equation

$$-j\omega B(x, y, z) = \nabla \times E(x, y, z), \quad (3.11)$$

where

$$E(x, y, z) = E(0, y, z)e^{-jk_0x}. \quad (3.12)$$

The corresponding components of the total magnetic field $B[_{0l}(0, y, z)]_t$ are given by [4]:

$$
\begin{aligned}
&\left[B_{0x}(0, y, z)\right]_t \\
&= \frac{2j\mu_0 I k_0}{\pi k_1^2} \left\{ \frac{(y - L_1)(z + d_1)}{\left[(z + d_1)^2 + (y - L_1)^2\right]^2} \right.\\
&\quad + e^{j2\pi/3} \left[\frac{(y + L_2)(z + d_2)}{\left[(z + d_2)^2 + (y + L_2)^2\right]^2}\right] \\
&\quad \left. + e^{-j2\pi/3} \left[\frac{(y - L_3)(z + d_3)}{\left[(z + d_3)^2 + (y - L_3)^2\right]^2}\right] \right\},
\end{aligned}
$$

$$(3.13)$$

$$
\begin{aligned}
&\left[B_{0y}(0, y, z)\right]_t \\
&= \frac{\mu_0 I}{2\pi} \left\{ \frac{4(z + d_1)(k_0 d)}{k_1^2 \left[(z + d_1)^2 + (y - L_1)^2\right]^2} \right.\\
&\quad \times \left[1 - \frac{2\left[(z + d_1)^2 - (y - L_1)^2\right]}{(z + d_1)^2 + (y - L_1)^2}\right] \\
&\quad - \frac{(z - d_1)}{(z - d_1)^2 + (y - L_1)^2} + \frac{(z + d_1)}{(z + d_1)^2 + (y - L_1)^2} \\
&\quad + e^{j2\pi/3} \left\{ \frac{4(z + d_2)(k_0 d)}{k_1^2 \left[(z + d_2)^2 + (y + L_2)^2\right]^2} \right.
\end{aligned}
$$

$$\times \left[1 - \frac{2\left[(z+d_2)^2 - (y-L_2)^2\right]}{(z+d_2)^2 + (y-L_2)^2} \right]$$

$$- \frac{(z-d_2)}{(z-d_2)^2 + (y-L_2)^2} + \frac{(z+d_2)}{(z+d_2)^2 + (y-L_2)^2}$$

$$+ e^{-j2\pi/3} \left\{ \frac{4(z+d_3)(k_0 d)}{k_1^2\left[(z+d_3)^2 + (y+L_3)^2\right]^2} \right.$$

$$\times \left[1 - \frac{2\left[(z+d_3)^2 - (y-L_3)^2\right]}{(z+d_3)^2 + (y-L_3)^2} \right]$$

$$\left. - \frac{(z-d_3)}{(z-d_3)^2 + (y-L_3)^2} + \frac{(z+d_3)}{(z+d_3)^2 + (y-L_3)^2} \right\},$$

$$(3.14)$$

$$[B_{0z}(0,y,z)]_t = \frac{\mu_0 I}{2\pi} \left\{ \frac{y-L_1}{(z-d_1)^2 + (y-L_1)^2} \right.$$

$$- \frac{(y-L_1)(1-2k_0^2/k_1^2)}{(z+d_1)^2 + (y-L_1)^2}$$

$$+ \frac{4(y-L_1)(k_0 d)}{k_1^2\left[(z+d_1)^2 + (y-L_1)^2\right]}$$

$$\times \left[1 + \frac{2\left[(z+d_1)^2 - (y-L_1)^2\right]}{(z+d_1)^2 + (y-L_1)^2} \right]$$

$$+ e^{j2\pi/3} \left\{ \frac{y+L_2}{(z-d_2)^2 + (y+L_2)^2} \right.$$

$$- \frac{(y+L_2)(1-2k_0^2/k_1^2)}{(z+d_2)^2 + (y+L_2)^2}$$

$$+ \frac{4(y+L_2)(k_0 d)}{k_1^2\left[(z+d_2)^2 + (y+L_2)^2\right]^2}$$

$$\times \left[1 + \frac{2\left[(z+d_2)^2 - (y+L_2)^2\right]}{(z+d_2)^2 + (y+L_2)^2} \right] \right\}$$

$$+ e^{-j2\pi/3} \left\{ \frac{y-L_3}{(z-d_3)^2 + (y-L_3)^2} \right.$$

$$- \frac{(y-L_3)(1-2k_0^2/k_1^2)}{(z+d_3)^2 + (y-L_3)^2}$$

$$+ \frac{4(y-L_3)(k_0 d)}{k_1^2\left[(z+d_3)^2 + (y-L_3)^2\right]^2}$$

$$\times \left[1 + \frac{2\left[(z+d_3)^2 - (y-L_3)^2\right]}{(z+d_3)^2 + (y-L_3)^2} \right] \right\} \right\}.$$

$$(3.15)$$

The component $B_{[0x}(0,y,z)]_t$ is negligible compared to the values of $B_{[0y}(0,y,z)]_t$ and $B_{[0z}(0,y,z)]_t$, respectively, as it is approximately $3 \cdot 10^3$ times smaller.

Additional approximation can be adopted due to the fact that the conductors are located away from the ground, i.e., the image terms, assuming the ground to be perfectly conducting, can be neglected yielding the following expressions:

$$[B_{0y}(0,y,z)]_t = -\frac{\mu_0 I}{2\pi} \left[\frac{(z-d_1)}{(z-d_1)^2 + (y-L_1)^2} \right.$$

$$+ \frac{(z-d_2)}{(z-d_2)^2 + (y+L_2)^2} e^{j2\pi/3}$$

$$\left. + \frac{(z-d_3)}{(z-d_3)^2 + (y-L_3)^2} e^{-j2\pi/3} \right],$$

$$(3.16)$$

$$[B_{0z}(0,y,z)]_t = \frac{\mu_0 I}{2\pi} \left[\frac{(y-L_1)}{(z-d_1)^2 + (y-L_1)^2} \right.$$

$$+ \frac{(y+L_2)}{(z-d_2)^2 + (y+L_2)^2} e^{j2\pi/3}$$

$$\left. + \frac{(y-L_3)}{(z-d_3)^2 + (y-L_3)^2} e^{-j2\pi/3} \right].$$

$$(3.17)$$

The computational example pertains to the most common type of long-distance 110 kV power line consisting of 3 conductors at different heights in the planes $z = d_1$, $z = d_2$, and $z = d_3$ above the surface ($z = 0$) of the earth. The conductors are located at the vertices of a triangle in the plane $x = 0$.

A model of the tower with all relevant dimensions is shown in Fig. 3.8. The properties of a three-wire, three-phase 110 kV power line are as follows: mean distance between two towers is $l_s = 300$ m; diameter of the conductor is $D = 17.3$ mm; cross-section of the aluminum is $S_{Al} = 148.9$ mm^2; cross-section of the steel is $S_{steel} = 25.4$ mm^2; mean value of the measured current $I = 90$ A; maximum value of the measured current $I = 130$ A.

Figs. 3.9 and 3.10 show B_y and B_z components of the magnetic flux density generated by the power line with current $I = 90$ A, for various values of z.

Fig. 3.11 shows the total magnetic flux density B generated by the power line depicted in Fig. 3.2, with current $I = 90$ A, for various values of z.

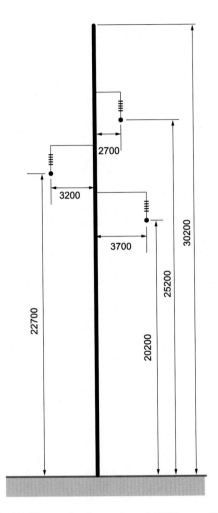

FIG. 3.8 Three-wire, three-phase 110 kV power line.

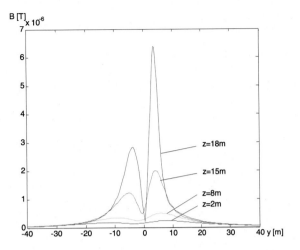

FIG. 3.9 B_y component of the magnetic flux density.

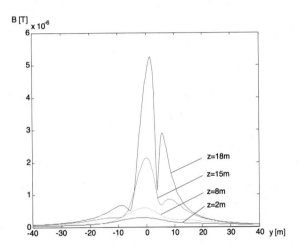

FIG. 3.10 B_z component of the magnetic flux density.

Fig. 3.12 shows the total magnetic field in the vicinity of a three-wire, three-phase power line in the cross-section plane of the configuration.

There are other types of configuration of three-phase, three-wire power lines such as equally spaced horizontally and equally spaced vertically.

3.1.2 Fields Generated by Substation Transformers

Contrary to the power line configurations which can be represented by long straight conductors, the geometries of conductor configurations in substations require curved wire models. For the assessment of electric fields, one first deals with a solution of scalar potential integral equation (SPIE), while the magnetic field can be obtained from Biot–Savart's law.

3.1.2.1 The Electric Field

Electric scalar potential at an arbitrary point $P(x, z)$ due to a conductor segment carrying linear charge density, as shown in Fig. 3.13, is defined as [5]

$$\varphi(x, z) = \frac{1}{4\pi\varepsilon} \int_{-L}^{+L} \frac{\rho_l(x')\,dx'}{R}, \qquad (3.18)$$

where ρ_l denotes the line charge density, $2L$ stands for the wire length, while R is the distance between a point on the straight wire and an arbitrary point P.

When the potential φ_l along the straight wire is known, the integral expression (3.18) becomes an integral equation

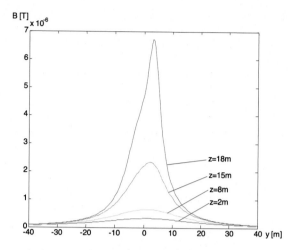

FIG. 3.11 Total magnetic flux density B due to the three-wire, three-phase 110 kV power line.

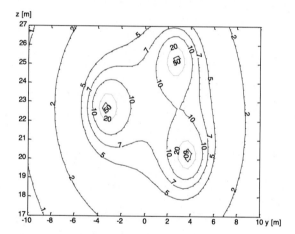

FIG. 3.12 Magnetic field H [A/m] in the vicinity of a three-phase power line.

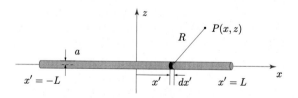

FIG. 3.13 Straight conductor geometry.

$$\varphi_l = \frac{1}{4\pi\varepsilon} \int\limits_{-L}^{+L} \frac{\rho_l(x')\,dx'}{\sqrt{(x-x')^2+a^2}}, \qquad (3.19)$$

where a denotes the wire radius.

Having performed a discretization of the substation conductors, a system of equations for unknown charges along each segment is obtained.

The integral equation (3.19) then transforms into a corresponding matrix equation [5]

$$\begin{bmatrix} P_{11} & P_{12} & \cdots & P_{1n} \\ P_{21} & P_{22} & \cdots & P_{2n} \\ \vdots & \vdots & \ddots & \vdots \\ P_{n1} & P_{n2} & \cdots & P_{nn} \end{bmatrix} \cdot \begin{bmatrix} q_1 \\ q_2 \\ \vdots \\ q_n \end{bmatrix} = \begin{bmatrix} \varphi_1 \\ \varphi_2 \\ \vdots \\ \varphi_n \end{bmatrix}, \qquad (3.20)$$

where $\varphi_1, \varphi_2, \ldots, \varphi_n$ stand for the conductor segment potentials, $q_1, q_2, \ldots, q_n$ denote the segment charges, while $P_{11}, P_{12}, \ldots, P_{nn}$ are Maxwell coefficients. The details pertaining to the Maxwell coefficients can be found in [7].

Having obtained the unknown charges along the conductor configuration, the electric field components in rectangular coordinates at an arbitrary point (x, y, z), generated by the ith segment are given by [5]:

$$E_{xi} = -\frac{\partial\varphi}{\partial x}$$
$$= \frac{q_i}{4\pi\varepsilon_0 L_i}\left[\frac{1}{\sqrt{(L_i-x)^2+W^2}} - \frac{1}{\sqrt{x^2+W^2}}\right], \qquad (3.21)$$

$$E_{yi} = -\frac{\partial\varphi}{\partial y}$$
$$= \frac{q_i}{4\pi\varepsilon_0 L_i}\frac{y}{W^2}\left[\frac{L_i-x}{\sqrt{(L_i-x)^2+W^2}} + \frac{x}{\sqrt{x^2+W^2}}\right], \qquad (3.22)$$

$$E_{zi} = -\frac{\partial\varphi}{\partial z}$$
$$= \frac{q_i}{4\pi\varepsilon_0 L_i}\frac{z}{W^2}\left[\frac{L_i-x}{\sqrt{(L_i-x)^2+W^2}} + \frac{x}{\sqrt{x^2+W^2}}\right], \qquad (3.23)$$

where q_i denotes the charge of the ith segment, L_i stands for the length of the ith segment, and

$$W^2 = y^2 + z^2. \qquad (3.24)$$

The total field components are assembled from each segment. Such a segment is shown in Fig. 3.14.

A computational example is related to the 110/10 kV/kV transmission substation of GIS (Gas-Insulated Substation) whose simplified two-dimensional layout is shown in Fig. 3.15. The 50 Hz electric field is calculated at height $z = 1$ m above ground. As there is no real

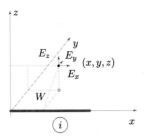

FIG. 3.14 Electric field components due to the ith segment.

possibility of a public exposure within the transmission substation, and a professional exposure is strictly limited to duration, the calculations are performed outside the fence of the substation.

The equipment with grounded shields (power cables), sheaths (GIS buses), or metallic casings (transformers, switch-gears) are neglected due to the shielding effect, i.e., the metallic enclosures are connected to the ground, thus generating negligible electric field. Consequently, overhead transmission lines and unshielded conductors are of interest for this particular substation. The calculation domains 1 to 5, in which higher field values are expected, are assigned as in Fig. 3.15.

The spatial distribution of electric field over domain 2, where the highest field value is captured, is shown in Fig. 3.16. It is worth noting that due to the shortness of the unshielded conductors, as well as their considerable distances from the substation fence, the main electric field sources are the overhead lines. This can be observed in Fig. 3.16 where the maximal electric

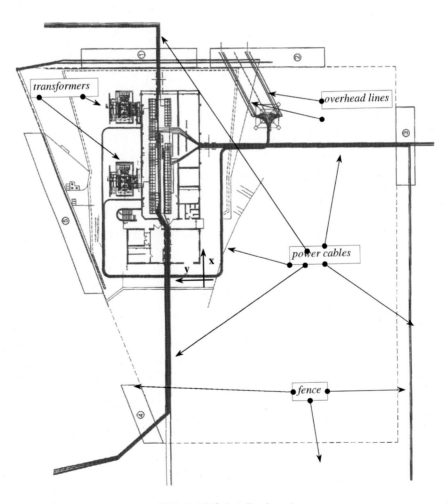

FIG. 3.15 Substation layout.

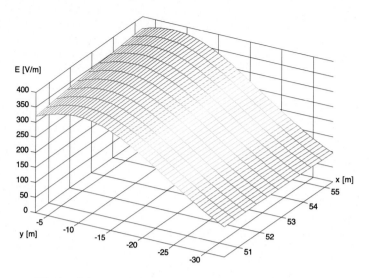

FIG. 3.16 Spatial distribution of the electric field over domain 2.

field is $E_{\max} = 380.70$ V/m. The point of maximum electric field is located just below the overhead line route.

Once the external electric field is known, the current density induced inside the human body can be computed. In addition, it is evident that the maximal value of the computed field in domain 2 (380.70 V/m) is appreciably lower from the exposure limits for both occupational (10 kV/m) and general population (5 kV/m), respectively proposed by ICNIRP guidelines [8,9].

3.1.2.2 The Magnetic Field

The magnetic flux density at an arbitrary point due to a current element shown in Fig. 3.17 is determined by the Biot–Savart's law:

$$d\vec{B} = \frac{\mu}{4\pi} \frac{i(t)\,\vec{dl} \times (\vec{r} - \vec{r}')}{|\vec{r} - \vec{r}'|^3} = \frac{\mu}{4\pi} \frac{i(t)\,\vec{dl} \times \vec{R}}{R^3}, \quad (3.25)$$

where μ is permeability, i denotes the current along the segment, and $R = |\vec{r} - \vec{r}'|$ is the distance from the source $i(t)\vec{dl}$ to the observation point P.

Performing some mathematical manipulation and integrating the contributions along the entire length of a conductor, we get

$$\vec{B} = \hat{e}_\varphi \frac{\mu\,i(t)}{4\pi\rho} \int_{\theta_1}^{\theta_2} \cos\theta\,d\theta = \hat{e}_\varphi \frac{\mu\,i(t)}{4\pi\rho} (\sin\theta_1 + \sin\theta_2),$$

$$(3.26)$$

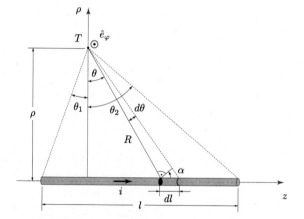

FIG. 3.17 Straight current element.

where ρ and θ are the variables in the cylindrical coordinate system.

The ELF magnetic field value at an arbitrary point can be assessed by assembling the contributions of all conductors divided in a certain number of straight segments. The kth straight segment carrying current i_k in Cartesian three-dimensional coordinate system is shown in Fig. 3.18.

Using the Biot–Savart's law, the magnetic field value at point C due to the considered conductor segment can be written as follows [6]:

$$B_k(t) = \frac{\mu\,i_k(t)}{4\pi\,R_{RS}} \left(\frac{R_{PS}}{R_{PR}} + \frac{R_{SQ}}{R_{QR}} \right), \quad (3.27)$$

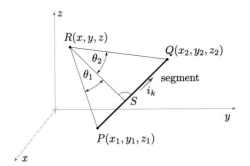

FIG. 3.18 Straight segment in Cartesian coordinate system.

where the corresponding distances R are assigned as in Fig. 3.18.

Total components of the magnetic flux density generated by N segments are assembled from the contributions of all segments. Therefore, the total value of the magnetic flux density at a given point of space can be expressed as

$$B(t) = \sqrt{\left(\sum_{i=1}^{N} B_{x,i}(t)\right)^2 + \left(\sum_{i=1}^{N} B_{y,i}(t)\right)^2 + \left(\sum_{i=1}^{N} B_{z,i}(t)\right)^2},$$

$$(3.28)$$

where $B_{x,i}(t)$, $B_{y,i}(t)$ and $B_{z,i}(t)$ are the components of the magnetic flux density due to the ith segment.

A computational example is related to the 110/10 kV/kV transmission substation of GIS (Gas-Insulated Substation) type. A simplified two-dimensional layout

of the substation is shown in Fig. 3.15. The calculation domains 1 to 5, in which higher field values are expected, are assigned as in Fig. 3.15. The spatial distribution of the magnetic field over domain 3, where the highest field value is captured, is shown in Fig. 3.19.

If the human body is exposed to an ELF magnetic field, the circular current density is induced inside the body due to the existence of the normal component of the magnetic flux density.

Once the magnetic flux density is determined, the internal current density can be calculated using the disk model of the human body.

3.1.3 Assessment of Circular Current Density Induced in the Body

Internal dosimetry of human exposure to ELF fields deals with internal electric and fields and current densities, respectively. According to ICNIRP 1998 [8], basic restrictions pertain to axial currents when a human is exposed to an electric field and circular current densities when the body is exposed to a magnetic field. On the other hand, ICNIRP 2010 [9] proposes induced electric field instead of the induced axial current density. The assessment methods for the axial currents in simplified body models are discussed in other chapters of this book. The internal circular current density due to the existence of the normal component of the magnetic flux density is presented in this section.

This internal current density can be assessed by using the disk model of the human body, shown in Fig. 3.20.

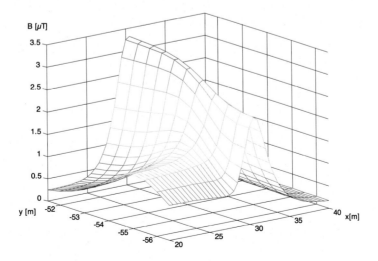

FIG. 3.19 Spatial distribution of the magnetic field over domain 3.

These analytical models of the body have been proposed in [10] to provide rapid estimation of the human exposure to an ELF magnetic field.

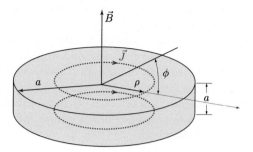

FIG. 3.20 Disk model of the human body.

The disk is homogeneous with radius a and conductivity σ. The analytical relation for the current density can be readily derived from Maxwell equations and it will be outlined in a few steps.

Starting from the differential form of Faraday's law:

$$\nabla \times \vec{E} = -\frac{\partial \vec{B}}{\partial t}, \qquad (3.29)$$

where $\vec{E}$ denotes the applied electric field, using the constitutive equation

$$\vec{J} = \sigma \vec{E}, \qquad (3.30)$$

and integrating over an arbitrary surface with related curve c, Eq. (3.29) becomes

$$\int_S \nabla \times \vec{J}\, d\vec{S} = -\int_S \sigma \frac{\partial \vec{B}}{\partial t}\, d\vec{S}. \qquad (3.31)$$

Applying Stokes theorem, for time harmonic fields, Eq. (3.29) can be written as follows:

$$\oint_c \vec{J}\, d\vec{s} = -j\omega\sigma \vec{B} \int_S d\vec{S}, \qquad (3.32)$$

i.e., it follows that

$$\int_0^{2\pi} J_\phi \rho\, d\phi = -j\omega\sigma B_z \int_0^\rho \int_0^{2\pi} \rho\, d\rho\, d\phi, \qquad (3.33)$$

where $\omega = 2\pi f$ is the angular frequency.

Taking into account the rotational symmetry, the analytical integration simply yields

$$J_\phi \cdot 2\pi\rho = -j\omega\sigma B_z \rho^2 \pi, \qquad (3.34)$$

and the value of the induced current density inside the disk is given by the following simple expression:

$$|J_\phi| = \sigma \pi \rho f \cdot B_z. \qquad (3.35)$$

Knowing the current density, it is also possible to calculate the total current flowing through the disk, as shown in Fig. 3.21.

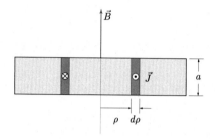

FIG. 3.21 Integration over the disk cross-section.

Thus, straightforward integration gives

$$I = \int_S \vec{J}\, d\vec{S} = \int_0^a \int_0^{2\pi} \sigma \pi f B_z \rho\, d\rho\, dz$$

$$= \sigma \pi f B_z \int_0^a \int_0^{2\pi} \rho\, d\rho\, dz = \sigma \pi f B_z \frac{a^3}{2}. \qquad (3.36)$$

Note that the thickness of the disk is not specified by the ICNIRP guidelines, as the total current is not a basic restriction. In [11], the thickness is assumed to be equal a (the disk radius).

Once the magnetic flux density is determined, the internal current density can be calculated using the disk model of the human body. The disk radius is $a = 0.14$ m while the conductivity is $\sigma = 0.5$ S/m.

The induced current densities and axial currents calculated for the case of maximal field values are presented in Table 3.1.

It is visible that the maximal value of the computed magnetic flux density in domain 3 (3.344 µT) is appreciably lower from the exposure limits for both occupational (500 µT) and general population (100 µT), respectively proposed by ICNIRP guidelines [8].

In addition, the maximal value of internal current density of 0.03677 mA/m^2 is also far below the exposure limits for both occupational (10 mA/m^2) and general population (2 mA/m^2), respectively proposed by ICNIRP [8].

TABLE 3.1
Maximal values of magnetic flux density and related internal current densities.

Domain	Magnetic flux density μT	Current density μA/m^2	Total axial current μA
1	2.701	29.7	0.29
2	0.945	10.39	0.102
3	3.344	36.77	0.36
4	2.552	28.06	0.275
5	0.747	8.21	0.085

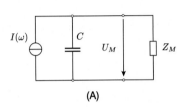
(A)

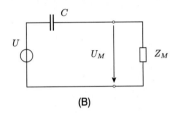
(B)

FIG. 3.22 Equivalent circuit for capacitive dipoles.

3.1.4 On the Basic Principles of Measurement of LF Fields

In some realistic calculation scenarios, the methods of theoretical dosimetry are not enough to provide information about accurate values of external field levels in a human environment. This section addresses some basic principles of electric and magnetics field measurements at low frequencies.

3.1.4.1 Measurement of LF Electric Fields

The most commonly used method to measure electric fields is the capacitive method, i.e., two electrodes measuring the displacement current are immersed into the electric field to be determined. Furthermore, contrary to the measurement of LF magnetic fields, the measurement of electric fields is more sensitive to the presence of the body which may distort the field, thus some precautions must be taken to provide accurate results. As the presence of a human operator in the field distorts the field so that the measured value is practically unrelated to the actual value, it is therefore necessary to isolate the sensor from the basic device.

Empirical facts require 7 m distance of a human from the test site.

A circuit of a sensor for measuring electric fields is presented in Fig. 3.22.

An alternative voltage signal occurs on the plate of a capacitor when illuminated by an alternating field. The capacitor behaves like a capacitive dipole as long as its dimensions are appreciably smaller than the wavelength of the measured signal. Such an assumption is valid for low frequency fields.

The equivalent circuits shown in Fig. 3.22 are convenient to calculate the output voltage of capacitive dipoles loaded with a test impedance Z. In the equivalent circuit of Fig. 3.22A, the dipole is represented as a current source with a parallel capacitance. The current source $I(\omega)$ represents the displacement current flowing into the sensor immersed in an electric field. The equivalent circuit of Fig. 3.22B is based on a voltage source with capacitive series impedance. The voltage source $U(\omega)$ is obtained as the product of the electric field strength and the effective length of the dipole. The parallel (or series) capacitance C is equal for both equivalent circuits to the capacitance of the dipole, which can be measured with an impedance or capacitance meter at low frequencies. The current flowing through capacitive sensor is shown in Fig. 3.23.

The displacement current flowing into the sensor can be written as an integral of the field over an electrode surface

$$I = j\omega\varepsilon \int_S \vec{E}\, d\vec{S}, \qquad (3.37)$$

where current I generates a voltage on the electrodes or on the input impedance of the measurement circuitry connected in parallel to the electrodes. The corresponding input impedance is determined by a high-

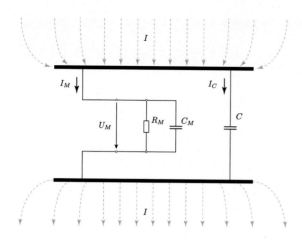

FIG. 3.23 Flow of current through a capacitive sensor.

impedance source follower. The input capacitance C_M of the source follower is parallel to the plate capacitance. The voltage U_M is induced on the sensor plates and, thus, at the input to the measurement circuitry is given by

$$U_M = I_M \cdot Z_M = I \cdot Z = \frac{j\omega\varepsilon \int_S \vec{E}\, d\vec{S}}{\frac{1}{R_M} + j\omega(C + C_M)}. \quad (3.38)$$

Given the assumed test impedance Z_M, the sensor represents a first-order high-pass filter with corresponding limit frequency

$$f_g = \frac{1}{2\pi R_M (C + C_M)}. \quad (3.39)$$

At frequencies well above the limit frequency, the input admittance of the measurement circuitry can be neglected with respect to the susceptance of the capacitive elements, i.e.,

$$\frac{1}{R_M} \ll \omega(C + C_M). \quad (3.40)$$

So it follows that

$$U_M = \frac{\varepsilon \int_S \vec{E}\, d\vec{S}}{C + C_M}. \quad (3.41)$$

Thus, the measurement of the electric field strength is independent of the frequency above the lower limit frequency. The ohmic resistance of the measurement

circuitry must be maximized to extend the lower limit frequency downwards and produce a sensor that is as broadband as possible. Typical sensors are cubically shaped.

The sensor has three plates arranged perpendicularly to one another, each of which forms a capacitance with respect to the core of the cube. The three field components are determined by measuring the displacement currents of the individual plates. These signals are sampled via three input amplifiers and multiplexer stages and processed by a signal processor. This makes it possible to break down the field components by spectrum and measure the fields selectively. To minimize distortion of the electric field, the data from the sensor are transmitted via optical fiber cables (max. length 20 m), thus making the sensor fully remote-controllable from the basic device.

3.1.4.2 Measurement of LF Magnetic Fields

Measurement of magnetic fields is based on the implementation of Faraday's law of induction (Maxwell's first equation in the integral form) stating that time-varying magnetic fields induce a voltage in a coil proportional to the field level:

$$e^{\text{ind}} = -\frac{\partial \phi}{\partial t} = -\frac{\partial}{\partial t} \left\{ \int_A \vec{B}\, d\vec{S} \right\}, \quad (3.42)$$

where e^{ind} denotes the induced electromagnetic force (EMF), B is the magnetic flux density, and S is the area of the loop.

Generally, the instruments for measuring magnetic fields consist of two parts: the probe, or field-sensing element, and the detector processing signal from the probe. Thus, the simplest meters measure the voltage induced in a coil of wire; see Fig. 3.24.

For a sinusoidal varying magnetic field B of frequency f, the EMF induced in the coil is then simply given by

$$e^{\text{ind}} = -2\pi f B_0 S \cos \omega t, \quad (3.43)$$

where f is the frequency of the field, $\omega = 2\pi f$, and B_0 is the component of magnetic induction perpendicular to the loop.

As the induced voltage has the opposite sign with respect to EMF, it follows that a magnetic field penetrating vertically through the coil induces the voltage in the coil

$$U_{\text{ind}} = n2\pi f B S, \quad (3.44)$$

where n denotes a number of turns, f stands for frequency, B is the root mean square (rms) value of mag-

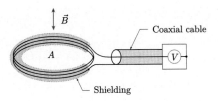

FIG. 3.24 Coil-type magnetic field meter.

netic induction, and S is the cross-sectional area of the coil. If the coil has high-impedance termination, the field strength can be determined directly by measuring the induced voltage as follows:

$$B = \frac{U_{\text{ind}}}{n2\pi f S},\qquad(3.45)$$

thus the induced voltage is dependent on the magnetic induction and on the frequency. It is necessary to compensate for the frequency response using an appropriate integrator element (as the frequency response compensation is improved, the influence of the frequency on the measurement accuracy decreases). The accuracy of the result is naturally highly dependent on how precisely the coil is situated at a right angle to the field. The measurement is regarded correct only if the magnetic field passes perfectly vertically through the coil.

Isotropic field probes, having three coils arranged perpendicularly to one another, as shown in Fig. 3.25, represent an improvement over one-dimensional magnetic field probes.

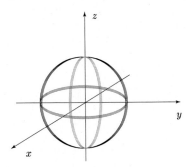

FIG. 3.25 Three-dimensional isotropic coil arrangement.

The benefit of such an arrangement is that the sensor always provides the same value for the equivalent field strength regardless of the orientation.

In general, a good sensor for magnetic field measurement should have a large effective area (to avoid effects due to small rotational fields), high sensitivity based on many winding turns, low capacitance of the windings, good shielding with respect to electric fields, low

isotropy error through precise positioning of the three coils, mechanical stability, etc.

3.1.4.3 Comparison of Calculated and Experimental Results

The first example deals with the measurement of a magnetic field from the three-phase 110 kV power line presented in [4].

A comparison of analytical results with the measurements available in [5] for the magnetic flux density B generated by the three-phase power line described in Sect. 3.1.1.2 (with current $I = 90$ A, conductor height of 11 m above the earth because of the sag) is given in Table 3.2.

It is obvious that the calculated/measured values of the magnetic flux density generated by the 110 kV power line at the height of $h = 2$ m above ground are well under the safety limits given by many of the national and international exposure limits.

Note that the highest difference does not exceed a few thousandths of μT. The model of the power line with straight conductors is a simplified one with average distance between towers of 300 m, and conductors are assumed to lay in the plane $z = $ const. over the earth. Minimal height of the conductor is 20.2 m above the earth.

Note that the calculated results are in a satisfactory agreement with the measured values.

The next example deals with power substations, i.e., with the measurement of the magnetic field in the proximity of a typical MV/LV power substation. The magnetic field level at the height of 1 m above the ground around two substations has been measured. The measurements intend to characterize the spatial variability of magnetic field levels around the power substation.

The measurements have been carried out in the vicinity of two characteristic MV/LV substations with nominal power of 630, 1000, 2 × 630, and 2 × 1000 kVA. As the professional exposure in MV/LV power substations is strictly limited to duration during the survey and maintenance, measurements have been undertaken within the interior of the substation. For each substation, 432 measurement points have been chosen, as shown in Fig. 3.26.

A simplified two-dimensional plot of substation 1 is shown in Fig. 3.27. The corresponding contour plots of measured magnetic flux density are presented in Fig. 3.28.

Furthermore, a simplified two-dimensional plot of substation 2 is shown in Fig. 3.29. The corresponding contour plots of measured magnetic flux density are presented in Fig. 3.30.

TABLE 3.2
Measured and calculated values of magnetic flux density B from the three-phase power line 11 m above the ground.

Distance from y-axis	Measured values at $I = 90$ A		Analytically calculated values at $I = 90$ A	
m	B (µT) at $z = 1$ m	B (µT) at $z = 2$ m	B (µT) at $z = 1$ m	B (µT) at $z = 2$ m
−30	0.08	0.08	0.129	0.132
−20	0.15	0.15	0.243	0.254
−15	0.22	0.23	0.353	0.377
−10	0.33	0.37	0.522	0.575
−5	0.54	0.61	0.736	0.842
0	0.76	0.95	0.890	1.046
5	0.77	0.94	0.832	0.973
10	0.56	0.60	0.608	0.680
15	0.35	0.37	0.405	0.435
20	0.22	0.23	0.273	0.287
30	0.10	0.11	0.140	0.144

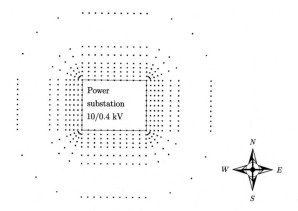

FIG. 3.26 The measurement points around the substation.

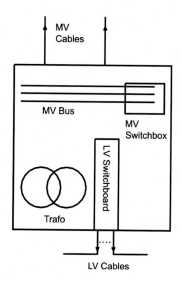

FIG. 3.27 MV/LV substation 1.

While measuring the magnetic field of substation 1, the transformer operated under 28% of the nominal load ($I_{avg} = 260$ A), whereas, in substation 2 the transformer operated under 24% of the average nominal load ($I_{avg} = 327$ A).

The measured values of the magnetic flux density can be extrapolated to the nominal transformer power by assuming that the magnetic flux density variation is proportional to the relative current variation.

The LV equipment (switchboard and cables) has the most significant influence on the magnetic field levels. The highest field level (18.3 µT) is measured in TS substation 1, while in TS substation 2 with the equipment placed in the center of the room, the influence of LV equipment is significantly lower. Furthermore, incoming and outgoing cables are coming from the cable duct and do not increase the level of the field.

In TS substation 1, at the western side, the cable that supplies the MV bus is attached to a wall and increases the value of the field appreciably.

Generally, the magnetic field levels from TS substation 2 are lower.

Therefore, the layout of the substation strongly influences the magnetic field, i.e., the rearrangement of the

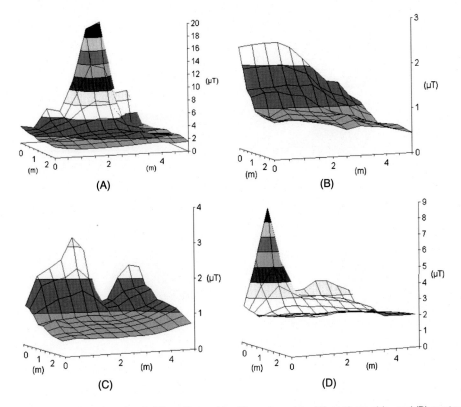

FIG. 3.28 TS 10/0,4 kV substation 1: (A) southern side, (B) eastern side, (C) northern side, and (D) western side.

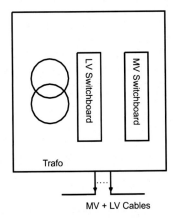

FIG. 3.29 MV/LV substation 2.

substations' layout can reduce the magnetic field levels. Furthermore, LV switchboard and cables produce the highest level of the magnetic field, while distribution transformers do not contribute significantly to the magnetic field. The magnetic field values fall off rapidly away from the substation.

3.2 ASSESSMENT OF HIGH FREQUENCY ELECTROMAGNETIC FIELDS

The presence of the non-ionizing radiation due to high frequency (HF) electromagnetic field sources in the environment has continuously caused an increasing public concern regarding possible health risk. Namely, during HF exposures the human body may absorb a significant amount of the radiated energy, as the dimensions of organs are comparable to the wavelength of the external field, thus making the tissue heating the dominant biological effect [8,12–14]. Consequently, HF dosimetry includes the assessment of the external field (incident field dosimetry), the rate of power deposition in tissue due to the exposure to HF radiation and the related temperature increase in tissue.

HF sources to electromagnetic interference (EMI) to be considered in this chapter are simple power line communication (PLC) systems, radio frequency identification (RFID) antitheft gate systems, and radio base station antennas. Some theoretical and experimental dosimetry procedures are presented. Also, details on rel-

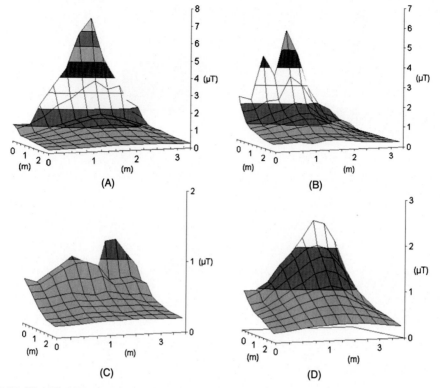

FIG. 3.30 TS 10/0,4 kV substation 2 (Podi 9): (A) southern side, (B) eastern side, (C) northern side, and (D) western side.

evant analytical and numerical techniques for theoretical dosimetry are given.

3.2.1 Fields Radiated by Power Line Communication (PLC) Systems

A power line communication (PLC) system aims to provide necessary communication means in terms of existing power line network and electrical installations in houses and buildings. A serious shortcoming of this technology is electromagnetic interference (EMI) aspect, as overhead power lines at the PLC frequency range (1 MHz to 30 MHz) behave as transmitting or receiving antennas.

The geometry of a simple PLC system is shown in Fig. 3.31. The system consists of two conductors placed in parallel above each other at a distance d. The conductors are suspended between two poles of equal height, thus having the shape of the catenary [15].

The geometry of a catenary is fully represented by such parameters as the distance between the points of suspension, L, the sag of the conductor, s, and the height of the suspension point, h, as shown in

Fig. 3.31. The imperfectly conducting ground is characterized with the electrical permeability ε_r and conductivity σ. The conductors are modeled as thin wire antennas excited by the voltage generator V_g at one end, and terminated by the load impedance Z_L at the other end.

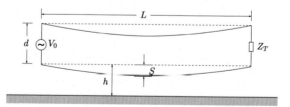

FIG. 3.31 Simple outdoor PLC circuit.

Theoretical incident dosimetry of a simple PLC system is based on the set of Pocklington integro-differential equations for the calculation of the current distribution and subsequently radiated field from power lines given by [16]

$$E_{sm}^{\text{exc}}(s) = \frac{j}{4\pi\omega\varepsilon_0} \sum_{n=1}^{N_w} \int_0^{L_n} \left\{ \left[k_1^2 \vec{e}_{sm} \vec{e}_{s_n'} - \frac{\partial^2}{\partial s_m \partial s_n'} \right] \right.$$

$$\times g_{0n}(s_m, s_n')$$

$$+ R_{\text{TM}} \left[k_1^2 \vec{e}_{sm} \vec{e}_{s_n^*} - \frac{\partial^2}{\partial s_m \partial s_n^*} \right] g_{in}(s_m, s_n^*)$$

$$+ (R_{\text{TE}} - R_{\text{TM}}) \vec{e}_{sm} \vec{e}_p \cdot \left[k_1^2 \vec{e}_p \vec{e}_{w*} - \frac{\partial^2}{\partial p \partial s^*} \right]$$

$$\left. \times g_i(s_m, s_n^*) \right\} I(s_n') ds' \bigg\}, \qquad (3.46)$$

where E^{exc} is the corresponding excitation function, N is the total number of wires, and $I_n(x')$ is the unknown current distribution induced on the nth wire, while $g_{0mn}(x, x')$ and $g_{imn}(s, s')$ are related Green functions:

$$g_{0mn}(s_m, s_n') = \frac{e^{-jk_1 R_{1mn}}}{R_{1mn}},$$

$$g_{imn}(s_m, s_n') = \frac{e^{-jk_1 R_{2mn}}}{R_{2mn}}, \qquad (3.47)$$

where indices 0 and i are related to the source and image wires, respectively, R_{1mn} and R_{2mn} are distances from the source point and corresponding image to the observation point of interest, respectively, and $k_1 = \omega\sqrt{\mu_0\varepsilon_0}$ is the propagation constant of free-space.

The influence of a lossy ground is taken into account via the Fresnel plane wave reflection coefficient for TM and TE polarization, respectively [16]:

$$R_{\text{TM}} = \frac{\underline{n}\cos\theta' - \sqrt{\underline{n} - \sin^2\theta'}}{\underline{n}\cos\theta' + \sqrt{\underline{n} - \sin^2\theta'}},$$

$$R_{\text{TE}} = \frac{\cos\theta' - \sqrt{\underline{n} - \sin^2\theta'}}{\cos\theta' + \sqrt{\underline{n} - \sin^2\theta'}}, \qquad (3.48)$$

where θ' is the angle of incidence, $\underline{n} = \varepsilon_{\text{eff}}/\varepsilon_0$ is the refraction index, and $\varepsilon_{\text{eff}} = \varepsilon_r\varepsilon_0 - j\sigma/\omega$ is the complex permittivity of the ground.

The integral equations (3.46) are numerically solved using the Galerkin–Bubnov scheme of the Boundary Element Method [17].

Once the equivalent current distribution over the wires is obtained, the values of the radiated electric and magnetic fields at an arbitrary point are readily obtained.

The total electric field irradiated by an arbitrary wire configuration is obtained from the following expression [16]:

$$\vec{E} = \frac{1}{j4\pi\omega\varepsilon_0} \sum_{n=1}^{N} \left\{ \left[k_1^2 \int_0^{L_n} \vec{e}_{s_n'} I(s_n') g_{0n}(\vec{r}, \vec{r}') ds_n' \right. \right.$$

$$+ \int_0^L \frac{\partial I(s_n')}{\partial s_n'} \nabla g_{0n}(\vec{r}, \vec{r}') ds_n' \bigg]$$

$$+ R_{\text{TM}} \left[k_1^2 \int_0^{L_n} \vec{e}_{s^*} I(s_n') g_{in}(\vec{r}, \vec{r}^*) dw' \right.$$

$$\left. - \int_0^{L_n} \frac{\partial I(s^*)}{\partial s^*} \nabla g_{in}(\vec{r}, \vec{r}^*) ds' \right]$$

$$+ (R_{\text{TE}} - R_{\text{TM}}) \left(\left[k_1^2 \int_0^{L_n} \vec{e}_{s^*} I(s_n') g_{in}(\vec{r}, \vec{r}^*) dw' \right. \right.$$

$$\left. \left. - \int_0^{L_n} \frac{\partial I(s^*)}{\partial s^*} \nabla g_{in}(\vec{r}, \vec{r}^*) ds' \right] \cdot \vec{e}_p \right) \vec{e}_p \right\}. \qquad (3.49)$$

The computational example pertains to the simple PLC circuit shown in Fig. 3.31. The distance between the poles is $L = 200$ m, with the radius of wires $a = 6.35$ mm. The wires are suspended on the poles at heights $h_1 = 10$ m and $h_2 = 11$ m above the ground, respectively. The sag of the conductor is assumed to be $s = 2$ m, and ground parameters are $\varepsilon_r = 13$ and $\sigma = 0.005$ S/m. The impressed power is $P = 1$ mW (average power required for the PLC system operation) and the operating frequency is changed from 1 to 30 MHz. The value of the terminating load Z_L is 50 Ω.

The calculated values of vertical component of the radiated electric field for the frequency range from 1 to 30 MHz and are shown in Fig. 3.32.

It is worth noting that the maximum levels of the calculated electric fields represent the worst case scenario for the assessment of the human exposure to a simple outdoor PLC system electromagnetic radiation (person standing directly under the power lines).

3.2.2 Fields in the Vicinity of RFID Loop Antennas

Radio Frequency Identification (RFID) is a term used for wireless description of an object by means of radio waves, with broad areas of applications: transport

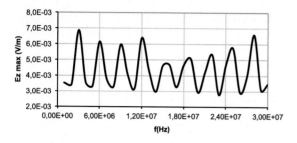

FIG. 3.32 Maximum value of E_z component below the conductor ($z = 1.75$ m).

ID through manufacturing processes, laundry identification, authentication/verification, employee ID, access control, automated airline baggage systems, shop antitheft protection, etc. [18]. An RFID system mainly consists of a transponder (RFID tag) and reader. There are two types of RFID system: passive and active. Passive RFID systems do not have the transmitter and just reflect the wave arriving from reader antenna (backscatter). Active systems have their own power supply and a transmitter, which is used for broadcasting the information stored on the microchip. Passive RFID tags consist of a microchip and antenna, which operates at low (124, 125, and 135 kHz), high (13.56 MHz) and ultra-high frequencies (860–960 MHz and 2.45 GHz) with range up to 10 m. Active RFID tags operate at 455 MHz, as well as 2.45 and 5.8 GHz, usually with range from 20 to 100 m. An RFID shop antitheft protection system is shown on Fig. 3.33.

FIG. 3.33 Antitheft store protection gate.

A numerical simulation of an RFID rectangular loop antenna operating at 13.56 MHz has been carried out in [18]. The loop antenna is a closed-circuit antenna that may have different forms, such as rectangle, square, triangle, ellipse, circle, etc. These antennas can generally be divided in two different classes: antennas with conductor length and loop dimensions small compared to wavelength ($< \lambda/10$), and antennas with conductor length and loop dimensions comparable to wavelength.

The measurement of the magnetic field generated from an RFID shop antitheft protection system, as shown in Fig. 3.34, has been reported in [18,19]. The input power is $P = 50$ mW per loop antenna. The wire radius is 0.9 mm, while the side length of the wire square loop is 46.5 cm.

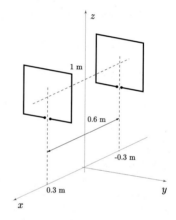

FIG. 3.34 Gate antenna system used measurement.

Fig. 3.35 shows the measured magnetic field distribution along the vertical line from $z = 0$ to $z = 2$ m, corresponding to a human standing beside the system, at a diagonal distance of approximately 1.8 m from the center of the antenna system ($x = 1.3$ m, $y = 1.3$ m), thus representing an employee exposure to the antitheft gate for a longer period.

The measured field value of 1.1 mA/m² is significantly below ICNIRP reference levels (160 mA/m² for workers and 73 mA/m² for general population).

3.2.3 Radiation From Base Station Antennas

The tremendous growth of the cellular telecommunication industry has resulted in an increasing number of various transmitting installations, particularly GSM base stations, and the related influence on human health has become a very hot subject in the last two decades. While the message or data-handling processes and computational capability are necessary aspects of a

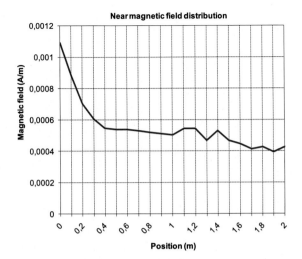

FIG. 3.35 Magnetic field intensity along the vertical line, $x = 1.3$ m, $y = 1.3$ m, from $z = 0$ to $z = 2$ m.

wireless communication system, the intensity and form of transmitted electromagnetic energy is of great interest to biological researchers. The estimation of the electromagnetic field generated from base station antennas regarding the possible health effects is usually carried out by taking into account the worst case scenario of the exposure.

The radiated power from a base station antenna may vary from less than a watt to hundreds of watts per channel (transmitter) or more, which depends on the particular location and type of the antenna used for a cellular communication system.

Base station antenna is usually mounted on a roof, or on a free-standing tower, or on the sides of a building, or on another structure that provides satisfactory height.

The low power antennas can be also mounted indoors, in offices, homes or airport terminals. High power antennas, on the other hand, are used at rural areas to relay calls between the user and the base stations, which are spaced too far apart.

The distribution of base stations with associated antenna powers is determined by coverage and capacity – two related factors of the wireless communication system.

Proper signal strength is necessary to cover the entire surface area and a sufficient capacity is needed to provide enough free channels to accommodate any user within the cell who might wish to use the system. As the number of users increases, more and more base stations must be installed closer together in order to ensure sufficient capacity. On the other hand, they operate at lower power levels, in order to provide comparable

signal intensity and to prevent undesired interaction among neighboring base stations.

Thus, in urban areas, cells are closer, but are operated at lower power levels than in rural areas, where the cells are larger. The amount of microwave power radiated from a base station is limited by national and international regulations. In rural or less-populated areas, cellular base stations may use several omnidirectional antennas, 3 to 5 m in length, which can be located from 30 to 100 m above the ground. An omnidirectional antenna radiates uniformly in all directions of the horizontal plane.

Base station antennas in rural areas are usually located on-free standing, tall, tapered poles, or on a towers of a metal-strut lattice construction. These antennas may also be located on existing structures (buildings, water tanks, high-voltage transmission line towers).

To summarize, the radiated power of the GSM base station antenna depends on the number of channels and temporary established links, quality and content of the information being transferred.

Nevertheless, the radiation of a GSM device is pulsed in nature, the control channel emits almost continuously so that the mean (measured) power is practically equal to the peak value and it is not necessary to take into account the pulse parameters.

Electromagnetic energy irradiated by an antenna system into space is directed in accordance to the radiation pattern. Consequently, the radiation on the street in the vicinity of the building, on which roof the antenna system is mounted (the roof-top antenna), is expected to be negligible.

The levels of public exposure to electromagnetic energy from any base station vary depending on antenna type, location and distance from the base station. The base station antennas are most commonly used in a sectorial arrangement and produce a so-called pie-shaped beam. This beam is wide in the horizontal direction and narrow in its vertical direction.

The most of irradiated energy is distributed within the main beam, wide in the horizontal plane as shown in Fig. 3.36A and relatively narrow in the vertical plane. Note that the vertical beam has mechanical or electrical tilt with angle of 5 to 10°, as depicted in Fig. 3.36B.

While most of the radiated energy is contained in the main beam, as one moves away from the antenna, the signal propagates and decreases in intensity as the inverse square of the distance from the antenna. Consequently, radiated fields are much weaker outside of the main beam than within it. The microwave exposure a person receives from a base station therefore depends on both the distance from the antenna and the angle

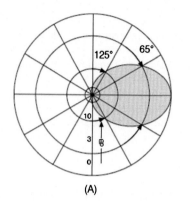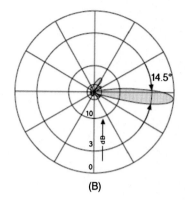

FIG. 3.36 (A) Horizontal pattern of GSM antenna, (B) Vertical pattern of GSM antenna.

below the direction of the main beam. At ground level, the signal is relatively weak near the base of an antenna tower, since the main beam is directed toward the horizon, and passes mostly overhead.

Exposure levels from cellular base stations can be determined by calculations and/or measurement. It is worth mentioning that the presence of reflecting and scattering structures, such as building walls, can have a profound influence on both the exposure and power deposition inside the human body. The microwave power density is lower inside buildings than outside, since a portion of the signal is absorbed when it passes through most building materials. For example, the attenuation factor of glass, wood or cement is between 10 and 100.

The term EIRP (equivalent isotropic radiated power) is an important concept to express the capabilities of RF transmission, and consequently a very important parameter of the base station antennas. EIRP could be derived from the power density of isotropic radiator.

For time-harmonic dependence and for the case of linear, homogeneous and isotropic media, the time averaged value of power density is given by

$$\vec{S}_{av} = \frac{1}{2\pi} \int_0^{2\pi} \vec{S}(\vec{r}, t)\, d(\omega t) = \frac{1}{2} \mathrm{Re}\left[\vec{E} \times \vec{H}^*\right]. \quad (3.50)$$

Factor 1/2 appears in (3.50) as E and H are expressed in terms of peak values and is to be omitted if root mean square values are considered.

The total average power (power flow) is then

$$P_{av} = \oint_S \frac{1}{2} \mathrm{Re}\left[\vec{E} \times \vec{H}^*\right] d\vec{A}, \quad (3.51)$$

where $d\vec{A}$ is the differential of the corresponding surface and represents the total radiated power by the radiation source.

The power density in the far field zone is

$$S_{av} = \left|\frac{1}{2}\mathrm{Re}\left[\vec{E} \times \vec{H}^*\right]\right| = \frac{1}{2}Z_0 H_\phi^2 = \frac{1}{2}\frac{E_\theta^2}{Z_0}, \quad (3.52)$$

where $Z_0 = 377\ \Omega$ is the free-space impedance.

For the case of a point source (isotropic radiator), the transmitter power is obtained by integrating the power density over imaginary sphere of radius r, that is,

$$P_t = \oint_S \vec{S}\, d\vec{A} = \int_0^{2\pi}\int_0^{\pi} S \cdot \hat{e}_\theta r^2 \sin\theta\, d\theta\, d\phi \cdot \hat{e}_r = S \cdot 4\pi r^2. \quad (3.53)$$

The power density then can be expressed as

$$S = \frac{P_t}{4\pi r^2}, \quad (3.54)$$

where P_t is the power at the antenna input, and r is the distance from the radiation source to the observation point.

For a directional antenna, the corresponding power density is defined as

$$S = \frac{P_t \cdot G_t}{4\pi r^2}, \quad (3.55)$$

where G_t is the gain of the transmitting antenna based on an isotropic radiator in the direction of maximal radiation.

EIRP is then given by the product

$$\text{EIRP} = P_t \cdot G_t, \tag{3.56}$$

and if the attenuation of the system (ohmic losses) is taken into account then (3.56) becomes

$$\text{EIRP} = \frac{P_t \cdot G_t}{L}, \tag{3.57}$$

or, in a decibel notation (with EIRP expressed in dBm),

$$\text{EIRP } [\text{dBm}] = P_t \, [\text{dB}] + G_t \, [\text{dBi}] - L \, [dB], \tag{3.58}$$

where G_t is the antenna gain in dBi, P_t is the maximal transmitted power (dB), L (dB) represents the ohmic losses between the transmitter and antenna (cable loss + combiner loss + feeder loss), and N is the number of transmitters.

Quantities from (3.58) are presented in Fig. 3.37.

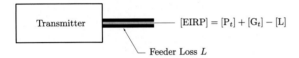

FIG. 3.37 Basic elements of a wireless communication system.

The choice of corresponding method for the calculation of the field radiated by a base station antenna system depends on various parameters, such as the distance from the radiation source.

Depending on the distance from the antenna, the observation point is located in the near or far field, respectively, as indicated in Table 3.3.

TABLE 3.3
The near and far field of an antenna (r is the distance from an antenna, L means antenna length, and λ is the wavelength).

near *induction* field	near *radiated* field	far field
$r \leq \dfrac{\lambda}{4}$	$\dfrac{\lambda}{4} \leq r \leq \dfrac{2D^2}{\lambda}$	$r \geq \dfrac{2D^2}{\lambda}$

The near field consists of radiated (Fresnel) and reactive (induction) fields. In the zone of inductive field, most of the energy is not radiated but oscillates from the source to the observation point. In radiative zone of the near field, the energy is propagated from the antenna, but not in the form of a plane wave. In the far field zone (Fraunhofer zone) of the antenna, the electromagnetic energy is radiated in the form of a plane wave.

3.2.3.1 Near Field Analysis: Assessment of Power Density

Base station antennas can be represented by an array of collinear half-wave resonant dipoles, as depicted in Fig. 3.38.

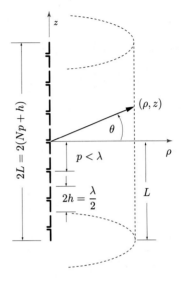

FIG. 3.38 A colinear array of resonant dipoles.

On the basis of cylindrical reference frame, the total average power P_{rad} radiated by an axial array can be written as [20,21]

$$P_{\text{tot}} = N P_{\text{rad}} = \oint_S \vec{S} \, d\vec{A} = \int_0^{2\pi} \int_{-L}^{+L} S \cdot \hat{e}_\theta^* \, \rho \, d\phi \, dz \cdot \hat{e}_\rho$$

$$= S \cdot 2\pi\rho \cdot 2L, \tag{3.59}$$

where N is the number of carriers (channels), ρ represents the radial distance from the antenna, and $2L$ is the entire antenna length.

The average power density is then [20,21]

$$S = \frac{N P_{\text{rad}}}{2\pi\rho 2L}. \tag{3.60}$$

Taking into account sectorial coverage with signal, the average power density is given by [22]

$$S = \frac{N P_{\text{rad}}}{\pi\rho 2L} \frac{180}{\overline{\phi}}, \tag{3.61}$$

where $2\overline{\phi}$ stands for -3 dB azimuth beam-width.

3.2.3.2 Far Field Analysis: Calculation of Power Density and Electric Field

Provided the observation point is in the far field, the base station antenna represents a point radiation source (Hertz dipole), as shown in Fig. 3.39.

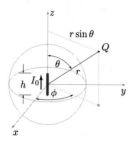

FIG. 3.39 Hertz dipole.

Combining (3.52) and (3.55), but for rms values, it follows that

$$\frac{E^2}{Z_0} = \frac{P_t G}{4\pi r^2}, \tag{3.62}$$

and the electric field is given by

$$E = \frac{\sqrt{30 P_t G}}{r}. \tag{3.63}$$

For the radio base station antenna system with N channels, using the EIRP concept, one obtains

$$S = \frac{N \cdot P_t G_t}{4\pi r^2} = \frac{N \cdot \text{EIRP}}{4\pi r^2}, \tag{3.64}$$

and the related rms field value is then given by

$$E = \frac{\sqrt{30 N \cdot P_t G}}{r} = \frac{\sqrt{30 N \cdot \text{EIRP}}}{r}. \tag{3.65}$$

Next step is an extension of expression (3.65) to the case of an antenna above the ground and to account for the observation point not being located in the direction of maximum radiation in the horizontal and vertical plane, respectively.

The magnitude of the far-field can by determined by using the geometrical optics method. The total field can be expressed as a superposition of the incident and reflected field components

$$\vec{E}^{\text{tot}} = \vec{E}^{\text{inc}} + \vec{E}^{\text{ref}}. \tag{3.66}$$

Considering the worst case scenario, i.e., by assuming the incident and reflected field to be in phase, vector

expression (3.66) simplifies into algebraic one, namely

$$E^{\text{tot}} = E^{\text{inc}} + E^{\text{ref}}, \tag{3.67}$$

where the corresponding incident and reflected field components are given by [23,24]

$$E^{\text{inc}} = \frac{\sqrt{30 N P_{\text{rad}} G(\phi, \theta)}}{R} e^{-j\beta R} \tag{3.68}$$

and

$$E^{\text{ref}} = \Gamma_R(\phi^*, \theta^*) \frac{\sqrt{30 N P_{\text{rad}} G(\phi^*, \theta^*)}}{R^*} e^{-j\beta R^*}, \tag{3.69}$$

where Γ_R is the corresponding reflection coefficient, $G(\phi, \theta)$ is the radiation pattern for a particular antenna, R is the distance from the source to observation point, while (*) denotes the image quantities.

Furthermore, if one uses the concept of effective isotropic radiated power (EIRP) and antenna gain G (assuming the direction of maximum radiation), the total field above a perfect electrically conducting (PEC) ground can then be written as follows:

$$E^{\text{tot}} = 2\frac{\sqrt{30 N P_{\text{rad}} G}}{R} = 2\frac{\sqrt{30 N \cdot \text{EIRP}}}{R}, \tag{3.70}$$

where factor 2 denotes the total reflection from PEC ground.

Finally, if the observation point is not in the maximum direction, the relative (numerical) gain $F(\phi, \theta)$ should be used, and for the total field and power density one has [23,24]:

$$E = 2\frac{\sqrt{30 N \cdot \text{EIRP}}}{R} F(\phi, \theta), \tag{3.71}$$

$$S = \frac{N \cdot \text{EIRP}}{\pi R^2} F^2(\phi, \theta). \tag{3.72}$$

The presence of reflected and scattering objects, such as building walls, may significantly affect the field levels. For example, the attenuation factor for materials such as glass, wood or concrete varies from 10 to 100 dB.

Thus, an overestimated prediction of the field at a given point is likely to be obtained from Eqs. (3.71) and (3.72).

In such cases, taking into account all these reflection phenomena, an approximation for predicting ground level field strength and power density is undertaken empirically by assuming a maximum 1.6-fold increase in field strength, resulting in an increase of power density of 2.56 (1.6 × 1.6).

TABLE 3.4
Parameters of the GSM base station mounted on a free standing tower.

Operating frequency	GSM downlink band (935–960 MHz)
Number of sectors	3
Directions of main lobes	sector A, 65°; sector B, 200°; sector C, 290°
Elevation of main lobes	sector A, −3°; sector B, 3°; sector C, 2°
Antenna type	sector A/B/C, Celwave AP907016, 2 antennas (per sector)
Number of channels	sector A/B/C, 2
Maximal permissible EIRP per channel	sector A/B/C, 59.5 dBm

Consequently, Eq. (3.72) is then corrected as

$$S = \frac{2.56 \mathrm{EIRP}}{4\pi R^2} F^2(\phi, \theta) = \frac{0.64 \mathrm{EIRP}}{\pi R^2} F^2(\phi, \theta). \quad (3.73)$$

The installation of base station antennas often raises concerns about their influence on human health, mostly for the people living in the close vicinity of such sites. Therefore, the circumstances where people could be exposed to field levels higher than exposure limits should be examined and the safety distances should be established.

The safety distance $d_{\min}$ from a base station antenna can be readily derived from the relation for the power density

$$S = N \frac{P_t 10^{\frac{G_t - L}{10}}}{4\pi R^2}, \quad (3.74)$$

where G_t is the antenna gain in dB, P_t is the maximal transmitted power, L represents the ohmic losses between the transmitter and the antenna, and N is the number of transmitters.

The safety distance from an antenna system is then given by

$$d_{\min} = \sqrt{N \frac{P_t 10^{\frac{G_t - L}{10}}}{4 S_{\max}}}, \quad (3.75)$$

where $S_{\max}$ denotes the maximum allowed power density for the considered frequency range.

3.2.3.3 Some Computational Examples

Consider a GSM radiation source represented by radio base station antenna system mounted on a 46 m high free standing tower. Technical data are given in Table 3.4.

The least distance from the source to the location of interest is 600 m. The location is defined with azimuth

from 150° to 200°, while the elevation angle with respect to horizon is from −7° to −8° measured from the source. The total field calculated at the 600 m distance is $E_{\mathrm{tot}} = 0.46$ V/m, which is far below the ICNIRP referent level at GSM frequency ($E_{\lim} = 42$ V/m) and Croatian exposure limit ($E_{\lim} = 16.82$ V/m) for general population.

Fig. 3.40 shows the distribution of the electric field in the vicinity of the antenna for the worst case scenario (points of interest are within the main beam). It is visible that at distances greater than 30 m from the antenna system the field levels fall below the reference level(s).

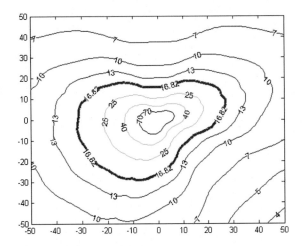

FIG. 3.40 Electric field distribution in the vicinity of an antenna system.

The next example deals with a radio base station antenna system mounted on a roof top at the height of 17 m above ground. Technical data are given in Table 3.5.

Fig. 3.41 shows the distribution of the radiated field in the vicinity of the antenna system for the worst

TABLE 3.5
Parameters of the GSM base station antenna system.

Operating frequency	GSM downlink band (935–960 MHz)
Number of sectors	3
Directions of main lobes	sector A, 5°; sector B, 125°; sector C, 245°
Elevation of main lobes	sector A, 0°; sector B, 2°; sector C, 2°
Antenna type	sector A/B/C, Kathrein 742264, 4 antennas (per sector)
Number of channels	sector A/B/C, 2
Maximal permissible EIRP per channel	sector A/B/C, 59.5 dBm

case scenario. At a distance of 20 m from the antenna, the electric field values are below referent level of 16.82 V/m.

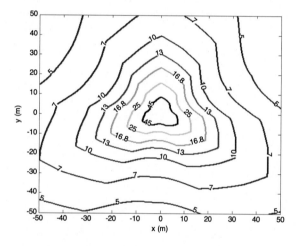

FIG. 3.41 Electric field distribution in the vicinity of the antenna system.

The last computational example deals with radiation of a base station at UMTS frequency range. Technical data are given in Table 3.6.

Fig. 3.42 shows the distribution of the electric field in the vicinity of UMTS antenna system for the worst case scenario (radius of 50 m in the vicinity of the antenna in the main beam).

It is visible that the electric field of UMTS base station falls below exposure limit ($E = 24.4$ V/m) [25] at distances less than 10 m from the antenna.

Calculated values of the electric field at the level of the antenna system within the main beam represent worst case scenario, while in practice the base stations are placed on a roof top, or on a free standing tower.

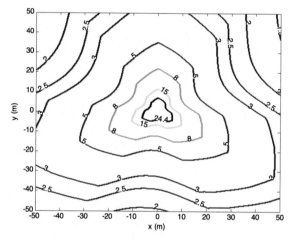

FIG. 3.42 Electric field distribution in the vicinity of the UMTS antenna system.

Therefore, it is not realistic for human beings to be within the main beam.

Moreover, if one were to take into account the attenuation due to propagation through concrete (~10 dB), the field levels at considered points would be even lower.

3.2.3.4 Some Measured Results

Consider a GSM base station mounted on a roof top at a height of 17 m above the ground, as shown on Fig. 3.43. Technical data are given in Table 3.7.

Using calibrated dipoles and spectrum analyzer, selective measurements at corresponding frequencies of BCCH channels yield the total field $E = 2.33$ V/m.

Finally, the last example deals with the roof-top antenna system in a sectorial arrangement mounted at a height of 48 m above ground, as shown on Fig. 3.44. The radiated power per channel is $P = 955$ W at $f = 935$ MHz. Technical data are given in Table 3.8.

TABLE 3.6
Parameters of the UMTS base station antenna system.

Operating frequency	UMTS downlink band (2110–2170 MHz)
Number of sectors	3
Directions of main lobes	sector A, 5°; sector B, 125°; sector C, 245°
Elevation of main lobes	sector A, 0°; sector B, 2°; sector C, 2°
Antenna types	sector A/B/C, Kathrein 742264, 1 antenna (per sector)
Number of channels	sector A/B/C, 4
Maximal permissible EIRP per channel	sector A/B/C, 54 dBm

TABLE 3.7
Parameters of the roof-top GSM base station.

Operating frequency	GSM downlink band (935–960 MHz)
Number of sectors	3
Directions of main lobes	sector A, 0°; sector B, 120°; sector C, 230°
Elevation of main lobes	sector A, 0°; sector B, 0°; sector C, 3°
Antenna types	sector A/B/C, Celwave APXV 906514, 2 antennas (per sector)
Number of channels	sector A/B/C, 3
Maximal permissible EIRP per channel	sector A/B/C, 58 dBm

TABLE 3.8
Parameters of the GSM base station antenna system.

Operating frequency	GSM downlink band (935–960 MHz)
Number of sectors	3
Directions of main lobes	sector A, 30°; sector B, 150°; sector C, 270°
Elevation of main lobes	sector A/B, 0°; sector C, 3°
Antenna types	sector A/B/C, Celwave APXV 906514, 2 antennas (per sector)
Number of channels	sector A/B, 4; sector C, 3
Maximal permissible EIRP per channel	sector A/B/C, 59.8 dBm

The electric field measured in a flat at the 15th floor below the antenna system at an approximated distance of 8 m from the radiation source is $E = 0.23$ V/m, which is rather negligible compared to the ICNIRP reference level $E_{\text{lim}} = 42$ V/m.

3.2.3.5 Far Field Analysis: Presence of a Lossy Ground and Layered Medium

This section deals with the simple analytical relations for the incident field dosimetry for the antenna radiating in the presence of inhomogeneous media based on the different schemes of ray tracing algorithm (arising from the geometrical optics approach [23,24] implementation. Six different approximations (used by different legislations) for the assessment of the field radiated by base station antenna are presented; the antenna insulated in free-space (FS), the antenna above perfect ground (PG), and the antenna above a lossy half-space. The half-space model uses the Fresnel reflection coefficient and the modified image theory (MIT) approximation [25,26]. Finally, the model of a layered medium featuring MIT approximation is used [27]. The results obtained by different models are compared to the results computed via Numerical Electromagnetics Code

FIG. 3.43 Antenna system of GSM base station mounted on a roof top.

FIG. 3.44 A view to the 2 sectors of base station antenna system.

(NEC) [28]. Several variables have been varied to examine the accuracy of a given approximation. All calculations have been carried out for the far field zone.

The free-space approximation is used by some dosimetry guidelines and neglects effects due to reflections and scattering, thus taking into account only the incident field; see Fig. 3.45.

As already discussed in (3.2.3.2), the electric field at some point in free-space in the far-field zone is given by

$$E^{tot} = E^{inc} = \frac{\sqrt{30N \cdot EIRP}}{R}. \tag{3.76}$$

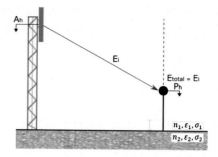

FIG. 3.45 Free-space approximation.

Eq. (3.76) gives the absolute value of the total electric field, thus taking into account the incident field only.

As previously mentioned, a simple way to account for the effect of the reflected wave on the total electric field is to assume the ground to be perfectly conducting; see Fig. 3.46.

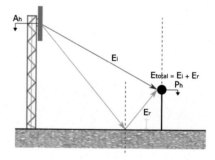

FIG. 3.46 Perfect ground approximation.

Therefore, the total electric field is

$$E^{tot} = E^{inc} + E^{ref} \tag{3.77}$$

where

$$E^{ref} = \frac{\sqrt{30N \cdot EIRP}}{R^*}, \tag{3.78}$$

with R^* being the distance a wave is traveling from the image antenna to the calculation point above perfect (PEC) ground.

Below analytical models, in which the presence of a lossy half-space is taken into account, are presented, see Fig. 3.47.

First, the half-space model featuring Fresnel's approach is analyzed.

The total field composed of incident and reflected fields is given by (3.77), and the reflected field can be

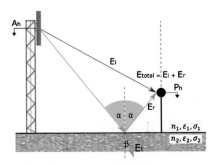

FIG. 3.47 Antenna above a lossy half-space.

calculated using Fresnel reflection coefficient (RC) approximation and given by

$$E^{\text{ref}} = \Gamma^{\text{Fr}} \cdot \frac{\sqrt{30N \cdot EIRP}}{R^*}. \tag{3.79}$$

The reflection coefficient arising from RC approach is

$$\Gamma^{\text{Fr}} = \frac{Z_1 \cos \alpha - Z_2 \cos \beta}{Z_1 \cos \alpha + Z_2 \cos \beta}, \tag{3.80}$$

where α and β is the angle of incidence and transmission, respectively, while Z_1 is the free-space impedance (377 Ω), and Z_2 is given by

$$Z_2 = \frac{j\omega\mu}{\gamma}. \tag{3.81}$$

Furthermore, the reflected electric field stemming from MIT approach is given by

$$E^{\text{ref}} = \Gamma^{\text{MIT}} \cdot \frac{\sqrt{30N \cdot EIRP}}{R^*}, \tag{3.82}$$

where the reflection coefficient arising from MIT approximation coefficient is

$$\Gamma^{\text{Fr}} = \frac{\varepsilon_{\text{eff}} - \varepsilon_0}{\varepsilon_{\text{eff}} + \varepsilon_0}, \tag{3.83}$$

where

$$\varepsilon_{\text{eff}} = \varepsilon_r \varepsilon_0 - j \frac{\sigma}{\omega}. \tag{3.84}$$

A computational example deals with 2 m long antenna configuration consisting of 8 dipoles and metal grid with the total radiated power of $P = 100$ W per channel. The operating frequency is $f = 936.8$ MHz. Simulations are carried out at calculation point height $P_h = 2$ m, by varying ground conductivity σ_2, relative permittivity ε_2, and antenna height A_h at calculation

point height ($P_h = 2$ m). The number of channels is $N = 1$.

Taking into account the far field zone condition,

$$r \geq \frac{2D^2}{\lambda}, \tag{3.85}$$

where r is a horizontal distance from the antenna, D is the largest antenna dimension and λ is wavelength. Calculating with $D = 2$ m and $\lambda = 0.32$ m, far field area is related to distances from 25 m. Thus, the electric field is calculated for distances between 25 and 400 m away from antenna with the step of 1 m (376 data in total per calculation).

Figs. 3.48 and 3.49 show the field versus distance for different approaches and compared with NEC.

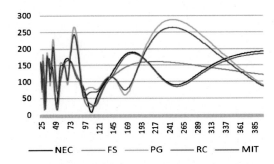

FIG. 3.48 Electric field computed via different approaches compared to NEC simulation in mV/m at specific distance from antenna ($\sigma = 0.01$ S/m, $\varepsilon_2 = 10$ and $A_h = 20$ m).

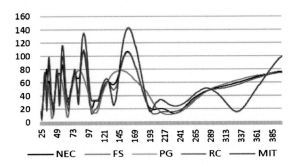

FIG. 3.49 Electric field computed via different approaches compared to NEC simulation in mV/m at specific distance from antenna ($\sigma = 0.01$ S/m, $\varepsilon_2 = 110$ and $A_h = 40$ m).

The PG and MIT methods are considered to be satisfactory approximations only for high ground conductivity values and provide similar results. The FS method provides a good approximation for higher relative permittivity values (above 10) and for higher A_h values, too. Finally, the RC approach provides rather satisfactory results in all cases.

Finally, the assessment of the electric field radiated by a base station antenna over a two-layered ground is analyzed. The approach used to assess the field reflected from the multilayered medium is based on the Modified Image Theory method (MIT) approach. Some basic concepts of MIT approximation are available in [29]. The geometry of the base station antenna radiating over a multilayered medium is shown in Fig. 3.50.

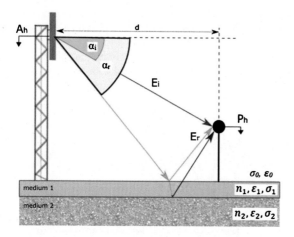

FIG. 3.50 The total electric field above a multilayered medium composed of the incident and reflected fields.

As shown in Fig. 3.50, the total electric field is composed of incident and reflected fields, and the reflected field from interface air-multilayer is given by [27]:

$$E^{ref} = \Gamma^{MIT}_{ml} \cdot \frac{\sqrt{30N \cdot EIRP}}{R^*}, \qquad (3.86)$$

where N is the total number of channels, P stands for the radiated power (W) and G denotes horizontal and vertical antenna gain (dB), respectively.

The reflected field is computed by using the simplified reflection coefficient arising from the use of the Modified Image Theory applied to the case of two-layered ground. Therefore, the reflected field from the interface is given by MIT reflection coefficient for two-layer ground given by [30]:

$$\Gamma^{MIT}_{ml} = \frac{R_{01} + R_{02}e^{-2\gamma l}}{1 + R_{01} \cdot R_{02}e^{-2\gamma l}}, \qquad (3.87)$$

where l is the given thickness of layer 2 and γ is the propagation constant of the medium, while R_{mn} by

which the reflection between the mth and nth layer is taken into account is given by

$$\Gamma^{Fr} = \frac{\varepsilon_{eff,m} - \varepsilon_{eff,n}}{\varepsilon_{eff,m} + \varepsilon_{eff,n}}, \qquad (3.88)$$

where $\varepsilon_{eff,m(n)}$ is the complex permittivity of the mth and nth layer, respectively.

A computational example is related to the antenna system with power $P = 100$ W, operating frequency $f = 936.8$ MHz and one active channel. The antenna is mounted 25 m above the ground. The horizontal antenna gain is supposed to be 0 dB while vertical antenna gain (α) is calculated according to radiation pattern and formula related to Fig. 3.50:

$$\alpha(°) = \arctan\left(\frac{A_h - P_h}{d}\right) - el_{tilt}, \qquad (3.89)$$

where A_h denotes an antenna height (m) and P_h calculation point height (m), while el_{tilt} denotes the electrical antenna tilt (for this purpose $el_{tilt} = 0°$). The total electric field is calculated only in the far zone (25 m away from the antenna). Therefore, the total electric field is calculated at the distances between 0 and 400 m from the antenna pillar at the height of 2 m above the ground.

The ground below antenna pillar consists of two different layers. The upper layer is presented as medium 1 (n_1, σ_1, ε_1), and the lower one as medium 2 (n_2, σ_2, ε_2). The basic physical characteristics of these two layers are presented in Table 3.9.

TABLE 3.9
Electrical properties of ground layers.

medium	σ [S/m]	ε_r
0	σ_0	ε_0
1	0.001	10
2	0.05	4

First, the impact of medium 1 thickness to the distribution of electric field above a multilayered soil is analyzed. The variation of thickness of medium 1 is shown Table 3.10.

Figs. 3.51 and 3.52 show the electric field distribution 2 m above the ground for various thickness of the first layer. Note that *Medium 1* denotes the field calculation with medium 1 included only (no medium 2 present), while *Medium 2* denotes the field calculation with medium 2 included only (no medium 1 present).

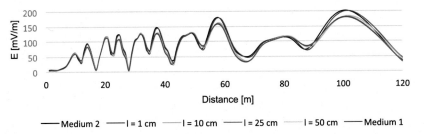

FIG. 3.51 Electric field for different values of the medium 1 thickness at distances up to 119 m away from antenna pillar.

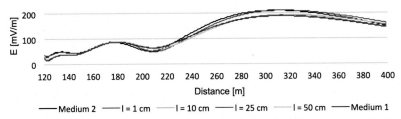

FIG. 3.52 Electric field for different values of the medium 1 thickness at distances between 120 and 399 m away from antenna pillar.

TABLE 3.10 Variation in thickness (l) of medium 1.			
I	II	III	IV
1 cm	10 cm	25 cm	50 cm

As could be seen from the obtained results, the curves have similar waveforms with slight amplitude differences, particularly at the local minimums and maximums at the distances of approximately up to 120 m away from antenna pillar (Fig. 3.53).

The field level curve for the $l = 1$ cm thick medium 1 almost perfectly follows the *Medium 2* curve, thus the influence of medium 1 could be neglected in this case. On the other hand, the increase of medium 1 thickness reduces the impact of medium 2 on the total field strength to a minimum. Thus, for the values of medium 1 thickness $l \geq 25$ cm, medium 2 soil can be ignored in field calculations as the average difference in comparison to *Medium 1* calculation is around 1.7%.

For the thickness of medium 1 of 1 cm, a maximum difference in the field value is 44% compared to *Medium 2* field calculation while the average difference is only 5%. Increasing the medium 1 thickness up to 50 cm, a maximum difference rises up to 68% with the average difference of approximately 12%.

The next set of results examines the impact of medium 1 conductivity to the electric field distribution.

Conductivity values are varied as shown in Table 3.11.

TABLE 3.11 Variation in conductivity of medium 1 in S/m.			
I	II	III	IV
0.05	0.1	1	10^3

The conductivity value of medium 2 remains the same ($\sigma = 0.001$ S/m), as well as relative permittivity ($\varepsilon_r = 10$). The results of field calculations are presented with respect to *Medium 2* calculation and shown in Figs. 3.53 and 3.54. The higher the conductivity of medium 1, the greater the difference in field level compared to *Medium 2* calculation. For the values of $\sigma_1 \leq 0.1$ S/m, a maximum field difference varies between 44% and 68%. Increasing σ_1 to 1 and 10^3 S/m, respectively, the highest field difference grows up to 96% and 114%, respectively.

Although the differences in field level at $\sigma_1 \leq 0.1$ S/m reach the value up to 68% of *Medium 2* calculation, the average differences are not crossing the 12% level. The same conclusion is valid for calculations with $\sigma_1 \geq 1$ S/m. However, the most significant growth in average field difference appears in the case with 1 cm thickness of medium 1 (between 5.2% and 10.9%) while the changes at other values of medium 1 thickness are much

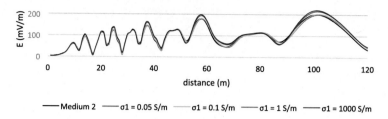

FIG. 3.53 Field curves for the medium 1 thickness $= 1$ cm between 0 and 120 m away from antenna pillar.

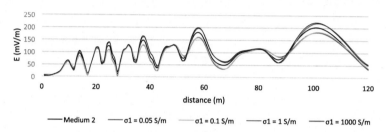

FIG. 3.54 Field curves for the medium 1 thickness $l = 25$ cm between 0 and 120 m away from antenna pillar.

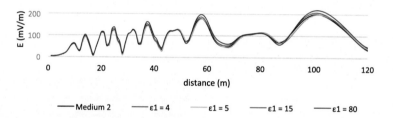

FIG. 3.55 Field curves with the medium 1 thickness $l = 1$ cm between 0 and 120 m away from antenna pillar.

lower (between 9.7% and 11.6%). The differences in field level occur at the points of local minima and maxima where the highest differences occur. The last set of results is related to the impact of the *Medium 1* permittivity on the field above a multilayered soil. Assuming the specific conductivity of medium 1 to be 0.05 S/m, the impact of relative permittivity of medium 1 on the total electric field is examined. Relative permittivity of medium 1 varied as shown in Table 3.12.

TABLE 3.12			
Variations in relative permittivity of medium 1.			
I	II	III	IV
4	5	15	80

The results for the electric field versus distance presented with respect to *Medium 2* calculation are shown in Figs. 3.55 and 3.56.

The highest differences in field level for medium 1 thickness $l \leq 10$ cm occur at the highest value of ε_1. For other medium 1 thicknesses ($l \geq 25$ cm), the highest field differences appear at the value of $\varepsilon_1 = 5$. The lowest average differences are present at the relative permittivity value of $\varepsilon_1 = 15$ independently of medium 1 thickness. Comparing the field levels for different values of relative permittivity ε_1, it is shown that a higher ε_1 results in a higher total field level. It is also interesting to notice the appearance of higher field level differences as the medium 1 thickness increases, which is also a proof of the impact of medium 1 thickness on total field strength. As in the case before, the higher field differences appear at points of local minima and maxima.

Generally, one concludes that the thicker the medium 1, the lower the impact of medium 2 on the field values. For the case of 1 cm thick medium 1, the impact is negligible. On the other hand, for the case of $l = 25$ cm thick medium 1, an impact of medium 2 on the total field strength is reduced and can be ignored.

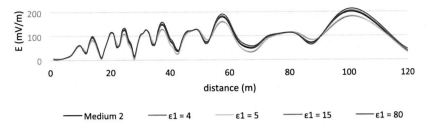

FIG. 3.56 Field curves with the medium 1 thickness $l = 25$ cm between 0 and 120 m away from antenna pillar.

This is obvious for medium 1 thickness of $l = 50$ cm or more, where the impact of medium 2 is brought to a minimum.

In conditions with lower values of specific conductivity of medium 1 ($\sigma_1 \leq 0.1$ S/m), a maximum difference in field level is between 44% and 68%, with the average difference between 5% and 12%.

Considering the impact of relative permittivity of medium 1 on the total electric field strength, it is shown that field differences can occur for any value of relative permittivity. It is worth noting that the influence of medium 1 can be ignored in field calculation if the thickness is lower than $l = 10$ cm. Similarly, medium 2 can be ignored if medium 1 is thicker than $l = 25$ cm.

3.2.3.6 Accurate Numerical Modeling of Radio Base Station Antenna Systems

An accurate and efficient simulation tool for the analysis of radiation from base station antenna is discussed in [31]. The base station antenna system is represented by the vertical antenna array in front of perfectly conducting (PEC) ground plane reflector.

The formulation of the problem is based on the set of coupled Pocklington integro-differential equations for a vertical antenna array. This set of coupled equations has been numerically treated by the indirect Galerkin–Bubnov scheme of the Boundary Element Method (GB-IBEM).

Knowing the current distribution along the antenna array, the corresponding radiated field is calculated.

Fig. 3.57 shows a vertical antenna array of M dipoles of length L_n, distance between centers of adjacent dipoles d_c, and horizontal distance between antennas and a perfectly conducting (PEC) ground plane reflector x_{dis}.

The currents induced along the antenna elements are governed by the set of coupled Pocklington integro-differential equations [31]:

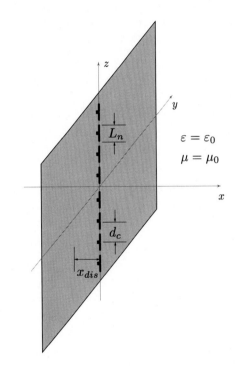

FIG. 3.57 Antenna array in front of reflector.

$$E_{zm}^{\text{inc}} = \frac{1}{j4\pi\omega\varepsilon_0} \sum_{n=1}^{M} \left[\frac{\partial^2}{\partial z^2} + k^2 \right]$$

$$\times \int_{-L/2}^{+L/2} I_n(z') G_{mn}(z, z')\, dz', \quad m = 1, 2, \ldots, M,$$

(3.90)

where $I_n(z')$ is the current distribution along the nth wire, and $G_{mn}(z, z')$ is the Green's function given by

$$G_{mn}(z, z') = g_{0mn}(z, z') - g_{imn}(z, z'), \quad (3.91)$$

where $g_{0mn}(z, z')$ is the free-space Green's function,

$$g_{0mn}(z, z') = \frac{e^{-jkR_{mn}}}{R_{mn}}, \qquad (3.92)$$

while $g_{imn}(z, z')$ is the Green's function due to the image wires,

$$g_{imn}(z, z') = \frac{e^{-jkR_{mn}^*}}{R_{mn}^*}. \qquad (3.93)$$

By solving equations (3.90) and obtaining the equivalent current distributions along the wires, the radiated electric field can be computed.

The three components of the electric field radiated by the vertical array from Fig. 3.57 are as follows [31]:

$$E_x = \frac{1}{j4\pi\omega\varepsilon_0} \sum_{n=1}^{M} \frac{\partial^2}{\partial x \partial z} \int_{-L/2}^{+L/2} G_{mn}(x, z') I_n(z') dz'$$

$$= \frac{1}{j4\pi\omega\varepsilon_0} \sum_{n=1}^{M} \int_{-L/2}^{+L/2} \frac{\partial I_n(z')}{\partial z'} \frac{\partial G_{mn}(x, z')}{\partial x} dz',$$

$$\qquad (3.94)$$

$$E_y = \frac{1}{j4\pi\omega\varepsilon_0} \sum_{n=1}^{M} \frac{\partial^2}{\partial x \partial z} \int_{-L/2}^{+L/2} G_{mn}(y, z') I_n(z') dz'$$

$$= \frac{1}{j4\pi\omega\varepsilon_0} \sum_{n=1}^{M} \int_{-L/2}^{+L/2} \frac{\partial I_n(z')}{\partial z'} \frac{\partial G_{mn}(y, z')}{\partial y} dz',$$

$$\qquad (3.95)$$

$$E_z = \frac{1}{j4\pi\omega\varepsilon_0} \sum_{n=1}^{M} \left[\frac{\partial^2}{\partial z^2} + k_1^2 \right]$$

$$\times \int_{-L/2}^{+L/2} G_{mn}(z, z') I_n(z') dz'$$

$$= \frac{1}{j4\pi\omega\varepsilon_0} \sum_{n=1}^{M} \left[\int_{-L/2}^{+L/2} \frac{\partial G_{mn}(z, z')}{\partial z} \frac{\partial I_n(z')}{\partial z'} dz' \right.$$

$$\left. + k^2 \int_{-L/2}^{+L/2} I_n(z') G_{mn}(z, z') dz' \right]. \qquad (3.96)$$

Numerical procedures involve solution of integral equations to obtain current distribution and subsequent evaluation of field integrals.

Through the GB-IBEM procedure, the set of integro-differential equations (3.90) is transformed into a linear equation system. Using the second-order Lagrange polynomials, i.e., discretizing the array geometry into quadratic boundary elements, the current distribution along the considered the ith wire segment is given by

$$I_{ni}(z') = I_{1i}^n f_1(z') + I_{2i}^n f_2(z') + I_{3i}^n f_3(z')$$

$$= I_{1i}^n \frac{z' - z_2}{z_1 - z_2} \frac{z - z_3}{z_1 - z_3} + I_{2i}^n \frac{z' - z_1}{z_2 - z_1} \frac{z' - z_3}{z_2 - z_3}$$

$$+ I_{3i}^n \frac{z' - z_1}{z_3 - z_2} \frac{z' - z_2}{z_3 - z_2}, \qquad (3.97)$$

while the current derivative is then

$$\frac{\partial I_{ni}(z')}{\partial z'} = \frac{2}{\Delta l^2} \{ 2 (I_{1i}^n - I_{2i}^n + I_{3i}^n) z' - [(z_3 + z_2) I_{1i}^n$$

$$- (z_3 + z_1) I_{2i}^n + (z_1 + z_2) I_{3i}^n] \}, \qquad (3.98)$$

where $\Delta l = z_3 - z_1$ denotes the segment length.

Using the weighted residual approach and performing certain mathematical manipulations, the set of algebraic equations arising from wire segmentation can be written in the form of matrix equation:

$$\sum_{n=1}^{M} \sum_{i=1}^{N_n} [Z]_{ji}^e \{I_n\}_i^e = \{V_m\}_j^e,$$

$$m = 1, 2, \ldots, M; \quad j = 1, 2, \ldots, N_m, \qquad (3.99)$$

where N_n is number of elements of the nth antenna, and N_m is the number of segments of the mth antenna.

Also $[Z]_{ji}$ is the mutual impedance matrix for the jth observed boundary element on the mth antenna and the ith source segment on the nth antenna:

$$[Z]_j^e = - \int_{\Delta l_j} \int_{\Delta l_i} \{D\}_j \{D'\}_i G_{mn}(z, z') dz'dz$$

$$+ k^2 \int_{\Delta l_j} \int_{\Delta l_i} \{f\}_j \{f'\}_i G_{mn}(z, z') dz'dz. \qquad (3.100)$$

Vectors $\{f\}$ and $\{f'\}$ contain shape functions $f_k(z)$ and $f_k(z')$, while vectors $\{D\}$ and $\{D'\}$ represent their derivatives. Vector $\{I\}_i$ contains the solution for current distribution in global nodes, and $\{V_m\}_j$ is a local voltage vector given by:

$$\{V_m\}_{ji}^e = -j4\pi\omega\varepsilon_0 \int_{\Delta l_j} E_{zm}^{inc}\{f\}_j \, dz, \qquad (3.101)$$

where index e suggests that operations are performed over a boundary element; Δl_i represents the ith element length (source element), and Δl_j represents the jth element length (observation element).

The excitation in the form of incident electric field is given by

$$E_{zm}^{inc} = \frac{V_g}{\Delta l_g}, \qquad (3.102)$$

where V_g is the impressed voltage across the feed-gap of width Δl_g.

The radiated field is calculated using the boundary element formalism, as well. The expressions for the field components are:

$$E_x = \frac{1}{j4\pi\omega\varepsilon_0} \sum_{n=1}^{M}\sum_{i=1}^{n} \left[\frac{4}{\Delta l_i^2}\left(I_{1i}^n - I_{2i}^n + I_{3i}^n\right) \right.$$

$$\times \int_{-\Delta l_i/2}^{+\Delta l_i/2} z' \frac{\partial G_{mn}(x,z')}{\partial x} dz'$$

$$- \frac{2}{\Delta l_i^2}\left[(z_3+z_2)I_{1i}^n - (z_3+z_1)I_{2i}^n\right.$$

$$\left. + (z_1+z_2)I_{3i}^n\right] \int_{-\Delta l_i/2}^{+\Delta l_i/2} \frac{\partial G_{mn}(x,z')}{\partial x}dz' \right], \qquad (3.103)$$

$$E_y = \frac{1}{j4\pi\omega\varepsilon_0} \sum_{n=1}^{M}\sum_{i=1}^{n} \left[\frac{4}{\Delta l_i^2}\left(I_{1i}^n - I_{2i}^n + I_{3i}^n\right) \right.$$

$$\times \int_{-\Delta l_i/2}^{+\Delta l_i/2} z' \frac{\partial G_{mn}(x,z')}{\partial y} dz'$$

$$- \frac{2}{\Delta l_i^2}\left[(z_3+z_2)I_{1i}^n - (z_3+z_1)I_{2i}^n\right.$$

$$\left. + (z_1+z_2)I_{3i}^n\right] \int_{-\Delta l_i/2}^{+\Delta l_i/2} \frac{\partial G_{mn}(y,z')}{\partial y}dz' \right] \qquad (3.104)$$

$$E_z = \frac{1}{j4\pi\omega\varepsilon_0} \sum_{n=1}^{M}\sum_{i=1}^{n} \left[-\frac{4}{\Delta l_i^2}\left(I_{1i}^n - I_{2i}^n + I_{3i}^n\right) \right.$$

$$\times \int_{-\Delta l_i/2}^{+\Delta l_i/2} z' \frac{\partial G_{mn}(z,z')}{\partial z} dz'$$

$$+ \frac{2}{\Delta l_i^2}\left[(z_3+z_2)I_{1i}^n - (z_3+z_1)I_{2i}^n + (z_1+z_2)I_{3i}^n\right]$$

$$\times \int_{-\Delta l_i/2}^{+\Delta l_i/2} \frac{\partial G_{mn}(z,z')}{\partial z}dz'$$

$$+ k^2\left(I_{1i}^n \frac{z'-z_2}{z_1-z_2}\frac{z-z_3}{z_1-z_3} + I_{2i}^n \frac{z'-z_1}{z_2-z_1}\frac{z'-z_3}{z_2-z_3}\right.$$

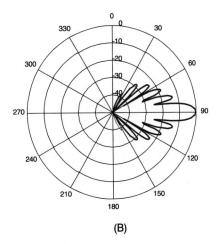

FIG. 3.58 (A) Horizontal pattern, (B) Vertical pattern.

$$+ I_{3i}^n \frac{z'-z_1}{z_3-z_2} \frac{z'-z_2}{z_3-z_2}\Bigg) \int\limits_{-\Delta l_i/2}^{+\Delta l_i/2} G_{mn}\left(z,z'\right) dz' \Bigg],$$

$$(3.105)$$

where I_{1i}, I_{2i}, I_{3i} are node currents on the ith element and nth antenna, respectively.

The computational example deals with the GSM sector antenna consisting of eight half-wave dipole antennas spaced by 0.75λ (between centers), with radius of 0.004λ. This array is distanced from reflector at $x = -0.176\lambda$.

For simplicity reasons, the reflector is modeled as an infinite, perfectly conducting plane. All dipoles are driven by the voltage generator placed at the center of each dipole. The total input power of all sources is chosen to be 30 W, which is a typical maximum power per GSM channel. The computation is carried out by discretizing the array configuration to 88 segments.

The horizontal and vertical radiation patterns are shown in Fig. 3.58.

Furthermore, the near field distributions have been calculated using both exact and approximate relations. Namely, the magnitude of the far-field radiated by the base station antenna system is determined analytically, by using the ray tracing algorithm based on the geometrical optics method, and also numerically.

A comparison of the results obtained via the different methods is shown in Figs. 3.59–3.61.

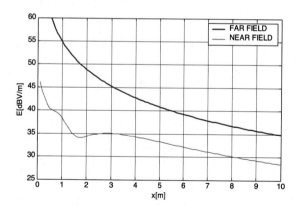

FIG. 3.59 Calculated electric field $\phi = 0°$, $z = 0$ m.

It is obvious that the use of analytical relations for the calculation of the radiated electric field in the near zone results in a significant overestimation (7–15 dB) only in the main lobe, e.g., for $z = 0$ (see Fig. 3.59). Beneath the main lobe, near field values could be higher than the values obtained by the far field pattern. This is

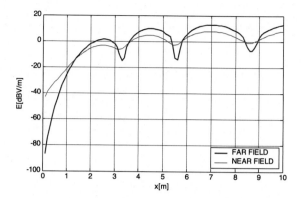

FIG. 3.60 Calculated electric field $\phi = 0°$, $z = -5$ m.

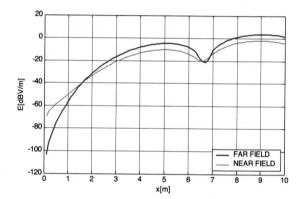

FIG. 3.61 Calculated electric field $\phi = 0°$, $z = -10$ m.

particularly visible for the nulls of the radiation pattern where the underestimation can reach the level of 12 dB (see Fig. 3.60).

REFERENCES

1. Ronold W.P. King, Tai T. Wu, The complete electromagnetic field of a three-phase transmission line over the Earth and its interaction with the human body, Journal of Applied Physics 78 (2) (1995) 668–683.
2. Ronold W.P. King, The electromagnetic field of a horizontal electric dipole in the presence of a three-layered region, Journal of Applied Physics 69 (12) (1991) 7987–7995.
3. Silvestar Šesnić, Dragan Poljak, Andres Peratta, Electromagnetic modeling of the human body exposed to Extremely Low Frequency (ELF) electromagnetic field, in: SoftCOM 2005 13th International Conference on Software, Telecommunications and Computer Networks, 2005, pp. 1–5.
4. S. Kraljevic, Dragan Poljak, V. Doric, A simplified method for the assessment of ELF magnetic fields from three-phase power lines, WIT Transactions on Modelling and Simulation 39 (2005) 551–562.

5. D. Poljak, N. Kovač, S. Kraljević, C.A. Brebbia, Simplified modeling of the human body exposed to power substation electric field using boundary element analysis, Modelling in Medicine and Biology VII 12 (2007) 213–222.

6. N. Kovač, D. Poljak, S. Kraljević, B. Jajac, Computation of maximal electric field value generated by a power substation, WIT Transaction on Modelling and Simulation 42 (2006) 165–174.

7. J.E.T. Villas, F.C. Maia, D. Mukhedkar, Vasco S. Da Costa, Computation of electric fields using ground grid performance equations, IEEE Transactions on Power Delivery 2 (3) (1987) 709–716.

8. International Commission on Non-Ionizing Radiation Protection (ICNIRP), Guidelines for limiting exposure to time-varying electric, magnetic and electromagnetic fields (up to 300 GHz), Health Physics 74 (4) (1998) 494–522.

9. International Commission on Non-Ionizing Radiation Protection (ICNIRP), Guidelines for limiting exposure to time-varying electric and magnetic fields (1 Hz to 100 kHz), Health Physics 99 (6) (2010) 818–836.

10. Exposure to electric or magnetic fields in the low and intermediate frequency range – methods for calculating the current density and internal electric field induced in the human body, Standard, International Electrotechnical Commission, Geneva: IEC, 2004.

11. D. Poljak, S. Sesnic, D. Cavka, M. Titlic, M. Mihalj, The human body exposed to a magnetotherapy device magnetic field, WIT Transactions on Biomedicine and Health 13 (2009) 203–211.

12. Riadh W.Y. Habash, Electromagnetic Fields and Radiation: Human Bioeffects and Safety, CRC Press, 2001.

13. Eleanor R. Adair, Ronald C. Petersen, Biological effects of radiofrequency/microwave radiation, IEEE Transactions on Microwave Theory and Techniques 50 (3) (2002) 953–962.

14. Dragan Poljak, Electromagnetic fields: environmental exposure, in: Encyclopedia of Environmental Health, Elsevier, 2011, pp. 259–268.

15. Vicko Dorić, Dragan Poljak, Khalil El Kamichi Drissi, Human exposure to outdoor PLC system, in: PIERS 2011 Marrakesh Progress in Electromagnetics Research Symposium, 2011, pp. 1602–1606.

16. Dragan Poljak, Khalil E. Drissi, Computational Methods in Electromagnetic Compatibility: Antenna Theory Approach Versus Transmission Line Models, John Wiley & Sons, 2018.

17. Dragan Poljak, Advanced Modeling in Computational Electromagnetic Compatibility, Wiley-Interscience, New Jersey (NJ), 2007.

18. Damir Senić, Dragan Poljak, Antonio Šarolić, Simulation and measurements of electric and magnetic fields of an RFID loop antenna anti-theft gate system, Journal of Communications Software and Systems 6 (4) (2010) 133–140.

19. Ivana Zulim, Damir Senic, Dragan Poljak, Antonio Sarolic, Assessment of SAR in the human body exposed to an RFID loop antenna, in: 2012 20th International Conference on Software, Telecommunications and Computer Networks, SoftCOM, IEEE, 2012, pp. 1–5.

20. Quirino Balzano, Antonio Faraone, Human exposure to cellular base station antennas, in: 1999 IEEE International Symposium on Electromagnetic Compatibility, vol. 2, IEEE, 1999, pp. 924–927.

21. Quirino Balzano, Antonio Faraone, Estimation of the average power density in the assessment of the human exposure to cellular base station antennas, in: 15th International Conference on Applied Electromagnetics and Communications, 1999, ICECom 2003, 1999, pp. 79–81.

22. Basic standard for the calculation and measurement of electromagnetic field strength and SAR related to human exposure from radio base stations and fixed terminal stations for wireless telecommunication systems (110 MHz – 40 GHz), Standard, Hrvatski Zavod za Norme, Zagreb, HZN, 2005.

23. P. Bernardi, M. Cavagnaro, S. Pisa, E. Piuzzi, Human exposure in the vicinity of radio base station antennas, in: Proc. 4th Eur. Symp. Electromagnetic Compatibility, 2000, pp. 187–192.

24. A. Faraone, Human exposure to radio base station – a review, in: Proc. 4th Eur. Symp. Electromagnetic Compatibility, vol. 2, Brugge, 2000, pp. 479–496.

25. Marin Galić, Dragan Poljak, Vicko Dorić, Comparison of different analytical models to determine electric field radiated by a base station antenna, in: 2017 2nd International Multidisciplinary Conference on Computer and Energy Science, SpliTech, IEEE, 2017, pp. 1–5.

26. Marin Galić, Dragan Poljak, Vicko Dorić, Comparison of free space, perfect ground and Fresnel's equation models to determine electric field radiated by a base station antenna, in: 2017 25th International Conference on Software, Telecommunications and Computer Networks, SoftCOM, IEEE, 2017, pp. 1–6.

27. Marin Galic, Dragan Poljak, Vicko Doric, Analytical technique to determine the electric field above a two-layered medium, in: 2018 2nd URSI Atlantic Radio Science Meeting, AT-RASC, IEEE, 2018, pp. 1–4.

28. 4nec2, http://www.qsl.net/4nec2/.

29. T. Takashima, T. Nakae, R. Ishibashi, Calculation of complex fields in conducting media, IEEE Transactions on Electrical Insulation 15 (1) (1980) 1–7.

30. Dragan Poljak, Vicko Dorić, Mario Birkić, Khalil El Khamlichi Drissi, Sebastien Lallechere, Lara Pajewski, A simple analysis of dipole antenna radiation above a multilayered medium, in: 2017 9th International Workshop on Advanced Ground Penetrating Radar, IWAGPR, IEEE, 2017, pp. 1–6.

31. Dragan Poljak, Vicko Doric, Damir Vucicic, C.A. Brebbia, Boundary element modeling of radio base station antennas, Engineering Analysis with Boundary Elements 30 (6) (2006) 419–425.

CHAPTER 4

Simplified Models of the Human Body

The starting point in the analysis of interaction of humans with time-harmonic or transient electromagnetic fields is the knowledge of the induced currents and fields inside the human body [1–9]. A simplified approach to the assessment of human exposure to EM fields involves the representation of the human body with a geometry with a high degree of simplification (canonical geometry), such as parallelepiped [4,5], or cylinder [6–14].

The simplified approach based on the equivalent parasitic antenna representation of the body deals with Pocklington or Hallén integral equation formulation, in either frequency or time domain, respectively. The electrical properties of the body related to the LF and HF exposures are taken into account via the load term in the Pocklington equation for a thick cylinder.

By obtaining the current distribution along the body, one can readily calculate the induced current density, induced electric field, specific absorption rate (SAR), specific absorption (SA), or other parameters of interest. This provides simple and efficient procedures for a rapid estimation of these electromagnetic phenomena. Furthermore, the cylindrical body model can be analyzed using the transmission line approach, as well [15,16].

4.1 PARALLELEPIPED MODEL OF THE HUMAN BODY

The exposure of humans to HF electromagnetic radiation is quantified by the specific absorption rate (SAR), which is defined as the mass averaged rate of energy absorption in tissue:

$$\text{SAR} = \frac{d}{dt}\frac{dW}{dm} = \frac{d}{dt}\frac{dW}{\rho dV} \quad (4.1)$$

and is expressed in [W/kg].

The absorption of electromagnetic energy causes a temperature rise within a tissue, therefore SAR is also a measure of the local heating rate and can be expressed in the form

$$\text{SAR} = C\frac{dT}{dt}, \quad (4.2)$$

where T denotes the temperature and C is the specific heat capacity.

A parallelepiped model of the body [4,5] is shown in Fig. 4.1.

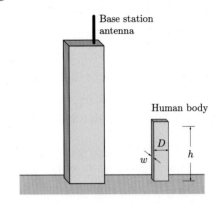

FIG. 4.1 Human body exposed to electromagnetic radiation.

If a human being is illuminated by the plane wave (e.g., base station antenna radiation), the analytical calculation of SAR can be carried out using the approximation formula [4,5]:

$$\text{SAR} = \frac{\sigma}{\rho}\frac{\mu\omega}{\sqrt{\sigma^2 + \varepsilon^2\omega^2}}(1 + \gamma_{pw})^2\frac{\left|E^{\text{inc}}\right|^2}{Z_0^2} \quad (4.3)$$

where E^{inc} is the root-mean-square (rms) value of the incident electric field, σ and ε are the electrical properties of the human body and γ_{pw} is the corresponding reflection coefficient given by [4,5]

$$\gamma_{pw} = \frac{2\left|\sqrt{\varepsilon'}\right|}{\left|\sqrt{\varepsilon'} + \sqrt{\varepsilon_0}\right|} - 1, \quad (4.4)$$

where ε' is the complex permittivity of the medium.

The free-space averaged SAR per gram tissue can be calculated, assuming the exponential decay of SAR (plain wave approximation) when penetrating through the lossy medium, as

$$\text{SAR}_{1g} = \frac{\delta_{\text{skin}}}{2d}\left(1 - e^{-\frac{2d}{\delta_{\text{skin}}}}\right)\text{SAR}_{\text{surf}} \quad (4.5)$$

where $d = 1$ cm and δ_{skin} is skin thickness,

Human Interaction with Electromagnetic Fields. https://doi.org/10.1016/B978-0-12-816443-3.00012-1

$$\delta_{\text{skin}} = \sqrt{\frac{2}{\omega\mu\sigma}}. \qquad (4.6)$$

The whole body averaged SAR for the parallelepiped model is

$$
\begin{aligned}
\text{SAR}_{\text{WB}} &= \frac{1}{HD} \int_0^H \int_0^D \text{SAR}_{\text{surf}}\, e^{-\frac{2x}{\delta_{\text{skin}}}} \, dx \, dz \\
&= \frac{\delta_{\text{skin}}}{2D} \left(1 - e^{-\frac{2D}{\delta_{\text{skin}}}}\right) \text{SAR}_{\text{surf}},
\end{aligned}
\qquad (4.7)
$$

where $D = 20$ cm is parallelepiped thickness.

4.2 CYLINDRICAL ANTENNA MODELS OF THE BODY

When exposed to electromagnetic radiation, the human body behaves as an imperfectly conducting straight wire antenna.

The currents and fields induced in human organs may give rise to thermal and non-thermal effects. When the body is exposed to low frequency (LF) fields, the thermal effects are found to be negligible, and possible non-thermal effects are often related to the behavior at the cell level [6–8]. In the case of HF exposure, the thermal effects are dominant due to resonance phenomena [9].

The mathematical model of a human being exposed to transient electromagnetic fields is based on the cylindrical representation of the human body and the corresponding space-time Hallén integral equation [13,14]. The Hallén integral equation is solved via the Galerkin–Bubnov scheme of the indirect boundary element method, and the equivalent space-time current distribution along the cylindrical body model is obtained. Besides the well-established concept of specific absorption (SA), the transient current flowing through the human body is post-processed in terms of certain measures of quantifying the transient response. These measures arise from circuit theory and they are *average* and *root-mean-square* value of time-varying current, *instantaneous power* dissipated in the body, and *total absorbed energy* in the body [14].

4.3 CYLINDRICAL MODELS – FREQUENCY DOMAIN ANALYSIS

At low frequencies the displacement currents can be neglected, allowing the electric and magnetic fields to be analyzed separately. The human body can be exposed to low voltage/high intensity systems, in which the principal radiation is generated from the magnetic field, or to high voltage/low intensity systems, in which the principal radiation is generated by the electric field. In the case of magnetic field exposure, the induced currents in the body form close loops, while in case of the electric field exposure the related currents induced in the body have the axial character.

This section is devoted to the assessment of human exposure to ELF electric fields used for power utilities. The current density is the main parameter for the evaluation of low frequency exposure hazard proposed by ICNIRP basic restrictions [3] and a key-point in any further analysis of possible adverse biological effects of ELF fields. The internal field is proposed as a basic restriction in [17].

The human body consists of various tissues and organs with different electrical properties, such as conductivity σ and relative permittivity ε_r. At low frequencies near the power frequency $f = 50$ or 60 Hz, the body is a good conductor with an average conductivity of approximately $\sigma \approx 0.5$ S/m.

Although the body cross-section changes in size and shape, this does not have a significant effect on the total current as long as the tissue material is predominantly conducting and the body can be represented as an equivalent cylindrical antenna model with a uniform cross-section and conductivity.

The maximum current flows from the body into the ground when it is assumed that a person is standing bare footed on a good conducting ground. Under the same circumstances, the cylindrical model can be extended to HF and transient exposures.

A person standing vertically on the earth and exposed to the electromagnetic field from extremely low (ELF) to GSM frequencies can be represented by an imperfectly conducting cylinder of length L and radius a, as it is shown in Fig. 4.2.

The current distribution along the human body is obtained as the solution of the Pocklington integro-differential equation for a thick loaded straight wire [10–12]. Knowing the current distribution provides the calculation of other parameters of interest.

4.3.1 Pocklington Equation Formulation for LF Exposures

This integro-differential equation can be derived starting from the time harmonic electric field $\vec{E}$ given by the magnetic potential $\vec{A}$ and the electric scalar potential φ, i.e.,

$$\vec{E} = -\nabla\varphi - j\omega\vec{A}. \qquad (4.8)$$

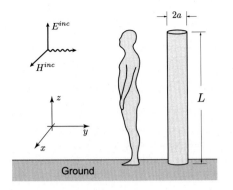

FIG. 4.2 The equivalent antenna model of the human body.

The vector and magnetic potentials are coupled through the previously defined Lorentz gauge:

$$\nabla \vec{A} = -j\omega\mu\varepsilon\varphi. \tag{4.9}$$

As only the axial component of the magnetic potential $\vec{A}$ along the cylinder exists, the combination of (4.8) and (4.9) results in

$$E_z = \frac{1}{j\omega\mu\varepsilon}\left(\frac{\partial^2 A_z}{\partial z^2} + k^2 A_z\right). \tag{4.10}$$

Applying the image theory, the magnetic vector potential A_z can be represented by the integral of the axial current $I(z)$ flowing along the equivalent dipole antenna, i.e.,

$$A_z = -\frac{\mu}{4\pi}\int_{-L}^{L}\frac{1}{2\pi}\int_{0}^{2\pi}\frac{e^{-jkR}}{R}I(z')\,dz'\,d\phi, \tag{4.11}$$

where R is the distance from the source point to the observation point and k is the wave number of the free-space.

The continuity conditions for the tangential electric field components can be written as

$$E_z^{inc} + E_z^{sct} = I(z)\,Z_L(z), \tag{4.12}$$

where E_z^{inc} is the excitation function in the form of incident field, E_z^{sct} is the scattered field due to the presence of the imperfectly conducting cylinder, $I(z)$ is the axial current distribution, and $Z_L(z)$ is the impedance per unit length of the cylinder.

Consequently, the tangential component of the electric field on the antenna surface is given by

$$E_z^{sct}(z, a) = \frac{1}{j4\pi\omega\varepsilon_0}\int_{-L}^{L}\left(\frac{\partial^2}{\partial z^2} + k^2\right)g_E(z, z')\,I(z')\,dz', \tag{4.13}$$

where ε_0 is the permittivity of the free space, ω is the applied frequency.

Combining (4.12) and (4.13), one obtains the Pocklington integro-differential equation which determines the current along a thick wire:

$$E_z^{inc}(z, a) = -\frac{1}{j4\pi\omega\varepsilon_0}\int_{-L}^{L}\left(\frac{\partial^2}{\partial z^2} + k^2\right)g_E(z, z')\,I(z')\,dz'$$
$$+ Z_L(z)\,I(z), \tag{4.14}$$

where $g_E(z, z')$ is the so-called "exact" kernel of the integral equation given by [3]:

$$g_E(z, z') = \frac{1}{2\pi}\int_{0}^{2\pi}\frac{e^{-jkR}}{R}\,d\phi, \tag{4.15}$$

in which R denotes the distance from the source point z to the observation point z', both located on the thick wire antenna surface.

The distance R is given by

$$R = \sqrt{(z - z')^2 + 4a^2\sin^2\frac{\phi}{2}}. \tag{4.16}$$

The conducting and dielectric properties of the body are taken into account by the properties of the impedance Z_L.

4.3.2 Numerical Solution of the Pocklington Equation

The Pocklington integral equation can be solved numerically using the boundary element formalism. The integral equation (4.14) can be, for convenience, written in an operator form as

$$K\,I = E, \tag{4.17}$$

where K is a linear operator, I is the unknown function to be found for a given excitation E.

The unknown current is then expanded into a finite sum of linearly independent basis functions f_i with unknown complex coefficients α_i, i.e.,

$$I \cong I_n = \sum_{i=1}^{n} \alpha_i f_i. \tag{4.18}$$

Substituting (4.18) into (4.17) yields

$$K I \cong K I_n = \sum_{i=1}^{n} \alpha_i K f_i. \tag{4.19}$$

Now a residual R_n can be defined as

$$R_n = K I_n - E. \tag{4.20}$$

Finally, according to the definition of the scalar product of functions in a Hilbert space, the error R_n can be weighted with respect to certain weighting functions W_j, as already discussed:

$$\int_{\Omega} R_n W_j^* \, d\Omega = 0, \quad j = 1, 2, \ldots, n, \tag{4.21}$$

where W_j^* is the complex conjugate of weighting function W_j.

The expression

$$\langle R_n, W_j \rangle = \int_{\Omega} R_n W_j^* \, d\Omega \tag{4.22}$$

is the scalar product of functions R and W, and Ω is the domain of interest.

Since the operator K is linear, one obtains a system of algebraic equations and, by choosing $W_j = f_j$, the Galerkin–Bubnov procedure yields

$$\sum_{i=1}^{n} \alpha_i \int_{\Omega} K(f_i) f_j \, d\Omega = \int_{\Omega} E f_j \, d\Omega, \quad j = 1, 2, \ldots, n. \tag{4.23}$$

Eq. (4.23) is the Galerkin formulation of the Pocklington integral equation (4.14). Using the integral equation kernel symmetry, taking into account the boundary conditions for the current at the free ends of the thin wires, and after integration by parts, Eq. (4.23) can be written as

$$\sum_{i=1}^{n} \alpha_i \frac{1}{j4\pi\omega\varepsilon} \left(-\int_{-L}^{L} \frac{df_j(z)}{dz} \int_{-L}^{L} \frac{df_i(z)}{dz'} g_E(z, z') \, dz' \, dz \right.$$

$$\left. + k_1^2 \int_{-L}^{L} f_i(z') g_E(z, z') \, dx \, dx' \right)$$

$$+ \int_{-L}^{L} Z_S(z) f_j(z) f_i(z) \, dz$$

$$= -\int_{-L}^{L} E_z^{inc}(z) f_j(z) \, dz, \quad j = 1, 2, \ldots, n. \tag{4.24}$$

Eq. (4.24) represents the weak Galerkin formulation of the integral equation (4.17).

The resulting matrix equation system that follows from discretization of (4.24) is given by

$$\sum_{i=1}^{M} [Z]_{ji} \{I\}_i = \{V\}_j, \quad j = 1, 2, \ldots, M, \tag{4.25}$$

where $[Z]_{ji}$ is the submatrix representing the interaction of the ith source boundary element and the jth observation element, i.e.,

$$[Z]_{ji} = -\frac{1}{j4\pi\omega\varepsilon} \left(\int_{\Delta l_j} \{D\}_j \int_{\Delta l_i} \{D'\}_i^T g_E(z, z') \, dz' \, dz \right.$$

$$\left. + k^2 \int_{\Delta l_j} \{f\}_j \int_{\Delta l_i} \{f\}_i^T g_E(z, z') \, dz' \, dz \right)$$

$$+ \int_{\Delta l_j} Z_L(z) \{f\}_j \{f\}_i^T \, dz. \tag{4.26}$$

Vector $\{I\}$ contains the unknown coefficients of the solution and represents the local voltage vector. Matrices $\{f\}$ and $\{f'\}$ contain shape functions, while $\{D\}$ and $\{D'\}$ contain their derivatives, where M is the total number of segments, and Δl_i, Δl_j are the lengths of the ith and jth segment, respectively.

Functions $f_k(x)$ are required to be differentiable at least once. A convenient choice for the shape functions is the family of Lagrange's polynomials given by

$$L_i(z) = \prod_{j=1}^{m} \frac{z - z_j}{z_i - z_j}, \quad j \neq i. \tag{4.27}$$

Lagrange's polynomials provide accurate results and are easily integrable in closed form. The local right-hand side vector $\{V\}_j$ for the jth observation boundary element can be written as

$$\{V\}_j = \int_{\Delta l_j} E_z^{inc} \{f\}_j \, dz \tag{4.28}$$

and represents the local voltage vector.

The linear approximation is used, as this choice provides accurate and stable results [3]. The evaluation of the right-hand side vector can be carried out in closed form if a constant incident electric field is assumed along the wire, i.e.,

$$E_z^{inc} = E_0. \qquad (4.29)$$

Thus, using the linear basis functions, it follows that

$$V_{ij} = \int_{-\Delta l/2}^{\Delta l/2} E_0 \frac{z_{j+1} - z}{\Delta l} dz = E_0 \frac{\Delta l}{2}, \qquad (4.30)$$

$$V_{2j} = \int_{-\Delta l/2}^{\Delta l/2} E_0 \frac{z - z_j}{\Delta l_g} dz = E_0 \frac{\Delta l}{2}. \qquad (4.31)$$

It is worth mentioning that the excitation function in the form of an incident field E_0 is usually obtained from the measurement.

In the ELF region (50/60 Hz), the cylinder representing the human body can be considered as a conducting medium whose impedance per unit length is [3]

$$Z_L(z) = \frac{1}{a^2 \pi \sigma} + Z_c, \qquad (4.32)$$

where $Z_c = 1/j\omega C$, and C is the capacitance between the soles of the feet and their image in the earth. If the foot-soles are bare and in direct contact with the moist earth (well-grounded body) then the capacity can be neglected, i.e., $Z_c = 0$.

Once the axial current is computed from the numerical solution, it is possible to calculate the induced current density, i.e.,

$$J_z(z) = \frac{I_z(z)}{a^2 \pi}, \qquad (4.33)$$

as well as the induced electric field

$$E_z(z) = \frac{J_z(z)}{\sigma}. \qquad (4.34)$$

The macroscopic average electric field discussed so far could be subsequently used for the calculation of the corresponding local electric fields induced in different organs [6–8].

4.3.3 Analytical Modeling of the Human Body – Hallén Equation for LF and HF Exposures

Contrary to the Pocklington integral equation approach presented so far, the total axial current induced in the human body, when it is approximated by a parasitic

cylindrical antenna with half-length L and mean radius a, can be obtained by analytically solving the Hallén integral equation [6–9].

The Hallén integral equation can be derived enforcing the boundary condition for the tangential electric field components:

$$E_z^{inc}(\rho, z)|_{\rho=a} + E_z^{sct}(\rho, z)|_{\rho=a} = E_z^{tot}(\rho, z)|_{\rho=a}, \qquad (4.35)$$

where E_z^{inc}, E_z^{sct} and E_z^{tot} are the incident, scattered and total fields, respectively.

Total and scattered fields are given by following expressions:

$$E_z^{tot} = Z I_z(z'), \qquad (4.36)$$

$$E_z^{sct} = -\frac{j\omega}{k^2} \left(\frac{\partial^2 A_z}{\partial z^2} + k^2 A_z \right). \qquad (4.37)$$

Therefore, combining (4.36)–(4.37), we obtain

$$\frac{\partial^2 A_z}{\partial z^2} + k^2 A_z = \frac{jk^2}{\omega} \left(Z_L I_z(z') - E_z^{inc} \right), \qquad (4.38)$$

where Z_L is the impedance per unit length of the imperfectly conducting cylinder (the body).

The solution of the differential equation (4.38) is given by [6,7]

$$A_z(z) = -\frac{j}{c} \left[K \cos kz + E_z^{inc} \right. $$
$$\left. - Z_L \int_0^z I_z(z) \sin k(z - s) ds \right], \qquad (4.39)$$

where c denotes the velocity of light and K is unknown constant.

The magnetic vector potential can also be represented by the following integral:

$$A_z(z) = -\frac{\mu_0}{4\pi} \int_0^z I_z(z') \frac{e^{-jkR}}{R} dz',$$

$$R = \sqrt{(z - z')^2 + a^2}. \qquad (4.40)$$

Combining the relations (4.39) and (4.40), the Hallén integral equation is obtained in the form [3,6,7]

$$\int_0^z I_z(z') \frac{e^{-jkR}}{R} \, dz' = -j\frac{4\pi}{Z_0}\left[K\cos kz + \frac{1}{k}E_z^{\text{inc}} \right.$$

$$\left. - Z_L \int_0^z I_z(s)\sin k(z-s)\,ds \right].$$

$$(4.41)$$

The unknown constant K is easily determined from the wire end conditions $I_z(\pm L)=0$.

The corresponding expression for Z_L associated with the frequency range of interest can be found elsewhere, e.g., in [6,7].

For the LF part of the electromagnetic spectrum, the solution of Eq. (4.41) is [6,7]

$$I_z(z) = j2\pi \frac{kL^2}{\psi_1 Z_0} E_z^{\text{inc}}\left[1-\left(\frac{z}{L}\right)^2\right],\qquad (4.42)$$

where parameter ψ_1 is given by [3,6,7]

$$\psi_1 = 2\ln\frac{2L}{a} - 3.\qquad (4.43)$$

The solution of Eq. (4.41) for higher frequencies is [3,9]

$$I_z(z) = j2\pi \frac{kL^2}{\psi_2 Z_0} E_z^{\text{inc}}\left[1-\left(\frac{z}{L}\right)^2\right]\left[1 - j4\pi\frac{Z_L}{Z_0 k\psi_2}\right],$$

$$(4.44)$$

where parameter ψ_2 is given by [18]

$$\psi_2 = \frac{1}{1-\cos kL}\{C_a(L,0) - C_a(L,L)$$
$$- [E_a(L,0) - E_a(L,L)]\cos kL\},\qquad (4.45)$$

where

$$C_a(h,z) = \int_0^z \cos ks\left(\frac{e^{-jkR_1}}{R_1} + \frac{e^{-jkR_2}}{R_2}\right) ds, \quad (4.46)$$

$$E_a(h,z) = \int_0^z \left(\frac{e^{-jkR_1}}{R_1} + \frac{e^{-jkR_2}}{R_2}\right) ds, \qquad (4.47)$$

$$R_1 = \sqrt{(z-s)^2 + a^2}, \quad R_2 = \sqrt{(z+s)^2 + a^2}. \quad (4.48)$$

Note that the human body is assumed to be well-grounded.

When a person stands, he/she is exposed to an axial electric field E_{2z}^{inc} and a transverse magnetic field B_{2y}^{inc}. The integral equation for the total current $I_{1z}(z)$ induced in a conductor with the half-length L and radius a when exposed to an incident electric field E_{2z}^{inc} parallel to the cylinder is

$$\int_{-L}^L I_{1z}(s)\frac{e^{-jk_2 r}}{r}\,ds = -\frac{j4\pi}{\xi_0}\left(C\cos k_2 z + \frac{1}{2}V\sin k_2\,|z| \right.$$

$$\left. + U^{\text{inc}} - Z^i P_z\right),\qquad (4.49)$$

where $\xi_0 = 120\pi\ \Omega$, $V = -I_{1z}(0)\left[Z_0 Z_L/(Z_0 + Z_L)\right]$, with Z_L being a load impedance at $z=0$. For simplicity it is assumed that $Z_L = 0$, which corresponds to a barefoot man on a highly conducting deck; Z_0 is the impedance of the cylinder, $U^{\text{inc}} = E_{2z}^{\text{inc}}/k_2$, and Z^i is the impedance per unit length of the cylinder. Also, $r = [(z-s)^2 + a^2]^{1/2}$, $k_2 = \omega/c$ is the wave number of air, and

$$P_{-z} = P_z = \int_0^z I_{1z}(s)\sin k_2(z-s)\,ds. \qquad (4.50)$$

The integral equation (4.49) has no exact solution, while approximate analytical solutions can be obtained in different ways [9,18].

The formula for the total axial current $I_{1z}(z)$ induced in the body when exposed to E_{2z}^{inc} is

$$I_{1z}(z) = \frac{j4\pi}{\xi_0\psi_u}\frac{E_{2z}^{\text{inc}}}{k_2}\left[\frac{(\cos k_2 z - \cos k_2 L)}{\cos k_2 L}\right.$$

$$\left. - \frac{Z_L}{(Z_0+Z_L)}\frac{(1-\cos k_2 L)}{\cos k_2 L}\frac{\sin k_2(L-z)}{\sin k_2 L}\right],$$

$$(4.51)$$

where

$$\psi_u = \psi_u(0)$$
$$= \frac{C_a(L,0) - C_a(L,L) - (E_a(L,0) - E_a(L,L))\cos k_2 L}{1-\cos k_2 L},$$

$$(4.52)$$

$$C_a(L,z) = \int_0^L \cos k_2 s\left(\frac{e^{-jk_2 r_1}}{r_1} + \frac{e^{-jk_2 r_2}}{r_2}\right) ds, \quad (4.53)$$

$$E_a(L,z) = \int_0^L \left(\frac{e^{-jk_2 r_1}}{r_1} + \frac{e^{-jk_2 r_2}}{r_2}\right) ds, \qquad (4.54)$$

$$r_1 = r = \left[(z-s)^2 + a^2 \right]^{1/2}, \quad (4.55)$$

$$r_2 = \left[(z+s)^2 + a^2 \right]^{1/2}. \quad (4.56)$$

Here, $k_2 = \omega/c$ is the wave number of air and L is the length of a person standing on a conducting plane like the metal deck of a ship.

The current density $J_{1z}(z)$ and electric field $E_{1z}(z)$ in the body are given by [3]

$$J_{1z}(z) = \sigma_1 E_{1z}(z) = \frac{I_{1z}(z)}{A(z)}, \quad (4.57)$$

where A represents the cross-sectional area of the soles of the feet.

4.3.4 Multiple Wire Model of the Body

If the arms are not in close contact with the body but raised to an arbitrary angle, a multiple wire model of the body and the surface integral equation formulation has to be used.

Thus, the effects of the arms being raised to various angles to the vertical body axis are modeled by attaching wires of radius $a = 0.05$ m and length $L = 0.8$ m to the cylindrical antenna [3]. The wires representing the arms are attached at a height of 1.4 m from the ground. The equivalent antenna model with the arms raised is shown in Fig. 4.3.

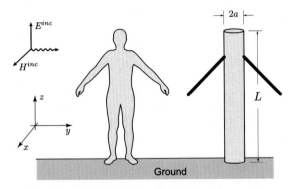

FIG. 4.3 Equivalent antenna model of the body with the arms raised.

The incident field $\vec{E}^{\text{inc}}$ tangential to the body surface is given by

$$\vec{E}_{\text{tan}} = (j\omega \vec{A} + \nabla \varphi)_{\text{tan}}, \quad (4.58)$$

where the vector potential $\vec{A}$ and scalar potential φ are:

$$\vec{A}(\vec{r}) = \frac{\mu}{4\pi} \iint_S \vec{J}(\vec{r}') \frac{e^{-jk|\vec{r}-\vec{r}'|}}{|\vec{r}-\vec{r}'|} dS(\vec{r}'), \quad (4.59)$$

$$\varphi(\vec{r}) = -\frac{1}{j4\pi\omega\varepsilon} \iint_S \nabla_S' \vec{J}(\vec{r}') \frac{e^{-jk|\vec{r}-\vec{r}'|}}{|\vec{r}-\vec{r}'|} dS(\vec{r}'), \quad (4.60)$$

where $\vec{J}$ is the current density.

The axial current distribution is obtained by integrating the current density over the cross-section of the body.

When the axial current is determined, it is possible to calculate the electric field, and the power density or specific absorption rate induced in the body [3]. Moreover, the macroscopic average electric field can be subsequently used for the calculation of the corresponding local electric fields induced in different organs.

The integral equation (4.58) is solved using the method of moments (MoM) [3].

4.3.5 Numerical Solution Via Method of Moments (MoM)

The vector basis functions formulated for triangular surface patches are used to approximate the current density. Note that these elements are reduced to linear elements when wires attached to a cylinder are modeled. The basis functions are of the form:

$$\vec{f}_n(\vec{r}) = \begin{cases} \dfrac{l_n}{2A_n^+} \rho_n^+, & \vec{r} \in T_n^+, \\ \dfrac{l_n}{2A_n^-} \rho_n^-, & \vec{r} \in T_n^-, \\ 0, & \text{otherwise,} \end{cases} \quad (4.61)$$

defined over the triangular patches $T_n^{\pm}$ of areas $A_n^{\pm}$ on either side of the nth edge of length l_n. For the wire elements, the basis function reduces to

$$\vec{f}_n(\vec{r}) = \frac{\rho_n^{\pm}}{h_n^{\pm}}, \quad (4.62)$$

where $h_n^{\pm}$ is the length of each wire segment on either side of the nth node.

The boundary condition in Eq. (4.58) is enforced by testing with a weighting function $\vec{f}_m$ as follows:

$$\langle \vec{E}, \vec{f}_m \rangle = j\omega \langle \vec{A}, \vec{f}_m \rangle + \langle \nabla \varphi, \vec{f}_m \rangle. \quad (4.63)$$

Galerkin's method in which the weighting function is the same as the basis function is used for the testing.

The line integral of the vector potential over the subdomain region resulting from the testing is approximated by replacing it with its value at the centroid of the triangular element. For a wire segment, the centroid is the midpoint of the linear element. The surface divergence in the scalar potential term can be simplified by exploiting the property of the triangular surface basis and the vector identity $\nabla(\varphi \vec{f}) = \varphi \nabla \vec{f} + \nabla \varphi \vec{f}$.

The obtained expressions are [3]:

$$\left\langle \vec{E}, \vec{f}_m \right\rangle = \frac{l_m}{2A_m^+} \iint_{T_m^+} \vec{E}(\vec{r})\rho_m^+(\vec{r}) \, dS$$

$$+ \frac{l_m}{2A_m^-} \iint_{T_m^-} \vec{E}(\vec{r})\rho_m^-(\vec{r}) \, dS, \qquad (4.64)$$

$$\left\langle \vec{A}, \vec{f}_m \right\rangle = \frac{l_m}{2A_m^+} \iint_{T_m^+} \vec{A}(\vec{r})\rho_m^+(\vec{r}) \, dS$$

$$+ \frac{l_m}{2A_m^-} \iint_{T_m^-} \vec{A}(\vec{r})\rho_m^-(\vec{r}) \, dS, \qquad (4.65)$$

$$\left\langle \nabla\varphi, \vec{f}_m \right\rangle = l_m \left[\varphi(\vec{r}_m^{c+}) - \varphi(\vec{r}_m^{c-}) \right]. \qquad (4.66)$$

The current $\vec{J}$ is expanded using the basis functions and takes the form

$$\vec{J} = \sum_{n=1}^{N} J_n \vec{f}_n(\vec{r}). \qquad (4.67)$$

The integral equation is then turned into a system of linear equations of order $N \times N$ and with the form shown below:

$$[Z_{mn}]\{J_n\} = \{V_m\}. \qquad (4.68)$$

The elements of $[Z_{mn}]$, known as the impedance matrix, are given by

$$Z_{mn} = l_m \left[j\omega \left(A_{mn}^+ \frac{\rho_m^{c+}}{2} + A_{mn}^- \frac{\rho_m^{c-}}{2} \right) + \varphi_{mn}^- - \varphi_{mn}^+ \right], \qquad (4.69)$$

while $\{J_n\}$ is a column vector of the current coefficients, which is required to be solved for, and $\{V_m\}$ is another column vector, the elements of which are the incident field, and given by

$$V_m^c = \frac{l_m}{2A_m^+} \iint_{s_m^+} \vec{E}^{\text{inc}}(\vec{r})\rho_m^+(\vec{r}) \, dS$$

$$+ \frac{l_m}{2A_m^-} \iint_{s_m^-} \vec{E}^{\text{inc}}(\vec{r})\rho_m^-(\vec{r}) \, dS. \qquad (4.70)$$

The expressions for the vector and scalar potentials are given by:

$$A_{mn}^\pm = \frac{\mu}{4\pi} \iint_{s_n^+ + s_n^-} \vec{f}_n(\vec{r}')G(\vec{r}_m^{c\pm}, \vec{r}') \, dS(\vec{r}'), \qquad (4.71)$$

$$\varphi_{mn}^\pm = -\frac{1}{j4\pi\omega\varepsilon} \iint_{s_n^+ + s_n^-} \nabla_s' \vec{f}_n(\vec{r}')G(\vec{r}_m^{c\pm}, \vec{r}') \, dS(\vec{r}'), \qquad (4.72)$$

where the free-space Green's function is

$$G(r, r') = \frac{e^{-jk|r-r'|}}{|r - r'|}. \qquad (4.73)$$

Note that the triangular patch surface domain has been replaced by the linear segments for the case of wires, $s_n^\pm$ are the segments attached to the nth edge.

Computations of the elements of the impedance matrix $[Z_{mn}]$ require the evaluation of the integrals for the vector and scalar potentials in (4.71) and (4.72). Note that the Green's function (4.73) is singular at the segment where the observation point coincides with the source point.

The singular integral

$$\int_{s_n^\pm} \int_{-\pi}^{\pi} \frac{e^{-jkr_m}}{r_m} \, d\varphi \, dS, \qquad (4.74)$$

where r_m is the distance from the observation point to the source point, is rewritten as

$$\int_{s_n^\pm} \int_{-\pi}^{\pi} \left(\frac{1}{r_m} + \frac{e^{-jkr_m} - 1}{r_m} \right) d\varphi \, dS. \qquad (4.75)$$

The first term can be evaluated as an analytic function. The second term is not singular and can be evaluated by the standard numerical integration routines.

4.3.6 Computational Examples

4.3.6.1 LF Exposures

The first example deals with a well-grounded human body exposed to the 60 Hz overhead power line electric field. The configuration of a power line used in many parts of the world, including USA, is shown in Fig. 4.4.

The conductors are spaced horizontally at a distance $s = 3$ m from each other and located at a height $d = 15$ m from the ground [3,6–8]. The person stands at a

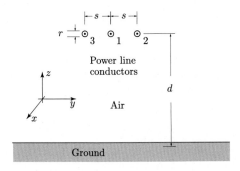

FIG. 4.4 Geometry of a three-wire power line.

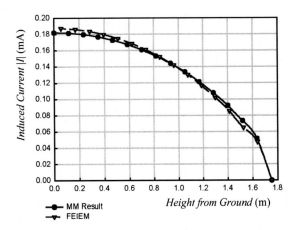

FIG. 4.5 Axial current induced in the body due to the incident field (10 kV/m, 60 Hz).

certain point y from the reference origin facing the positive y direction.

Although the body is exposed to all six components of the field, it is shown [3] that $E_{0z}(0, y, z)$, the component parallel to the upright cylindrical body axis, is the largest within the practical ranges of body height z and distance y from the origin. In this case the component of the electric field tangential to the body is assumed to be $E = 10$ kV/m.

The foot current results obtained by numerical solution of Pocklington equation via GB-IBEM are compared with the results computed via other methods in Table 4.1.

TABLE 4.1
Foot current on antenna model of the human body $E_z^{\mathrm{inc}} = 10$ kV/m, 60 Hz.

Foot current	BEM	FDTD [9]	Analytical [1,7]
$I_z(0)$ [µA]	191	179	183

The current density induced by the external field in individual organs inside the body may be obtained from the axial current distribution. Note that tissue conductivity varies from $\sigma = 0.01$ S/m (bone) to 0.6 S/m (muscle).

The corresponding axial current distribution along the body is shown in Fig. 4.5. The results obtained using GB-IBEM are compared with results obtained by solving the surface integral equation using the Method of Moments [3].

At $f = 60$ Hz, the body physical dimensions are electrically very small, thus resulting in MoM solver computation error [3]. Therefore, the computation has been carried out for $f = 30$ kHz and scaled to $f = 60$ Hz. Similarly, the finite difference time domain (FDTD) method results used for the comparison purpose are undertaken

at a few MHz and then scaled. In particular, FDTD calculation uses 45.024 cubical cells of 1.37 cm, which can be considered as computationally expensive.

This clearly demonstrates the efficiency of BEM (FEIEM), which uses only 31 linear elements for the entire cylinder discretization. Knowing the current distribution, one can easily determine the current density and the electric field induced in the body. The results are presented in Table 4.2.

TABLE 4.2
ELF exposure parameters.

$I_z(0)$ [µA]	$J_z(0)$ [mA/m²]	E_z [mV/m]
191	3.1	6.2

The basic restrictions for ELF exposure according to the International Commission on Non-Ionizing Radiation Protection (ICNIRP) Guidelines [19] are given in terms of permissible values for the induced current density inside the body.

It has to be underlined that the obtained value of 3.1 mA/m² is less than the limit of 10 mA/m² for occupational exposure, but exceeds the limit of 2 mA/m² for general public exposure.

The next example deals with the power line with maximum current of $I = 300$ A and operating voltage $V = 100$ kV at $f = 60$ Hz. The conductors are spaced horizontally at distance $s = 3$ m from each other and suspended at a height $d = 15$ m from the ground. The person is standing at a point $y = 7$ m from the reference origin facing the positive y direction. The body is exposed to all 6 components of the field. However, it was shown [3,6,7] that $E_{0z}(0, y, z)$, the component parallel to the upright cylindrical body axis, is the largest within the practical ranges of body height z and distance y from the origin. These conditions are given as:

$$0 \leq z \leq 2m,$$
$$0 \leq y \leq \infty.$$

Being parallel, the external field causes the largest field and current density in the body. The incident field is $E_z^{inc} = 530$ V/m, at 60 Hz, while the value of the foot current and the current density at the ground contact point calculated by using the MoM compared with boundary element method (BEM) [3] and an analytical method [3,6,7] are shown in Table 4.3.

TABLE 4.3
Foot current on antenna model of human body $E_z^{inc} = 530$ V/m, 60 Hz.

Result	MoM	BEM	Analytic
Foot current $I_z(0)$ [μA]	j9.64	j10.11	j9.8
Current density [μA/m^2]	j157	j169	j163

The foot current is also calculated with the incident field at 10 kV/m and 60 Hz. The result for this case compared with measurement and also for several other methods is given in Table 4.4.

The current density induced by the external field in individual organs inside the body can be obtained from the axial current distribution at various heights.

At the low power frequency of 60 Hz, the body physical dimensions are very small electrically, resulting in the MoM solver giving computation error. This is a common problem faced by the MoM, and is also known to occur with the NEC. The computation has to be performed in double precision at a higher frequency of 30 kHz and scaled to 60 Hz. Similarly the FDTD result used for the comparison above is also performed at several MHz and scaled. The FDTD uses 45.024 cubical cells of 1.37 cm size for the modeling, which is computationally very expensive.

In the next example, the human body is exposed to an electric field of 60 Hz from a power line, $E_z^{inc} = 1081.2$ V/m. The numerical results obtained using the GB-BEM are compared with King–Sandler analytical results [6] and presented in Table 4.5. It can be seen that there is good agreement between the present solution and the analytical one, with max discrepancy of 5–10%.

Fig. 4.6 shows the distribution of the induced current along the human body exposed to the incident electric field $E_z^{inc} = 104$ V/m at the frequency of $f = 60$ Hz. The present results are compared to the analytical results obtained by King and Sandler [6] and against the FDTD results obtained by Gandhi and Chen [20]. It can be seen that good agreement is obtained.

Relatively satisfactory agreement between the results obtained via rather different approaches is obtained. From the kink in the curve when the reduced kernel is used, the necessity of applying the exact kernel, which provides stable numerical results, is obvious.

In Table 4.6, the present solution values for the induced current in the foot are compared with the results obtained by the following empirical formulae [21]:

$$I_{fc} = 0.108h^2 f E \quad [\text{mA}], \tag{4.76}$$

TABLE 4.4
Foot current on antenna model of human body $E_z^{inc} = 10$ kV/m, 60 Hz.

Result	MoM	Measured [16]	BEM [6,9]	FDTD [4]	Analytical [1,19]
$I_z(0)$ [μA]	182	165	190	179	≈183

TABLE 4.5
Comparison of the analytical and numerical results.

Physical parameter	GB-BEM	Analytic [1]	Discrepancy [%]
$I_{(z=0)}$ [μA]	7.85	8.22	4.5
$J_{(z=0)}$ [μA/m^2]	13.08	13.7	4.5
$E_{(z=0)}$ [μV/m]	261.67	274.6	4.7
$P_{(z=0)}$ [nW/m^3]	17.12	18.85	9.2

TABLE 4.6
Comparison of numerical results obtained for $E_z^{inc} = 8.7$ kV/m, $f = 50$ Hz.

Incident field	Foot current [GB-BEM] [μA]	Foot current [9] [μA]
$E_z^{inc} = 10$ kV/m, $f = 60$ Hz	191	198.45
$E_z^{inc} = 8$ kV/m, $f = 50$ Hz	142	143.9

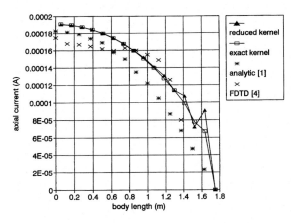

FIG. 4.6 Current distribution induced in the human body obtained for ELF exposure: $E_z^{inc} = 10$ kV/m, $f = 60$ Hz.

where frequency f is expressed in MHz.

Obviously, all the values computed from the empirical formulae are within 10% of the present values computed using the GB-BEM approach.

Table 4.7 shows the calculated results for ELF exposure, for $E_z^{inc} = 16.2$ kV/m, $f = 60$ Hz.

TABLE 4.7
Comparison of numerical results obtained by GB-IBEM and by analytical results available from [5], $E_z^{inc} = 16.2$ kV/m, $f = 60$ Hz.

Physical parameter	GB-IBEM	Analytic [5]
$I_z(0)$ [μA]	117.59	123.42
$J_z(0)$ [μA/m²]	0.196	0.2057
$E_z(0)$ [μV/m]	3.92	4.114
$P_z(0)$ [μW/m³]	3.84	4.23

According to [19], the induced current density is limited to less than 10 mA/m² for occupational exposure and 2 mA/m² for general public exposure.

The results presented so far are obtained by considering the equivalent antenna of the body with both arms resting close to the sides. The effects of the arms being raised to various angles to the vertical body axis are also modeled by attaching wires of radius $a = 0.05$ m and length $L = 0.8$ m to the cylindrical antenna [3]. The wires representing the arms are attached at a height of 1.4 m from the ground. The effect of the arms vertical position in respect to body is shown in Fig. 4.7.

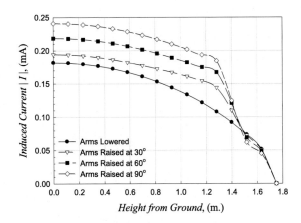

FIG. 4.7 Current induced with arms stretched out by incident field of 1 kV/m, 60 Hz.

Note that the outstretched arms increase the induced current on the body.

4.3.6.2 HF Exposures

The first example for HF sources deals with the human exposed to a field from a broadcast tower. The current distribution along the body due to the electromagnetic field in the vicinity of a broadcast tower operating at 700 kHz with the incident field $E_z^{inc} = 1$ V/m is shown in Fig. 4.8.

The next example deals with the exposure to a field from a missile boat antenna. The current distribution induced along the body in the vicinity of a monopole antenna on a navy missile boat operating at 5 MHz with the field $E_z^{inc} = 10$ V/m is shown in Fig. 4.9.

The current in this case is about two orders of magnitude larger than that due to the broadcast tower and the power lines.

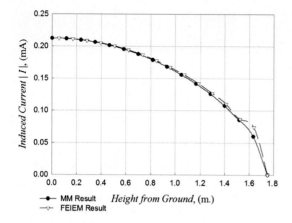

FIG. 4.8 Current induced by the field in vicinity of broadcast tower.

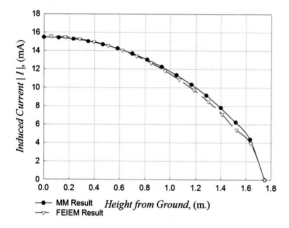

FIG. 4.9 Current induced by the field in vicinity of a missile boat antenna.

The last example in this section is related to the radiation from a shipboard antenna system. Almost all ships are equipped with radio-frequency equipment, which, in cooperation with objects on and above deck, forms the shipboard electromagnetic environment. Such an environment is characterized by high amplitude electromagnetic fields that might be harmful to humans. Naval personnel standing on the metal deck of the ship are exposed to the near fields of several vertical antennas operating at frequencies from 1 to 30 MHz used intermittently for radio communication over the surface of the sea to other ships or the shore. The electromagnetic field close to such an antenna is quite large and the induced current is significant.

For this analysis, a typical 100 W peak envelope power HF transmitter is chosen. The HF transmitter characteristics were: 100 W maximum output power, HF frequency band (3–30 MHz), 5 MHz transmitting frequency during measurement, SSB AM modulation, 6 m long vertical whip antenna located 16.5 m above the deck.

The ship of interest was located in a port, about 100 m away from other ships. The coast station was about 200 m from the ship. The shipboard electromagnetic environment consisted of:

- analyzed sources antennas;
- other emitters and antennas on the topside;
- metallic (perfectly conductive) objects on the deck;
- cabins of different size and shape and other objects;
- metallic (perfectly conductive) ship hull;
- sea (ground plane).

The accessible parts of the deck were covered by the grid of points – 36 points in all, each point 1 m high above the deck (Fig. 4.10). The number of points was limited by the duration of entire measurement and practical access on the deck.

The distance between two adjacent points was not greater than 2 m. The points were numerated and marked on the deck. The vertical component of HF electric field (E_z) was measured at each point. SAR and the current induced in the human body were obtained for the each value of measured electric field E_z. Calculations of the current distribution and current density and SAR were undertaken at $f = 5$ MHz, $\sigma = 0.5$ S/m, $\varepsilon_r = 50$, $k_2 = 0.105$ m^{-1}, $h = 1.75$ m, $Z_L = 0$ (corresponding to a barefoot man on a high conducting deck), $a = 0.14$ m. Figs. 4.11 and 4.12 show the axial current and current, and current density distribution along the human body, respectively.

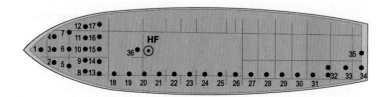

FIG. 4.10 Measurement points on the ship deck.

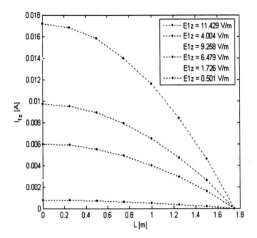

FIG. 4.11 Current distribution along the human body.

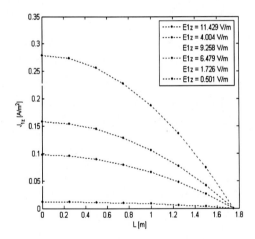

FIG. 4.12 Current density distribution along the human body.

Finally, the obtained maximum value of whole body averaged SAR is $SAR_{WB} = 0.079899$ mW/kg, which is found to be far below 0.08 W/kg (ICNIRP exposure limit for general public).

It is evident that the obtained values of SAR, current density and induced currents in the human body are found to be below appropriate limits.

4.4 TIME DOMAIN MODELING – EXPOSURE OF HUMANS TO TRANSIENT RADIATION: CYLINDRICAL MODEL OF THE HUMAN BODY

A direct time domain simulation of the human body exposed to a transient electromagnetic field can be under-

taken either using realistic, anatomically based models [22–24], or using the simplified human equivalent antenna model [13]. The coupling of transient electromagnetic fields to anatomically based body models has been usually analyzed via the finite-difference time-domain (FDTD) method.

The human equivalent antenna has been designed for experimental dosimetry, and is valid within the frequency range from 50 Hz to 110 MHz [25]. It is worth noting that the dimensions of the human equivalent antenna are within the thin wire approximation and the effective frequency bandwidth of the pulsed electromagnetic waveforms corresponds to the frequency range of the human equivalent antenna.

The time domain human equivalent antenna model has been based on the Hallén integral equation and on dimensions available from [13,25]. A solution of this integral equation using the time domain Galerkin–Bubnov scheme of the boundary element method (GB-BEM) has been presented in [13].

As an extension of the work reported in [25], some convenient measures for post-processing the human body transient response have been proposed in [14].

Further, to the specific absorption (SA) concept for quantifying transient exposures, commonly used within the bioelectromagnetic community [19], additional measures of the body transient response promoted in [14] are given in terms of *average* and *root-mean-square* value a of time-varying current, *instantaneous power* dissipated in the body, and *total absorbed energy* by the body.

After determining the transient response of the human body by solving the Hallén equation, one can readily calculate a distribution of the average and root-mean-square values of the space-time varying current flowing through the body as a measure of the transient behavior of the induced current.

Electromagnetic energy absorbed by the body can be quantified using the circuit-theory quantities in terms of instantaneous power and total absorbed energy.

The main advantages of the proposed formulation, when compared to a more complex realistic models, are its simplicity and efficiency in getting the rapid estimation of the transient phenomena.

4.4.1 Time Domain Formulation

The time domain analysis of the transient electromagnetic field illuminating the well-grounded human body standing vertically on the perfectly conducting (PEC) ground, as shown in Fig. 4.2, is based on the human equivalent antenna concept [13,25].

The dimensions of the human equivalent antenna ($L = 1.8$ m, $a = 5$ cm) are within the thin wire approximation, and the effective bandwidth of the EMP frequency spectrum is 5 MHz. This bandwidth is also within the frequency range of the human equivalent antenna which is stated to be valid from 50 Hz to 110 MHz.

The space-time integro-differential equation for the current along the wire is obtained by enforcing the condition for the total tangential electric field component at the wire surface [3]:

$$E_z^{\text{inc}} + E_z^{\text{sct}} = R_L\, I, \tag{4.77}$$

where E_z^{inc} is the incident electric field and the scattered electric field E_z^{sct} is expressed in terms of the vector and scalar potentials:

$$\vec{E}_{|\text{tan}}^{\text{sct}} = \left(\frac{\partial \vec{A}}{\partial t} + \Delta\varphi \right)_{|\text{tan}}, \tag{4.78}$$

where the vector potential is defined by

$$\vec{A} = \frac{\mu}{4\pi} \int_{S'} \frac{\vec{J}(r', t - R/c)}{R}\, dS' \tag{4.79}$$

and the scalar potential is given by

$$\varphi = \frac{1}{4\pi\varepsilon} \int_{S'} \frac{\rho(r', t - R/c)}{R}\, dS', \tag{4.80}$$

where ρ_s and $\vec{J}$ are space-time dependent surface charge and surface current density, respectively, satisfying the continuity equation

$$\nabla \vec{J}_s = -\frac{\partial \rho_s}{\partial t}. \tag{4.81}$$

Finally, R_L is the resistance per unit length of the antenna given by

$$R_L = \frac{1}{a^2 \pi \sigma}. \tag{4.82}$$

According to the thin wire approximation and performing certain forward mathematical manipulations, the space-time integro-differential equation is obtained [3]:

$$-\varepsilon \frac{\partial E_z^{\text{inc}}}{\partial z} = \left[\frac{\partial^2}{\partial z^2} - \frac{1}{c} \frac{\partial^2}{\partial t^2} \right] \int_0^L \frac{I(z', t - R/c)}{4\pi R}\, dz'$$

$$- R_L \frac{\partial I(z, t)}{\partial t}. \tag{4.83}$$

Integrating Eq. (4.83), the corresponding Hallén integral equation counterpart is readily obtained [13]:

$$\int_0^L \frac{I(z', t - R/c)}{4\pi R}\, dz'$$

$$= F_0 \left(t - \frac{z}{c} \right) + F_L \left(t - \frac{L - z}{c} \right)$$

$$+ \frac{1}{2Z_0} \int_0^L E_z^{\text{inc}} \left(z', t - \frac{|z - z'|}{c} \right) dz'$$

$$- \frac{1}{2Z_0} \int_0^L R_L(z')\, I \left(z', t - \frac{|z - z'|}{c} \right) dz', \tag{4.84}$$

where $I(z', t - R/c)$ is the unknown space-time dependent current to be determined, c is the velocity of light, and Z_0 is the wave impedance of free-space. The unknown functions $F_0(t)$ and $F_L(t)$ account for the multiple reflections of the current wave from the wire ends.

Transient current induced in the person due to the EMP excitation is obtained as a solution of the time domain Hallén integral equation (4.84) by the time domain Galerkin–Bubnov scheme of the boundary element method [13].

When the axial current is determined, it is possible to calculate the electric field, the power density and the specific absorption (SA) induced in the body and further measures of the transient response [14].

4.4.2 Numerical Solution of the Time Domain Hallén Integral Equation

The integral equation (4.84) can be written in the operator form

$$L(I) = Y, \tag{4.85}$$

where L is a linear integral operator, and I is the unknown function to be determined for a given excitation Y.

The unknown solution for current is given in the form of a linear combination of the basis functions:

$$I(z', t') = \{f\}^T \{I\}, \tag{4.86}$$

where $\{f\}$ is the vector containing the basis functions and $\{I\}$ is the vector containing unknown time dependent coefficients of the solution.

In accordance to the weighted residual approach, Eq. (4.85) is multiplied by the test function W_j and integrated over the domain of interest.

Thus, the request for minimization of the interpolation error yields

$$\int_0^L [L(I) - Y] W_j \, dz = 0, \quad j = 1, 2, \ldots, N. \quad (4.87)$$

Applying the boundary element algorithm, the local matrix system for the ith source element interacting with the jth observation element is given as follows:

$$\int_{\Delta l_j} \int_{\Delta l_i} \{f\}_j \{f\}_i^T \frac{1}{4\pi R} \, dz' \, dz \{I\} \Big|_{t - \frac{R}{c}}$$

$$= \int_{\Delta l_j} F_0 \left(t - \frac{z}{c} \right) \{f\}_j \, dz$$

$$+ \int_{\Delta l_j} F_L \left(t - \frac{L - z}{c} \right) \{f\}_j \, dz$$

$$+ \frac{1}{2Z_0} \int_{\Delta l_j} \int_{\Delta l_i} E_z^{inc} \left(z, t - \frac{|Z - z|}{c} \right) \{f\}_j \, dz' \, dz$$

$$- \frac{1}{2Z_0} \int_{\Delta l_j} \int_{\Delta l_i} R_L(z') \{f\}_j \{f\}_i^T \, dz' \, dz. \quad (4.88)$$

In addition, the calculation procedure is more efficient if the known excitation is also interpolated over the wire segment.

Using the same shape functions as those used in the current evaluation procedure, we can write:

$$E_z^{inc}(z', t') = \{f\}^T \{E\}, \quad (4.89)$$

where $\{E\}$ is the time dependent vector containing known values of the transient excitation.

Hence, the matrix equation (4.89) becomes:

$$\int_{\Delta l_j} \int_{\Delta l_i} \{f\}_j \{f\}_i^T \frac{1}{4\pi R} \, dz' \, dz \{I\} \Big|_{t - \frac{R}{c}}$$

$$= \int_{\Delta l_j} F_0 \left(t - \frac{z}{c} \right) \{f\}_j \, dz$$

$$+ \int_{\Delta l_j} F_L \left(t - \frac{L - z}{c} \right) \{f\}_j \, dz$$

$$+ \frac{1}{2Z_0} \int_{\Delta l_j} \int_{\Delta l_i} \{f\}_j \{f\}_i^T \, dz' \, dz \{E\} \Big|_{t - \frac{|z - z'|}{c}}$$

$$- \frac{1}{2Z_0} \int_{\Delta l_j} \int_{\Delta l_i} R_L(z') \{f\}_j \{f\}_i^T \, dz' \, dz \{I\} \Big|_{t - \frac{|z - z'|}{c}}. \quad (4.90)$$

The time domain signals F_0 and F_L can be expressed in terms of auxiliary functions $K_0(t)$ and $K_L(t)$:

$$\int_{\Delta l_j} F_0 \left(t - \frac{z}{c} \right) \{f\}_j \, dz$$

$$= \int_{\Delta l_j} \sum_{n=0}^{\infty} K_0 \left(t - \frac{z}{c} - \frac{2nL}{c} \right) \{f\}_j \, dz$$

$$- \int_{\Delta l_j} \sum_{n=0}^{\infty} K_L \left(t - \frac{z}{c} - \frac{2n + 1}{c} L \right) \{f\}_j \, dz, \quad (4.91)$$

$$\int_{\Delta l_j} F_0 \left(t - \frac{L - z}{c} \right) \{f\}_j \, dz$$

$$= \int_{\Delta l_j} \sum_{n=0}^{\infty} K_0 \left(t - \frac{L - z}{c} - \frac{2nL}{c} \right) \{f\}_j \, dz$$

$$- \int_{\Delta l_j} \sum_{n=0}^{\infty} K_L \left(t - \frac{L - z}{c} - \frac{2n + 1}{c} L \right) \{f\}_j \, dz, \quad (4.92)$$

defined by relations:

$$K_0(t) = \int_{\Delta l_i} \{f\}_i^T \frac{1}{4\pi R_0} \, dz' \{I\} \Big|_{t - \frac{R_0}{c}}$$

$$- \frac{1}{2Z_0} \int_{\Delta l_i} \{f\}_i^T \, dz' \{E\} \Big|_{t - \frac{z'}{c}}$$

$$- \frac{1}{2Z_0} \int_{\Delta l_i} R_L(z') \{f\}_i^T \, dz' \{I\} \Big|_{t - \frac{z'}{c}}, \quad (4.93)$$

$$K_L(t) = \int_{\Delta l_i} \{f\}_i^T \frac{1}{4\pi R_L} \, dz' \{I\} \Big|_{t - \frac{R_L}{c}}$$

$$- \frac{1}{2Z_0} \int_{\Delta l_i} \{f\}_i^T \, dz' \{E\} \Big|_{t - \frac{L - z'}{c}}$$

$$- \frac{1}{2Z_0} \int_{\Delta l_i} R_L(z') \{f\}_i^T \, dz' \{I\} \Big|_{t - \frac{L - z'}{c}}. \quad (4.94)$$

Rearranging of the matrix equation (4.90) yields

$$[A]\{I\}\Big|_{t-\frac{R}{c}} = [B]\{E\}\Big|_{t-\frac{|z-z'|}{c}}$$

$$+[C]\left\{\sum_{n=0}^{\infty} I^n\right\}\Big|_{t-\frac{z}{c}-\frac{2nL}{c}-\frac{R_0}{c}}$$

$$-[B]\left\{\sum_{n=0}^{\infty} E^n\right\}\Big|_{t-\frac{z}{c}-\frac{2nL}{c}-\frac{z'}{c}}$$

$$+[R]\left\{\sum_{n=0}^{\infty} I^n\right\}\Big|_{t-\frac{z}{c}-\frac{2nL}{c}-\frac{z'}{c}}$$

$$-[D]\left\{\sum_{n=0}^{\infty} I^n\right\}\Big|_{t-\frac{z}{c}-\frac{2n+1}{c}L-\frac{R_L}{c}}$$

$$+[B]\left\{\sum_{n=0}^{\infty} E^n\right\}\Big|_{t-\frac{z}{c}-\frac{2n+1}{c}L-\frac{L-z'}{c}}$$

$$-[R]\left\{\sum_{n=0}^{\infty} I^n\right\}\Big|_{t-\frac{z}{c}-\frac{2n+1}{c}L-\frac{L-z'}{c}}$$

$$+[D]\left\{\sum_{n=0}^{\infty} I^n\right\}\Big|_{t-\frac{L-z}{c}-\frac{2nL}{c}-\frac{R_L}{c}}$$

$$-[B]\left\{\sum_{n=0}^{\infty} E^n\right\}\Big|_{t-\frac{L-z}{c}-\frac{2nL}{c}-\frac{L-z'}{c}}$$

$$+[R]\left\{\sum_{n=0}^{\infty} I^n\right\}\Big|_{t-\frac{L-z}{c}-\frac{2nL}{c}-\frac{L-z'}{c}}$$

$$-[C]\left\{\sum_{n=0}^{\infty} I^n\right\}\Big|_{t-\frac{L-z}{c}-\frac{2n+1}{c}L-\frac{R_0}{c}}$$

$$+[B]\left\{\sum_{n=0}^{\infty} E^n\right\}\Big|_{t-\frac{L-z}{c}-\frac{2n+1}{c}L-\frac{z'}{c}}$$

$$-[R]\left\{\sum_{n=0}^{\infty} I^n\right\}\Big|_{t-\frac{L-z}{c}-\frac{2n+1}{c}L-\frac{z'}{c}}, \quad (4.95)$$

where the interaction matrices $[A]$, $[C]$ and $[D]$ are of the form

$$[\ ] = \int_{\Delta l_j} \int_{\Delta l_i} G(z,z')\{f\}_j\{f\}_i^T \, dz' \, dz, \quad (4.96)$$

where $G(z,z')$ is the corresponding Green's function, $1/4\pi R$ in the $[A]$ matrix, where R is the distance from the source to the observation element, $1/4\pi R_0$ in the $[C]$ matrix and $1/4\pi R_L$, where R_0 and R_L are the distances from the source elements to the wire ends.

The $[B]$ matrix is given by the expression

$$[B] = \frac{1}{2Z_0} \int_{\Delta l_j} \int_{\Delta l_i} \{f\}_j\{f\}_i^T \, dz' \, dz, \quad (4.97)$$

and the resistance matrix is of the form

$$[R] = \frac{1}{2Z_0} \int_{\Delta l_j} \int_{\Delta l_i} R_L(z')\{f\}_j\{f\}_i^T \, dz' \, dz. \quad (4.98)$$

The last step in obtaining the space-time varying current distribution is the time discretization.

For convenience, the matrix system (4.95) can be written in the form

$$[A]\{I\}\Big|_{t-\frac{R}{c}} = \{g\}, \quad (4.99)$$

where g is the space-time dependent vector representing the entire right-hand side of Eq. (4.95).

The unknown time coefficients are interpolated by the time domain shape functions T^k over the time increment,

$$I_i(t') = \sum_{k=1}^{N_t} I_i^k T^k(t'), \quad (4.100)$$

where N_t is the total number of time segments.

Using the weighted residual approach in the time domain, it simply follows for the time increment that

$$\int_{t_k}^{t_k+\Delta t} \left([A]\{I\}\Big|_{t-\frac{R}{c}} - \{g\}\right) \theta_k \, dt = 0, \quad k=1,2,\dots,N_t,$$

$$(4.101)$$

where N_t is the total number of time samples, and θ_k denotes the set of time domain test functions.

Choosing the Dirac impulses as test functions, we get

$$[A]\{I\}\Big|_{t_k-\frac{R}{c}} = \{g\}\Big|_{\text{all discrete previous instants}}. \quad (4.102)$$

Performing the space-time discretization by satisfying the condition that within the one time increment the propagation on at least one space segment should be considered,

$$\Delta t \leq \frac{\Delta z}{c}, \qquad (4.103)$$

the following recurrence formula for the space-time dependent current is obtained:

$$I_j\Big|_{t_k} = \frac{-\sum_{i=1}^{N_g} \overline{a}_{ji} I_i\Big|_{t_k+\frac{r}{c}} + g_j\Big|_{\text{all discrete previous instants}}}{a_{jj}},$$
$$j = 1, 2, \ldots, N_g, \quad k = 1, 2, \ldots, N_t, \qquad (4.104)$$

where N_g is the total number of space nodes, and the horizontal line over matrix $[A]$ denotes the absence of diagonal terms.

What is necessary to prescribe are the boundary and initial conditions for the space-time varying current. Thus the stepping procedure starts by specifying the following boundary conditions at the wire ends:

$$I(0, t) = I(L, t) = 0, \qquad (4.105)$$

and the initial condition to be satisfied over the entire length of the wire at $t = 0$:

$$I(z, 0) = 0, \qquad (4.106)$$

i.e., the assumption is that the cylinder (the body) is not excited before $t = 0$.

4.4.3 Measures of the Transient Response

After obtaining the transient current flowing through the human body, it is possible to calculate additional measures of the body transient response.

The convenient parameters characterizing the human body transient response (arising from the basic theory of electric circuits), in addition to specific absorption concept, are the spatial distribution of *average* and *rms* values of a space-time varying current induced in the body, *instantaneous power* dissipated in the body, and *total energy absorbed* by the body.

4.4.3.1 Average Value of the Transient Current

The average value of the transient current is associated with a DC component in the spectrum of a particular transient waveform. The average value of a time varying current $i(t)$ is defined as follows:

$$I_{av} = \frac{1}{T_0} \int_0^{T_0} i(t)\, dt, \qquad (4.107)$$

where T_0 is the time interval of interest.

The distribution of average values of current is simply given by

$$I_{av}(x) = \frac{1}{T_0} \int_0^{T_0} i(x, t)\, dt. \qquad (4.108)$$

Obviously, for any waveform having approximately equal area above and below abscissa, the average value tends to be equal to zero. Therefore, from this parameter one can obtain a rapid estimation of the character of the given transient waveform properties.

When the space-time varying current along the cylinder at each space node z_i and time instant t_k is determined by solving the integral equation (4.95), the *average value* of this current can be computed from the following relation:

$$I_{av} = \frac{1}{T_0} \sum_{k=1}^{N_t} \int_{t_k}^{t_{k+1}} \left(\{T\}^T \{I\} \right)_i dt. \qquad (4.109)$$

Straightforward integration yields

$$I_{av}\Big|_{z_i} = \frac{\Delta t}{2T_0} \sum_{k=1}^{N_t} \left[\left(I_i^k \right) + \left(I_i^{k+1} \right) \right], \qquad (4.110)$$

where $\{T\}$ is the vector containing the time domain linear shape functions, and $\{I\}$ denotes the vector containing time-dependent values of the current.

4.4.3.2 Root-Mean-Square Value of the Transient Current

A time varying current delivers an average power to resistive load. The amount of delivered power strongly depends on the particular waveform.

A measure of comparing the power delivered by different waveforms is the root-mean-square (*rms*) or effective value of a transient current. The *rms* value of a time-varying current is a constant that is equal to the direct current value that would deliver the same average power to a given resistance R_L and that would produce the same heating effect on R_L.

Instantaneous power delivered to a resistance R_L by a transient current $i(t)$ is

$$p(t) = R_L i^2(t), \qquad (4.111)$$

while the corresponding average power P_{av} is determined by the integral relation:

$$P_{av} = \frac{1}{T_0} \int_0^{T_0} p(t)\,dt = \frac{1}{T_0} \int_0^{T_0} R_L\, i^2(t)\,dt = R_L\, I_{rms}^2,$$

$$(4.112)$$

from which the *rms* current is then

$$I_{rms} = \sqrt{\frac{1}{T_0} \int_0^{T_0} i^2(t)\,dt}.\qquad (4.113)$$

Consequently, the spatial distribution of the *rms* values of the space time current along the cylinder is given by

$$I_{rms}(x) = \sqrt{\frac{1}{T_0} \int_0^{T_0} i^2(x,t)\,dt}.\qquad (4.114)$$

When the current along the wire at each node and time instant is determined, the *rms* value of the wire current can be computed from the following relation:

$$I_{rms} = \sqrt{\frac{1}{T_0} \sum_{k=1}^{N_t} \int_{t_k}^{t_{k+1}} \left(\{T\}^T \{I\} \right)_i^2 dt},\qquad (4.115)$$

and, by performing a straightforward integration, we get

$$I_{rms}\Big|_{z_i} = \sqrt{\frac{\Delta t}{3T_0} \sum_{k=1}^{N_t} \left[\left(I_i^k\right)^2 + I_i^k\, I_i^{k+1} + \left(I_i^{k+1}\right)^2 \right]}.$$

$$(4.116)$$

Therefore, the *rms* value of the transient current flowing through the body appears to be a more interesting parameter from the bioelectromagnetics point of view as it is, by its definition, directly associated with the thermal effect of a time varying current flowing through a lossy material.

4.4.3.3 Instantaneous Power
Instantaneous power delivered to a certain resistance R_L or to some resistive medium having equivalent resistance R_L by a transient current is defined by relation (4.111). On the other hand, the absorbed power in the human body expressed by the field quantities is equivalent to the concept of instantaneous power arising from the circuit theory and it is usually defined as a volume

integral over power density, i.e.,

$$P_{rad}(t) = \int_V \sigma \left| \vec{E}(\vec{r},t) \right|^2 dV = \int_V \frac{\left| \vec{J}(\vec{r},t) \right|^2}{\sigma}\, dV.$$

$$(4.117)$$

Assuming the transient current to be approximately constant over the cylinder cross-section yields

$$i(z,t) = J(z,t)\, S,\qquad (4.118)$$

where $J(z,t)$ is the current density along the cylinder, and $S = a^2 \pi$ is the cylinder cross-section.

The instantaneous power can be obtained by spatially integrating the squared space-time varying current:

$$P_{rad}(t) = \frac{1}{\sigma S} \int_0^L i^2(z,t)\,dz.\qquad (4.119)$$

Provided the transient induced current is determined, the instantaneous power can be represented by the following relation:

$$P_{rad}(t) = \frac{1}{\sigma S} \sum_{i=1}^M \int_{\Delta l_i}^L \left(\{f\}_i^T \{I\}_i \right)^2 dz,\qquad (4.120)$$

where $\{f\}$ is the vector containing linear interpolation functions and $\{I\}$ stands for the induced current coefficients.

Straightforward integration yields

$$P_{rad}\Big|_{t_k} = \frac{1}{\sigma S} \frac{\Delta z}{3} \sum_{i=1}^M \left[\left(I_i^k\right)^2 + I_i^k\, I_{i+1}^k + \left(I_{i+1}^k\right)^2 \right],$$

$$(4.121)$$

where M denotes the total number of spatial segments along the cylinder.

4.4.3.4 Total Absorbed Energy
In accordance to the circuit theory, the total absorbed energy in the resistance or resistive material can be obtained by integrating the instantaneous power:

$$W_{tot}(t) = \int_0^t P_{rad}(t)\,dt.\qquad (4.122)$$

Using the numerical representation of the instantaneous power, we obtain

$$W_{\text{tot}}(t) = \sum_{k=1}^{N_t} \int_{t_k}^{t_k + \Delta t} \{T\}_k^T \{P\}_k \, dt, \qquad (4.123)$$

and then straightforward integration gives

$$W_{\text{tot}}\Big|_{t_k} = \frac{\Delta t}{2} \sum_{k=1}^{N_t} \left[\left(P^k + P^{k+1} \right) \right], \qquad (4.124)$$

where P^k and P^{k+1} stand for a discrete value of instantaneous power dissipated in the resistive medium at time instants t_k and t_{k+1}, respectively.

4.4.3.5 The Specific Absorption

After calculating the transient current flowing through the human body, the specific absorption, a principal measure of this human body transient response, can be also determined in terms of circuit theory concepts.

According to the general definition, the already mentioned total energy absorbed by the resistive material can be obtained by temporally integrating the radiation power $p_{\text{rad}}(t)$:

$$W_{\text{tot}} = \int_0^{T_0} p_{\text{rad}}(t) \, dt, \qquad (4.125)$$

where T_0 is the interval of interest.

The power dissipated in the human body is usually given by a volume integral over power density $\overline{P}_d$:

$$p_{\text{rad}}(t) = \int_V \overline{P}_d \, dV, \qquad (4.126)$$

where $\overline{P}_d$ is given by

$$\overline{P}_d = \sigma \left| \vec{E}(\vec{r}, t) \right|^2 = \frac{\left| \vec{J}(\vec{r}, t) \right|^2}{\sigma}, \qquad (4.127)$$

while $\vec{E}(\vec{r}, t)$ and $\vec{J}(\vec{r}, t)$ represent the electric field and current density induced inside the body, respectively.

Combining Eqs. (4.125) to (4.127) yields

$$W_{\text{tot}} = \int_0^{T_0} \int_V \sigma \left| \vec{E}(\vec{r}, t) \right|^2 dV \, dt = \int_0^{T_0} \int_V \frac{\left| \vec{J}(\vec{r}, t) \right|^2}{\sigma} \, dV \, dt. \qquad (4.128)$$

Specific absorption (SA) is defined as a quotient of the incremental energy (dW) absorbed and incremental mass (dm) contained in the volume (dV) of given density (ρ):

$$\text{SA} = \frac{dW}{dm} = \frac{dW}{\rho \, dV}. \qquad (4.129)$$

Substituting expression (4.128) into (4.129) results in

$$\text{SA} = \int_0^{T_0} \frac{1}{\rho} \frac{d}{dV} \left(\int_V \frac{\left| \vec{J}(\vec{r}, t) \right|^2}{\sigma} \, dV \right) dt, \qquad (4.130)$$

which after rearrangement becomes

$$\text{SA} = \int_0^{T_0} \frac{1}{\rho} \frac{\left| \vec{J}(\vec{r}, t) \right|^2}{\sigma} \, dt. \qquad (4.131)$$

Assuming the transient current distribution to be approximately constant over the cylinder cross-section gives

$$i(z, t) = J(z, t) \cdot S = J(z, t) \cdot a^2 \pi, \qquad (4.132)$$

where $J(z, t)$ denotes the space-time varying current density along the cylindrical body.

Finally, the SA can be expressed as

$$\text{SA}(z) = \frac{1}{\rho \sigma S^2} \int_0^{T_0} i^2(z, t) \, dt = \frac{1}{\rho \sigma \, (a^2 \pi)^2} \int_0^{T_0} i^2(z, t) \, dt. \qquad (4.133)$$

Note that the specific absorption defined by relation (4.133) and the spatial distribution of root-mean-square values of transient current given by

$$I_{\text{rms}}^2(z) = \frac{1}{T_0} \int_0^{T_0} i^2(z, t) \, dt \qquad (4.134)$$

are simply related as follows:

$$\text{SA}(z) = \frac{T_0}{\rho \sigma S^2} I_{\text{rms}}^2(z). \qquad (4.135)$$

Once the induced current is known, SA can be represented, using the boundary element formalism, by the following relation:

$$SA = \frac{1}{\rho\sigma \left(a^2\pi\right)^2} \frac{\Delta t}{3} \sum_{k=1}^{N_t} \int_{t_k}^{t_k+\Delta t} \{T\}_k^T \{I\}_k \, dt, \quad (4.136)$$

where vector $\{T\}$ contains the linear temporal interpolation functions, while vector $\{I\}$ represents the values of transient current distribution.

Performing straightforward integration leads to a simple formula

$$SA = \frac{1}{\rho\sigma \left(a^2\pi\right)^2} \frac{\Delta t}{3} \sum_{k=1}^{N_t} \left[\left(I_i^k\right)^2 + I_i^k I_{i+1}^k + \left(I_{i+1}^k\right)^2 \right],$$

$$(4.137)$$

where N_t denotes the total number of time increments and I_i^k denotes the ith node current at the kth time instant.

4.4.4 Numerical Results
Fig. 4.13 shows the transient current induced in the waist due to exposure to the standard double-exponential EMP waveform,

$$E_z^{\text{inc}}(t) = E_0 \left(e^{-at} - e^{-bt}\right), \quad (4.138)$$

where $E_0 = 1$ kV/m, $a = 4 \cdot 10^6$, $b = 4.76 \cdot 10^8$.

The results obtained from the time domain simulation are compared with the results calculated by using the frequency domain Method of Moments code and inverse Fourier transform [3]. The agreement seems to be satisfactory.

Fig. 4.14 shows the transient current induced in the foot due to exposure to the standard double-exponential EMP waveform (4.138) with $E_0 = 1.05$ V/m, $a = 4 \cdot 10^6$, and $b = 4.76 \cdot 10^8$.

The related distribution of the *average* and *rms* values along the body is shown in Figs. 4.15 and 4.16, respectively.

The next example deals with the transient current induced in the feet due to the Gaussian pulse exposure,

$$E_z^{\text{inc}}(t) = E_0 e^{-g^2(t-t_0)^2}, \quad (4.139)$$

with the following parameters: $E_0 = 1$ V/m, $g = 2 \cdot 10^9$ s^{-1} and $t_0 = 2$ ns.

As this pulse is a numerical equivalent of Dirac pulse, the obtained transient exposure shown in Fig. 4.17 can be regarded as the impulse response of the human body.

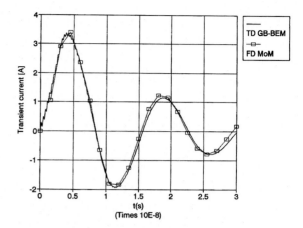

FIG. 4.13 Transient current induced in the waist exposed to the double-exponential EMP waveform.

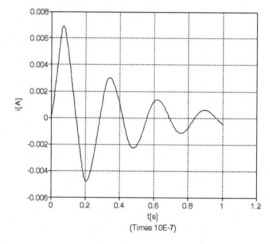

FIG. 4.14 Transient current induced in the foot exposed to the double-exponential EMP waveform.

Transient response to any other incident waveform then can be computed by performing a simple convolution.

While the spatial distribution of the average values of the space-time current flowing through the cylindrical body model is nearly zero for the case of a Gaussian pulse, more information regarding the heating effect due to the Gaussian pulse exposure can be obtained from the spatial distribution of the rms values of the transient current, shown in Fig. 4.18.

What can be observed from Fig. 4.18 is that the Gaussian pulse that induces the peak value of current around 1.5 mA in the feet corresponds to the equivalent DC current (approx. 0.4 mA). Thus, the transient current induced in the feet shown in Fig. 4.17 would produce

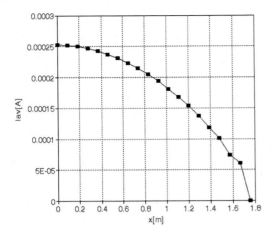

FIG. 4.15 Spatial distribution of the average values along the body for the EMP exposure.

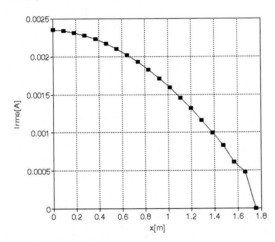

FIG. 4.16 Spatial distribution of the rms values along the body for the EMP exposure.

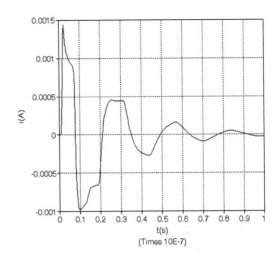

FIG. 4.17 Transient current induced in the feet due to the Gaussian pulse exposure.

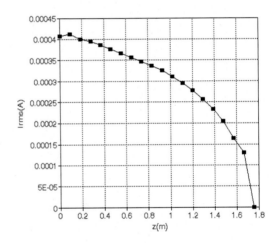

FIG. 4.18 Spatial distribution of average values of the transient current due to the Gaussian pulse exposure.

the same heating effect as the constant DC current of 0.4 mA.

Furthermore, the transient behavior of the instantaneous power dissipated in the body is shown in Fig. 4.19.

It can be noticed that the power dissipation, with the peak value slightly above 2.5 mW occurs early, within the first 50 ns.

The same conclusion can be drawn from Fig. 4.20 representing the total energy absorbed in the body versus time.

It is clearly visible that body does not absorb a significant amount of energy after the first 50 ns.

Therefore, a negligible amount of energy is absorbed by the human body exposed to pulsed field, due to a rather limited time duration (up to 100 ns) of the incident pulse.

Fig. 4.21 shows the corresponding spatial distribution of the specific absorption along the body for the case of Gaussian pulse (4.139) with $E_0 = 1$ V/m, $g = 2 \cdot 10^9$ s^{-1} and $t_0 = 2$ ns, temporal step being

$$E_z^{\text{inc}}(t) = u(t), \tag{4.140}$$

where $u(t)$ denotes the unit step, and standard double-exponential electromagnetic pulse (EMP) waveform (4.138) with $E_0 = 1.05$ V/m, $a = 4 \cdot 10^6$ s^{-1}, and $b = 4.76 \cdot 10^8$ s^{-1}.

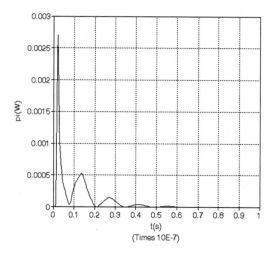

FIG. 4.19 Instantaneous power dissipated within the body due to the Gaussian pulse exposure.

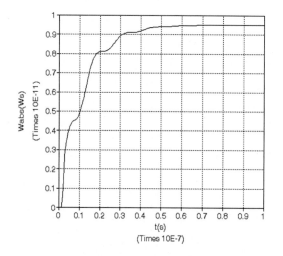

FIG. 4.20 Total energy absorbed in the body due to the Gaussian pulse exposure.

It is visible that the highest value of absorption is achieved in the case of EMP exposure, while the lowest value is reached in the case of Gaussian pulse exposure. The maximal values of SA (below 12 pJ/kg), compared to the limits of 28.8 J/kg for whole body average, and 576 J/kg for 1-g spatial peak per pulse proposed by an IEEE Standard, stay far below the given threshold. It is worth mentioning that these SA results are obtained for the case of unit pulse amplitudes.

However, even by scaling the results with realistic values of pulse amplitudes corresponding to the order of magnitude from 50 to 100 kV/m, and also by taking into account a rather thin radius of antenna body model, the

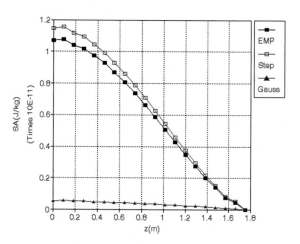

FIG. 4.21 Specific absorption for various transient exposure waveforms.

computed SA values, though considered rather underestimated, are still pretty negligible.

4.5 TRANSMISSION LINE MODELS OF THE HUMAN BODY

The representation of the human body in terms of one or more vertical conductors can be formulated via Transmission Line (TL) approach as reported in [15]. One can also use an enhanced TL theory which was outlined in [16]. Thus in [15] the human body is represented by one or more vertical conductors and modeled via the TL theory in the frequency domain. The related transient response is computed using the Inverse Fourier Transform (IFT). Per unit length parameters are calculated for finite heights and the physical characteristics of the human and soil are taken into account. The assessment of the space-time dependent current distribution induced in the body is carried out in the general case where the human is not directly in contact with the ground.

The results, related to various transient exposures, obtained with TL body representation are compared to the direct time domain results computed using the antenna body model, based on the Hallén equation and GB-IEM.

4.5.1 Theoretical Background

A person standing vertically on the ground and exposed to an incident transient electric field oriented tangential to the body can be represented by multiple imperfectly conducting conductors of length l and radius a, as shown in Fig. 4.22. The human body contains different tissues and organs with related conductivities σ_w and relative permittivities ε_{rw}.

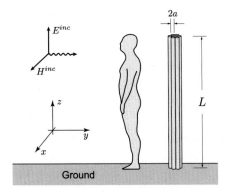

FIG. 4.22 Multi-conductor model of the human body.

The calculation of the current distribution within the human body is based on corresponding transmission line equations for multiple straight wires excited by a transient incident electric field given by [15]:

$$\frac{d[U(z, \omega)]}{dz} + [Z(\omega)][I(z, \omega)] = [E_z^e], \qquad (4.141)$$

$$\frac{d[I(z, \omega)]}{dz} + [Y(\omega)][U(z, \omega)] = 0, \qquad (4.142)$$

where $[U(z, \omega)]$, $[I(z, \omega)]$ are the total voltage and current along the multi-conductor system, respectively, while $[Z(\omega)]$ is the longitudinal per unit length impedance matrix of the vertical multi-conductor system given by

$$[Z] = j\omega[L] + [Z_w] + [Z_g]. \qquad (4.143)$$

Note that $[L]$ is the per-unit-length longitudinal self and mutual inductances matrix of the lossless vertical multi-conductor system above a perfect soil, $[Z_w]$ is the per-unit-length internal impedance matrix of conductors, $[Z_g]$ is the per-unit-length ground impedance matrix, $[Y(\omega)]$ is the transverse per unit length admittance matrix of the vertical multi-conductor system given by

$$[Y] = j\omega[C], \qquad (4.144)$$

where $[C]$ is the per-unit-length transverse self and mutual capacitances matrix of the lossless vertical multi-conductor system above a perfect soil and $[E_z^e]$ is the tangential incident electric field.

Fig. 4.23 shows the equivalent representation with the incident electric field being replaced by two current generators to each conductor's terminations in the frequency domain [15].

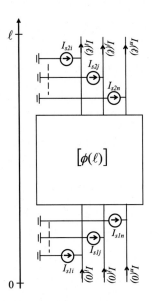

FIG. 4.23 Equivalent model of a vertical multi-conductor configuration excited by an external field.

The general solution of the transmission line coupling equations has been obtained in terms of the chain parameter matrix and the incident-field forcing functions [26]. The terminal conditions for the total voltage of each conductor are given by:

$$U(0) = -Z_0 I(0), \qquad (4.145)$$

$$U(l) = Z_l I(l), \qquad (4.146)$$

where Z_0 and Z_l are the conductor (human body) terminations.

The contact of the human body with the ground can be taken into account by means of reactive impedance Z_0:

$$Z_0 = \frac{1}{j\omega C_0}, \qquad (4.147)$$

where C_0 is the capacitance between the soles of the feet and their image in the earth.

If the foot soles are bare, i.e., in direct contact with moist earth, the body is assumed to be well grounded and the capacitance between the foot soles and their image in the earth is neglected.

The second end of the human is connected by a high impedance, i.e., the case $Z_l = \infty$ is assumed.

The corresponding transient response can now be evaluated using the Inverse Fourier Transform.

In [16], the analysis is based on the enhanced transmission line (TL) theory aiming to quantify the induced current inside the human body. A linear system equations, where the electromagnetic field excitation is represented by two equivalent current and voltage generators, is solved. Once the axial current is determined, it is possible to calculate the specific absorption rate (SAR).

The study [16] extends the work reported in [15], by representing the body in terms of wire-junction models to account for the influence of arms. The formulation is based on the transmission line (TL) equations in the frequency domain and valid for a range frequency of 50 Hz to 110 MHz.

The human body in a direct contact with the ground is considered, as shown in Fig. 4.24. The arms are represented by a set of wires attached to the thick cylinder. The dimensions are given in Table 4.8.

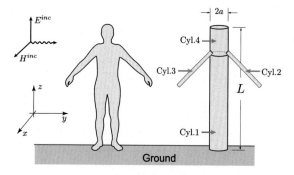

FIG. 4.24 Model of the human body with arms outstretched.

TABLE 4.8
Geometric parameters.

	Length L_i [m]	Radius a_i [m]
Cylinder 1 (body)	1.5	0.14
Cylinder 2 / Cylinder 3 (arms)	0.5	0.04
Cylinder 4 (head)	0.3	0.14

The per unit length parameters of the horizontal segments are calculated using the formalism of Ametani [27]. For vertical segments the formalism of Eldon J. Rogers et al. is used [28].

The set of equations for the coupling between a transmission line (n conductors) and an external electromagnetic field is given by [29]:

$$\frac{d\,[U(z,\omega)]}{dz} + [Z]\,[I(z,\omega)] = [U_F(z,\omega)], \quad (4.148)$$

$$\frac{d\,[I(z,\omega)]}{dz} + [Y]\,[U(z,\omega)] = [I_F(z,\omega)], \quad (4.149)$$

where $[U(z,\omega)]$ and $[I(z,\omega)]$ are n-dimensional complex vectors containing total voltage and current along the transmission line; $[Z]$ and $[Y]$ square matrices of size $n \times n$, impedance and admittance per unit length complex. The expressions for the linear longitudinal impedance $[Z]$ and admittance per unit length transverse $[Y]$ used in this study are documented in detail in [29], $[U_F(z,\omega)]$ and $[I_F(z,\omega)]$ are complex vectors by which an external electromagnetic field is taken into account, given by:

$$[U_F(z,\omega)] = -\frac{\partial}{\partial z}\,[E_T^e(z,\omega)] + [E_L^e(z,\omega)], \quad (4.150)$$

$$[I_F(z,\omega)] = -[Y]\,[E_T^e(z,\omega)], \quad (4.151)$$

where $[E_T^e(z,\omega)]$ and $[E_L^e(z,\omega)]$ are sources due to incident transverse electric field and incident longitudinal electric field, respectively.

It is worth noting that only the vertical incident electric field (longitudinal) to the human body is of interest.

4.5.2 Solution of the Transmission Line Equations in the Frequency Domain

TL Eqs. (4.148) and (4.149) can be written in state variable forms as a coupled first order, ordinary differential equations in matrix form:

$$\frac{d}{dz}\underbrace{\begin{bmatrix} [U(z)] \\ [I(z)] \end{bmatrix}}_{X(z)} = \underbrace{\begin{bmatrix} [0] & -[Z] \\ -[Y] & [0] \end{bmatrix}}_{A}\underbrace{\begin{bmatrix} [U(z)] \\ [I(z)] \end{bmatrix}}_{X(z)}$$
$$+ \begin{bmatrix} [U_F(z)] \\ [I_F(z)] \end{bmatrix}. \quad (4.152)$$

Using the state variable formalism [27–30], the solution of (4.152) is given by

$$X(z) = \phi(z - z_0)\,X(z_0) + \int_{z_0}^{z} [\phi(z - \tau)]\begin{bmatrix} [U_F(\tau)] \\ [I_F(\tau)] \end{bmatrix} d\tau, \quad (4.153)$$

with $X(z) = \begin{bmatrix} U(z) \\ I(z) \end{bmatrix}$ being the matrix of voltages and currents at any point z of the transmission line representing the body.

The chain parameter matrix $[\phi(z)]$ is defined as [16]:

$$[\phi(z)] = e^{A \cdot z} = \begin{bmatrix} [\phi_{11}(z)] & [\phi_{12}(z)] \\ [\phi_{21}(z)] & [\phi_{22}(z)] \end{bmatrix}. \quad (4.154)$$

Note that (4.153) provides the values of currents and voltages at an arbitrary point z of the line according to their value at the origin z_0.

As the values of voltages and currents at the extremities of the line (the human body) are of interest, (4.153) corresponding to a line that starts with $z_0 = 0$ and applied to the end $z_L = l$ becomes

$$X(l) = \phi(l) \cdot X(0) + \int_0^l [\phi(l - \tau)] \begin{bmatrix} [U_F(\tau)] \\ [I_F(\tau)] \end{bmatrix} d\tau.$$

$$(4.155)$$

The $n \times n$ submatrices of the chain parameter matrix are given [27]:

$$[\phi_{11}(l)] = +\frac{1}{2}[Y]^{-1}[T]\left(e^{[\gamma(l)]} + e^{-[\gamma(l)]}\right)[T]^{-1}[Y],$$

$$(4.156a)$$

$$[\phi_{12}(l)] = -\frac{1}{2}[Y]^{-1}[T][\gamma]\left(e^{[\gamma(l)]} - e^{-[\gamma(l)]}\right)[T]^{-1},$$

$$(4.156b)$$

$$[\phi_{21}(l)] = -\frac{1}{2}[T]\left(e^{[\gamma(l)]} - e^{-[\gamma(l)]}\right)[\gamma]^{-1}[T]^{-1}[Y],$$

$$(4.156c)$$

$$[\phi_{22}(l)] = +\frac{1}{2}[T]\left(e^{[\gamma(l)]} + e^{-[\gamma(l)]}\right)[T]^{-1},$$

$$(4.156d)$$

where $[T]$ is a matrix of size $n \times n$, $[\gamma]$ represents the diagonal matrix of propagation constants squared with $[\gamma^2] = [T]^{-1}[Y][Z][T]$.

From (4.155) the terms $U(0)$, $I(0)$, $U(l)$ and $I(l)$ can be expressed as

$$\begin{bmatrix} [U(l)] \\ [I(l)] \end{bmatrix} = \begin{bmatrix} [\phi_{11}(l)] & [\phi_{12}(l)] \\ [\phi_{21}(l)] & [\phi_{22}(l)] \end{bmatrix} \begin{bmatrix} [U(0)] \\ [I(0)] \end{bmatrix}$$
$$+ \begin{bmatrix} [U_{FT}(l)] \\ [I_{FT}(l)] \end{bmatrix}, \qquad (4.157)$$

which can be written as follows:

$$[I_{2n}] \cdot \begin{bmatrix} [U(l)] \\ [I(l)] \end{bmatrix} - \begin{bmatrix} [\phi_{11}(l)] & [\phi_{12}(l)] \\ [\phi_{21}(l)] & [\phi_{22}(l)] \end{bmatrix} \begin{bmatrix} [U(0)] \\ [I(0)] \end{bmatrix}$$
$$= \begin{bmatrix} [U_{FT}(l)] \\ [I_{FT}(l)] \end{bmatrix}, \qquad (4.158)$$

where $[I_{2n}]$ is the identity matrix of order $2n$ (n being the number of conductors by line) and the total forcing

functions are given in [27]:

$$[U_{FT}(l)] = \int_0^l ([\phi_{11}(l - \tau)][U_F(\tau)]$$
$$+ [\phi_{12}(l - \tau)][I_F(\tau)]) d\tau, \qquad (4.159)$$

$$[I_{FT}(l)] = \int_0^l ([\phi_{21}(l - \tau)][U_F(\tau)]$$
$$+ [\phi_{22}(l - \tau)][I_F(\tau)]) d\tau. \qquad (4.160)$$

For the case where the body is exposed to an external electromagnetic field, i.e., when $[U_F(\tau)] = [E_z^e]$ and $[I_F(\tau)] = [0]$, (4.159) and (4.160) become:

$$[U_{FT}(l)] = \int_0^l ([\phi_{11}(l - \tau)][E_z^e(\tau)]) d\tau, \qquad (4.161)$$

$$[I_{FT}(l)] = \int_0^l ([\phi_{21}(l - \tau)][E_z^e(\tau)]) d\tau. \qquad (4.162)$$

Provided the general solution of the transmission line coupling equations in terms of the chain parameter matrix and the external-field excitation is obtained, one then incorporates the appropriate boundary conditions. Thus, at both ends of the line, one has the following relationships:

$$[U(0)] = -[Z_0][I(0)], \qquad (4.163)$$
$$[U(l)] = [Z_L][I(l)], \qquad (4.164)$$

where $[Z_0]$ and $[Z_L]$ are the impedance matrices termination of the line.

In [16], two specific ends representing the feet and head are considered. The human is in contact with the ground through with shoes (foot soles) taken into account in terms of an equivalent capacitor which we represent by a capacitor C [16]:

$$Z_0 = \frac{1}{j\omega C}, \quad C = \frac{A\varepsilon_0\varepsilon_r}{d}, \quad A = \pi a^2, \qquad (4.165)$$

where C is the capacitance between the sole of the foot and its image in the ground, ε_0 is the permittivity of free-space, ε_r is the relative permittivity of the material constituting the sole of the shoe, A is the contact surface, and d is the distance between the sole of the shoe and its image in the ground.

The other end of the equivalent cylinder is connected by infinite impedance, i.e., $Z_L = \infty$, namely cylinders 2, 3 and 4 in the case of the human body with the arms outstretched.

An extended TL formalism to account for the multiple wire model, in the frequency valid for radial and the multiple wire representation, is related to the solution of a general matrix equation:

$$[A][X] = [B]. \qquad (4.166)$$

Matrix $[A]$ is the topological matrix of a radial network. Note that the human being is modeled as a radial network of transmission lines, where vector $[X]$ contains the unknown currents and voltages at each node, while vector $[B]$ includes the external field effects.

Hence, one should transform the propagation relations (4.158) for all the lines representing the body and the relations in the all nodes (between different organs) to matrix equation (4.166).

Matrix $[A]$ is composed of two submatrices as follows:

$$[A] = \begin{bmatrix} [A_1] \\ [A_2] \end{bmatrix}. \qquad (4.167)$$

Submatrix $[A_1]$ accounts for the propagation in all line segments which form the human body (trunk and arms) and is constructed from Eq. (4.158).

Submatrix $[A_2]$ accounts for the electrical relations (current and voltage) on all nodes of the human body topology [27]. Thus, applying the Kirchhoff law in the mth node, the following equation is obtained [27]:

$$\sum_{i=1}^{N} ([Y_i^m][U_i^m] + [Z_i^m][I_i^m]) = [0], \qquad (4.168)$$

where $[Z_i^m]$ and $[Y_i^m]$ are respectively the impedance and admittance matrices resulting of the application of Kirchhoff's laws in the mth node and containing numerical values $0, 1, -1$, impedance and admittance values according to the human body topology.

Finally, solving the matrix system (4.166) yields the currents and voltages at all nodes of the human body topology.

4.5.3 Computational Examples

The results obtained by two different cylindrical models are considered in this section, namely the single cylinder model, and the cylindrical model with the thin wires attached.

4.5.3.1 Single Cylinder Model

The human body represented by a cylinder of length $L = 1.75$ m and radius $a = 0.14$ m, with the base and the top of the cylinder terminated by impedances Z_L and Z_0, is depicted in Fig. 4.25.

The body is assumed to be well-grounded, thus neglecting the capacitance between the soles of the feet and their image in the earth ($Z_0 = 0\,\Omega$), while the average value of the conductivity is chosen to be $\sigma = 0.5$ S/m (conductivity at 60 Hz).

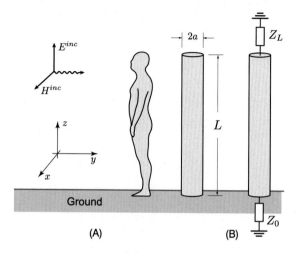

FIG. 4.25 (A) Cylindrical model of the human body exposed to vertical electric field; (B) Impedances Z_L and Z_0 on both ends of the cylinder.

The first set of numerical results is related to a validation of TL theory with time domain antenna model [13]. The three types of transient electric field excitation are considered in this work: Gaussian pulse, temporal step function, and EMP.

Figs. 4.26 to 4.28 show respectively the transient current induced in the feet due to EMP exposure (4.138) with $E_0 = 1.05$ V/m, $\alpha = 4 \cdot 10^6$ s^{-1}, $\beta = 4.76 \cdot 10^8$ s^{-1}, the temporal step exposure (4.140) and the Gaussian pulse exposure (4.139), where $E_0 = 1$ V/m, $g = 2 \cdot 10^9$ s^{-1}, and $t_0 = 2$ ns.

A satisfactory agreement between the proposed TL model and the numerical results obtained from the solution of the time domain Hallén equation by the Galerkin–Bubnov boundary element method (GB-BEM) [3] is achieved [15]. Small discrepancies appear due to the implementation of the Inverse Fast Fourier Transform algorithm.

Furthermore, the effects due to the soil conductivity and the capacitance effect are investigated. Fig. 4.29 shows the variation of the transient current induced in the feet due to the EMP excitation for different values of conductivity.

It can be seen that certain differences occur especially if the perfect conductivity of the body is assumed.

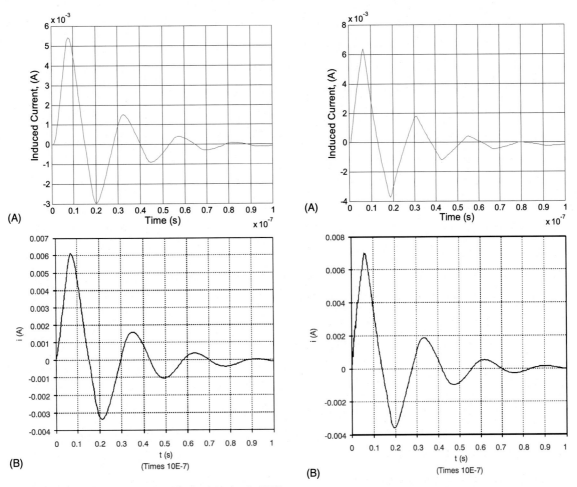

(A)

(B) (Times 10E-7)

FIG. 4.26 Transient current induced in the feet due to EMP exposure [15]: (A) TL model, (B) Antenna model.

Contrary to the previous examples in which the foot soles are considered to be bare, i.e., the body is in direct contact with the earth, thus neglecting the capacitance between the foot soles and their image in the earth, Fig. 4.30 shows the effect of variable capacitance of the foot soles to the transient response.

It is visible that the assumption of direct contact of the body with the earth implies higher values of the induced transient current. Obviously, the transient response of the body is more affected by the capacitance than by soil conductivity.

4.5.3.2 Human Body With the Arms Outstretched

This subsection deals with the study of the human body with the arms outstretched exposed to the double exponential pulse. Figs. 4.31 and 4.32 represent the variation

(A)

(B) (Times 10E-7)

FIG. 4.27 Transient current induced in the feet due to temporal step exposure [15]: (A) TL model, (B) Antenna model.

of transient current induced in the feet, the free end of the left and right arms, and the head. In this case, the current flows through the path of the least resistance.

Furthermore, the human body is exposed to transient electric field in the form of Gaussian pulse (4.139) with $E_0 = 1$ V/m, $g = 2 \cdot 10^9$ s^{-1}, and $t_0 = 2$ ns.

Fig. 4.33 shows the transient current induced in the feet due to the Gaussian pulse exposure. The transient current induced in the free end of the left/right arms and head is presented in Fig. 4.34.

4.5.3.3 Human Exposure to High Frequency (HF) Radiation

SAR (W/kg) at any point in the human head is defined as [1]

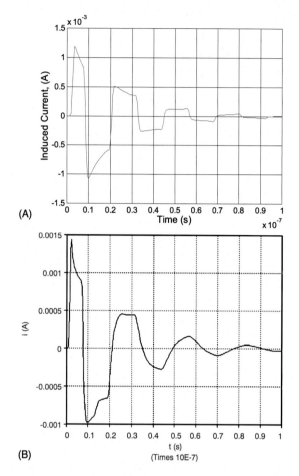

(A)

(B)

FIG. 4.28 Transient current induced in the feet due to Gaussian pulse exposure [15]: (A) TL model, (B) Antenna model.

$$SAR = \frac{\sigma |E|^2}{2\rho}, \qquad (4.169)$$

where σ and ρ are electric conductivity and density of the biological tissue, respectively, $|E|$ is the maximal value of the electric field induced in the human body.

The current density induced in the body can be expressed in terms of the axial current I_z as follows [16]:

$$J_z(r, z) = \frac{I_z(z)}{a^2\pi}\left(\frac{ka}{2}\right)\frac{J_0(j^{-1/2}k \cdot r)}{J_1(j^{-1/2}k \cdot a)}, \qquad (4.170)$$

where J_0 and J_1 are the Bessel functions, k is the free-space phase constant.

The induced electrical field is given by [16]

$$E_z(z) = \frac{J_z(r, z)}{\sigma + j\omega\varepsilon}. \qquad (4.171)$$

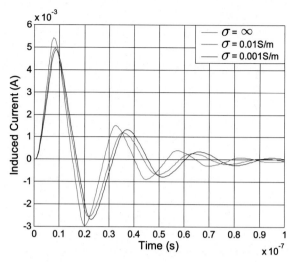

FIG. 4.29 Transient current induced in the feet due to the double-exponential pulse exposure for different value of conductivity.

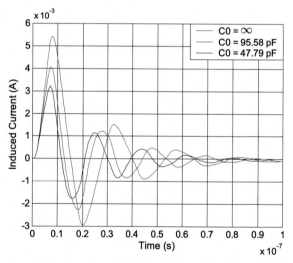

FIG. 4.30 Transient current induced in the feet of grounded and ungrounded human due to the double exponential pulse exposure for different value of capacitance.

In this case, a human having arms in contact with the sides, standing on the earth, exposed to HF radiation, where the value of the uniform electrical field is 1 V/m at a frequency of 30 MHz, is represented by a cylinder of the entire length L and radius a. The average values of the conductivity and permittivity of the human body, respectively, are assumed to be $\sigma = 0.6$ S/m and $\varepsilon_r = 60$ [16].

TABLE 4.9 Various parameters used in the eye thermal model.			
	$J_z(a, 0)$ [A/m^2]	$E_{(z=0)}$ [V/m]	SAR$_{(z=0)}$ [W/kg]
$f = 30$ MHz, $a = 0.14$ m	0.1318	0.2534	$1.6693 \cdot 10^{-5}$

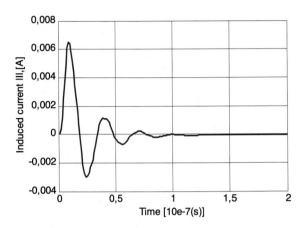

FIG. 4.31 Transient current induced in the feet due to the EMP exposure.

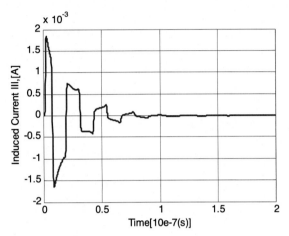

FIG. 4.33 Transient current induced in the feet due to the Gaussian pulse exposure.

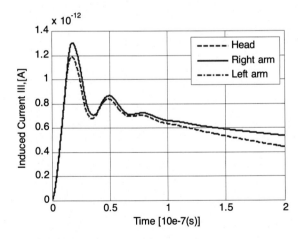

FIG. 4.32 Transient current induced in the right arm, left arm and head due to the EMP exposure.

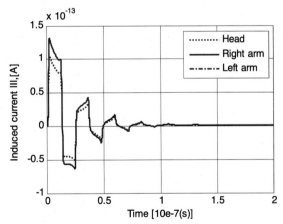

FIG. 4.34 Transient current induced in the right arm, left arm and head due to the Gaussian pulse exposure.

Figs. 4.35 and 4.36 show respectively the induced current density and electric field calculated inside human body due to HF radiation.

The maximum coupling conditions (uniform field along the body) are adopted to quantify the worst conditions of the maximum perturbation induced in the human body.

Fig. 4.37 shows the induced values of SAR inside the body. The maximum values of the induced current density, induced electric field and specific absorption rate at the feet are presented in Table 4.9.

The results presented in Figs. 4.36 and 4.37 show that the induced electric field decreases gradually, as one approaches the axis of the equivalent cylinder. Namely,

TABLE 4.10
Electrical parameters for a few frequencies.

Frequency [MHz]	Electrical conductivity σ [S/m]	Relative permittivity
5	0.54	150
15	0.56	100
20	0.57	80
30	0.6	60
35	0.66	53
40	0.7	50

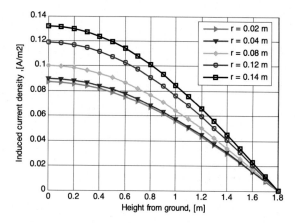

FIG. 4.35 Current density inside the human body $E_z = 1$ V/m, $f = 30$ MHz.

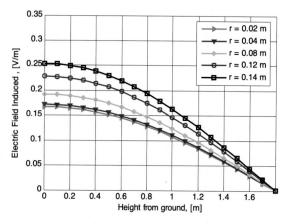

FIG. 4.36 Calculated electric field inside the human body $E_z = 1$ V/m, $f = 30$ MHz.

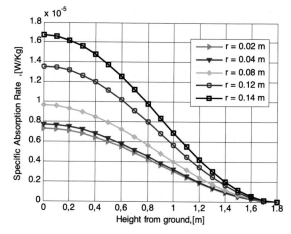

FIG. 4.37 Calculated specific absorption rate (SAR) inside the human body, $E_z = 1$ V/m, $f = 30$ MHz.

The obtained results are compared to the exposure limits proposed by ICNIRP (SAR = 0.4 W/kg for general population, SAR = 0.08 W/kg for occupational). The SAR values do not exceed the basic restrictions ($SAR_{max} = 1.6693 \cdot 10^{-5}$ W/Kg) and as such, we conclude that the heating effect is negligible.

Finally, the behavior of SAR versus frequency in the range from 5 to 40 MHz is analyzed.

The body model consists of a material with parameters simulating a muscle tissue: the frequency dependent permittivity and electrical conductivity (see Table 4.10) [16].

The incident electric field amplitude is $E_z = 1$ V/m, the maximum values of SAR at the feet obtained for $J(r = a, 0)$ are shown in Table 4.11.

According to Table 4.11, we remark that at range frequency 5–40 MHz, the specific absorption rate (SAR) inside the human body increases rapidly with frequency. The main conclusion is that the SAR_{max} for different frequency never exceeds the limit of 0.08 W/kg

at high frequencies, the induced current moves to the periphery of the cylinder due to skin effect. The outer layers of the body, behaves like a "shielding".

TABLE 4.11
Maximum values of the SAR inside the human body.

Frequency [MHz]	SAR_{max} [W/kg] for J $(r = a, 0)$
5	$6.0333 \cdot 10^{-10}$
15	$9.0103 \cdot 10^{-8}$
20	$4.3463 \cdot 10^{-7}$
30	$1.6693 \cdot 10^{-5}$
35	$2.4612 \cdot 10^{-4}$
40	$2.4223 \cdot 10^{-3}$

defined by ICNIRP, moreover, the SAR depends on the frequency.

REFERENCES

1. Riadh W.Y. Habash, Electromagnetic Fields and Radiation: Human Bioeffects and Safety, CRC Press, Boca Raton, Florida, 2001.
2. Om P. Gandhi, Some numerical methods for dosimetry: extremely low frequencies to microwave frequencies, Radio Science 30 (1) (1995) 161–177.
3. Dragan Poljak, Human Exposure to Electromagnetic Fields, WIT Press, Southampton, Boston, 2004, pp. 1–183.
4. Niels Kuster, Quirino Balzano, Energy absorption mechanism by biological bodies in the near field of dipole antennas above 300 MHz, IEEE Transactions on Vehicular Technology 41 (1) (1992) 17–23.
5. Ronold W.P. King, Sheldon S. Sandler, Electric fields and currents induced in organs of the human body when exposed to ELF and VLF electromagnetic fields, Radio Science 31 (5) (1996) 1153–1167.
6. Ronold W.P. King, Models and methods for determining induced ELF and VLF electromagnetic fields in the human body: a critical study, in: Proceedings of the IEEE 1997 23rd Northeast Bioengineering Conference, 1997, IEEE, 1997, pp. 48–49.
7. Ronold W.P. King, Fields and currents in the organs of the human body when exposed to power lines and VLF transmitters, IEEE Transactions on Biomedical Engineering 45 (4) (1998) 520–530.
8. Ronold W.P. King, The electric field induced in the human body when exposed to electromagnetic fields at 1–30 MHz on shipboard, IEEE Transactions on Biomedical Engineering 46 (6) (1999) 747–751.
9. Ronold W.P. King, Electric current and electric field induced in the human body when exposed to an incident electric field near the resonant frequency, IEEE Transactions on Microwave Theory and Techniques 48 (9) (2000) 1537–1543.
10. Dragan Poljak, Vesna Roje, Currents induced in human body exposed to the power line electromagnetic field, in:

11. Dragan Poljak, Youssef F. Rashed, The boundary element modelling of the human body exposed to the ELF electromagnetic fields, Engineering Analysis with Boundary Elements 26 (10) (2002) 871–875.
12. Dragan Poljak, Choy Yoong Tham, Niksa Kovac, The assessment of human exposure to low frequency and high frequency electromagnetic fields using the boundary element analysis, Engineering Analysis with Boundary Elements 27 (10) (2003) 999–1007.
13. Dragan Poljak, Choy Yoong Tham, Om Gandhi, Antonio Sarolic, Human equivalent antenna model for transient electromagnetic radiation exposure, IEEE Transactions on Electromagnetic Compatibility 45 (1) (2003) 141–145.
14. Dragan Poljak, Postprocessing of the human body response to transient electromagnetic fields, Progress in Electromagnetics Research 49 (2004) 219–238.
15. S. Mezoued, B. Nekhoul, Dragan Poljak, K. El Khamlichi Drissi, K. Kerroum, Human exposure to transient electromagnetic fields using simplified body models, Engineering Analysis with Boundary Elements 34 (1) (2010) 23–29.
16. A. Laissaoui, B. Nekhoul, S. Mezoued, Dragan Poljak, Assessment of the human exposure to transient and time-harmonic fields using the enhanced transmission line theory approach, Automatika 58 (4) (2017) 355–362.
17. International Commission on Non-Ionizing Radiation Protection (ICNIRP), Guidelines for limiting exposure to time-varying electric and magnetic fields (1 Hz to 100 kHz), Health Physics 99 (6) (2010) 818–836.
18. Ivana Zulim, Dragan Poljak, Antonio Sarolic, Assessment of human exposure to high frequency electromagnetic fields using simplified models of human body, in: 17th International Conference on Software, Telecommunications & Computer Networks, 2009, SoftCOM 2009, IEEE, 2009, pp. 10–14.
19. International Commission on Non-Ionizing Radiation Protection (ICNIRP), Guidelines for limiting exposure to time-varying electric, magnetic and electromagnetic fields (up to 300 GHz), Health Physics 74 (4) (1998) 494–522.
20. Om P. Gandhi, Jin-Yuan Chen, Numerical dosimetry at power-line frequencies using anatomically based models, Bioelectromagnetics 13 (S1) (1992) 43–60.
21. Don W. Deno, Currents induced in the human body by high voltage transmission line electric field – measurement and calculation of distribution and dose, IEEE Transactions on Power Apparatus and Systems 96 (5) (1977) 1517–1527.
22. J-Y. Chen, Om P. Gandhi, Currents induced in an anatomically based model of a human for exposure to vertically polarized electromagnetic pulses, IEEE Transactions on Microwave Theory and Techniques 39 (1) (1991) 31–39.
23. Jin-Yuan Chen, Cynthia M. Furse, Om P. Gandhi, A simple convolution procedure for calculating currents induced in the human body for exposure to electromagnetic pulses, IEEE Transactions on Microwave Theory and Techniques 42 (7) (1994) 1172–1175.

24. Om P. Gandhi, Cynthia M. Furse, Currents induced in the human body for exposure to ultrawideband electromagnetic pulses, IEEE Transactions on Electromagnetic Compatibility 39 (2) (1997) 174–180.

25. Om P. Gandhi, Edward E. Aslan, Human-equivalent antenna for electromagnetic fields, US Patent 5,394,164, 1995.

26. Dragan Poljak, Khalil El Khamlichi Drissi, Computational Methods in Electromagnetic Compatibility: Antenna Theory Approach Versus Transmission Line Models, John Wiley & Sons, New York, 2018.

27. Akihiro Ametani, Y. Kasai, J. Sawada, A. Mochizuki, T. Yamada, Frequency-dependent impedance of vertical conductors and a multiconductor tower model, IEE Proceedings-Generation, Transmission and Distribution 141 (4) (1994) 339–345.

28. Eldon J. Rogers, John F. White, Mutual coupling between finite lengths of parallel or angled horizontal Earth return conductors, IEEE Transactions on Power Delivery 4 (1) (1989) 103–113.

29. C. Taylor, R. Satterwhite, C. Harrison, The response of a terminated two-wire transmission line excited by a nonuniform electromagnetic field, IEEE Transactions on Antennas and Propagation 13 (6) (1965) 987–989.

30. R. Clayton Paul, A SPICE model for multiconductor transmission lines excited by an incident electromagnetic field, IEEE Transactions on Electromagnetic Compatibility 36 (4) (1994) 342–354.

CHAPTER 5

Realistic Models for Static and Low Frequency (LF) Dosimetry

Various power installations in modern society have been continuously causing public concern regarding possible health hazards due to exposure to fields generated by these sources.

In the last decades the topic has been extensively investigated and comprehensive views to the subject have been reported in many review papers; see, e.g., [1–4]. Nevertheless, the human body is a tremendously complex structure to study as measurement of induced currents and fields in the body in realistic scenarios is not possible on living humans. Consequently, measurements are carried out on laboratory animals or on phantoms having some electrical parameters corresponding to humans; see, e.g., [4]. However, as it is extremely difficult to extrapolate findings arising from experiments on phantoms and/or animal studies, many theoretical dosimetry procedures for the human exposure to low frequency fields, pertaining to the use of highly sophisticated numerical methods [5–8], are being developed for the analysis of bioelectromagnetic phenomena.

Generally, computational models used in low frequency dosimetry involve realistic human body models (or some of its parts, organs, etc.) with relatively high resolution (sophisticated discretization schemes) – mostly based on Magnetic Resonance Imaging (MRI) [1]. Note that these models usually have a rather high computational cost.

This chapter aims to review some numerical procedures based on the application of Finite Element Method (FEM) and Boundary Element Method (BEM) used in static and low frequency (LF) dosimetry.

The study of electrostatic and LF exposures is based on the calculation of induced currents and electric fields obtained by numerically solving the corresponding governing equations. The electrostatic exposures are formulated in terms of Laplace equation while the formulation for the low frequency exposures is based on the quasi-static approximation and the related Laplace equation form of the continuity equation [9,10].

Computational examples presented throughout this work are related to the human head exposed to an electrostatic field from a video display unit (VDU), as well as whole body and pregnant woman/foetus exposed to

high voltage (HV) extremely low frequency (ELF) electric fields generated by overhead power lines [9,10].

Thus, sophisticated simulation tools based on realistic numerical modeling are necessary for an accurate prediction of the internal fields [9,10]. A number of anatomically based computational models comprising cubical cells are related to the application of the Finite Difference Methods (FDM) methods. The Finite Element Method (FEM), Boundary Element Method (BEM) and Method of Moments (MoM) are generally used to a somewhat lesser extent [11].

It is worth noting that boundary integral equation methods, such as BEM or MoM, are considered to be promising efficient approaches for static and ELF dosimetry. As documented in ICNIRP [12,13], the staircasing error [8] is significant not only for the current density, but also for the internal electric field, i.e., 99th percentile value of E-field is suggested. However, such a measure is not quite appropriate for localized exposures. Therefore, BEM oriented approach seems rather important aiming to verify conventional Finite Difference (FD) based approaches.

5.1 PARAMETERS FOR QUANTIFYING LF EXPOSURES

While exposure to high frequency (HF) fields, due to the resonance effect, results in related tissue heating, LF fields may cause excitation of sensory, nerve and muscle cells with the thermal effects being negligible.

Note that, according to ICNIRP 1998 guidelines [12], the current density was a principal parameter for the estimation of LF exposure effects, while the 2010 ICNIRP guidelines [13] propose the induced electric field instead of the induced current density. However, there is a substantial amount of the numerical results for the current density in the relevant literature, therefore, for the comparison purpose this work pertains to the assessment of current density, as well.

The internal current density is generated due to an external electric or magnetic field. The current density J induced in the body due to electric fields is axial in nature and given by the following constitutive equation

Human Interaction with Electromagnetic Fields. https://doi.org/10.1016/B978-0-12-816443-3.00013-3

(differential form of Ohm's law):

$$J = \sigma E, \tag{5.1}$$

where σ is tissue conductivity while E is the corresponding internal electrical field.

The internal current density generated by the magnetic field has circular character and is given by

$$J = \sigma \pi r f B, \tag{5.2}$$

where B is the corresponding magnetic induction normal to the human body, f is the operating frequency, and r is the radius of the loop. Note that (5.2) is derived in Chapter 3.

5.2 HUMAN HEAD EXPOSED TO ELECTROSTATIC FIELD

Video display units (VDUs) based on cathode ray tube (CRT) are sources of many types of radiation, e.g., X-ray, ultraviolet, infrared, electromagnetic, etc. On the one hand, the radiation levels, such as levels of X-ray or optical radiation, high ($\sim$MHz) and low ($\sim$kHz) frequency fields stay below exposure limits [14,15], however, electrostatic and low frequency fields might be associated with some skin diseases, suppression of melatonin, or induction of phosphenes in the eyes, despite of the fact that there is no strong evidence of adverse health effects from domestic levels of ELF electromagnetic fields [14,15].

This section deals with human exposure to electrostatic fields generated by some VDUs (possibly associated with skin rashes, [5,16,17]) and presents a three-dimensional, anatomically based model of the head exposed to electrostatic field from a VDU via FEM and BEM, contrary to the usual approach featuring the use of FDM [5]. The formulation is based on the Laplace equation for a scalar potential [7,18]. The results are obtained for different face geometries.

Neglecting the charge density in the space between the head and display, a 3D electrostatic field between a VDU and the head is given by Laplace equation for a scalar potential φ [7,18]. This equation can be readily derived as outlined below.

The electrostatic field satisfies the reduced form of the first Maxwell curl equation:

$$\nabla \times \vec{E} = 0, \tag{5.3}$$

from which the electric field can be expressed as follows:

$$\vec{E} = -\nabla\varphi. \tag{5.4}$$

Taking into account Gauss law in a differential form:

$$\nabla \cdot \left(\varepsilon \vec{E} \right) = \rho, \tag{5.5}$$

and combining with (5.4), Poisson equation is obtained:

$$\nabla \cdot (\varepsilon \nabla \varphi) = -\rho. \tag{5.6}$$

For a linear, homogeneous and isotropic and source-free medium, (5.6) simplifies into Laplace equation

$$\nabla^2 \varphi = 0. \tag{5.7}$$

Laplace equation (5.7) is accompanied with the following set of associated boundary conditions:

$$\varphi = \varphi_s \quad \text{on the display}, \tag{5.8}$$
$$\varphi = \varphi_h \quad \text{on the head}, \tag{5.9}$$
$$\nabla\varphi \cdot \hat{n} = 0 \quad \text{on the far field boundaries}. \tag{5.10}$$

As depicted in Fig. 5.1, the Dirichlet's boundary conditions (5.8) and (5.9) are prescribed on the face and display, respectively, while Neumann conditions (5.10) are imposed on the rest of the outer boundary.

The head is assumed to be perfectly conducting, thus representing an equipotential surface potential φ_h.

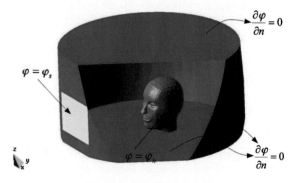

FIG. 5.1 Geometry and associated boundary conditions for the head in front of a VDU.

The parameters l_s, d_s, φ_s and φ_h represent the distance between display and nose tip, the size of display (diagonal display size given in inches), the electrostatic potential on the display and the head, respectively. The electric potential on a CRT monitor is taken to be between 1 and 15 kV. The potential of the display is assumed to be rather high (15 kV) – representing the worst-case scenario. The conditions for a human in front of the display are defined as follows: $l_s = 40$ cm, $d_s = 17''$, $\varphi_s = 15$ kV and $\varphi_h = 0$ kV. The monitor is of

4 : 3 format type, i.e., the width of the screen is 34.3 cm (13.6″) and height is 25.7 cm (10.2″). The radius of the computational domain is around 0.5 m, while the height is around 0.6 m, i.e., the head dimensions are 21 cm × 16.5 cm [19].

The eyebrows potential is assumed to be the same potential as for the face. Other parameters, such as temperature, humidity and conductivity of the screen glass surface, are simply neglected, and the air humidity is considered to be very low (dry air).

Fig. 5.2 shows two types of face to be analyzed.

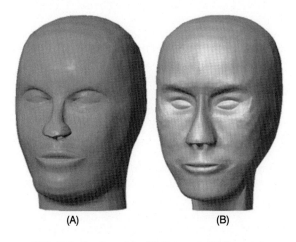

(A)   (B)

FIG. 5.2 Head models: (A) Person 1, (B) Person 2.

Note that meshing has been carried out with GID [20].

Laplace equation (5.7) is solved by means of FEM and BEM, and both methods are outlined below.

5.2.1 Finite Element Solution

A weighted residual approach applied to (5.7) results in the following integral [18,21–24]:

$$\int_{\Omega} \nabla^2 \varphi W_j \, d\Omega = 0, \tag{5.11}$$

where W_j denotes the set of test functions.

The weak formulation of the problem yields

$$\int_{\Omega} \nabla \varphi \cdot \nabla f_j \, d\Omega = \int_{\Gamma} \frac{\partial \varphi}{\partial n} f_j \, d\Gamma. \tag{5.12}$$

The potential over the finite element is expressed in terms of a linear combination of 4 shape functions [18]

as follows:

$$\varphi^e = \sum_{i=1}^{4} \alpha_i f_i, \tag{5.13}$$

where α_i represent unknown coefficients of the solution, while f_i are shape functions.

Expression (5.13) in matrix notation becomes

$$\varphi^e = \{f\}^T \{\alpha\}. \tag{5.14}$$

For three-dimensional problems, the shape functions f_i are given by

$$f_i(x, y, z) = \frac{1}{D} (V_i + a_i x + b_i y + c_i z), \quad i = 1, 2, 3, 4. \tag{5.15}$$

Parameters a_i, b_i, c_i can be found elsewhere, e.g., in [21]. The global matrix of the FEM system is assembled from local finite element matrices $[a]^e_{ji}$.

The potential gradient, which in rectangular coordinates is defined as

$$\nabla \varphi = \frac{\partial \varphi}{\partial x} \hat{e}_x + \frac{\partial \varphi}{\partial y} \hat{e}_y + \frac{\partial \varphi}{\partial z} \hat{e}_z, \tag{5.16}$$

in terms of shape functions can be written as follows:

$$\nabla \varphi = \begin{bmatrix} \dfrac{\partial \phi}{\partial x} \\[2mm] \dfrac{\partial \phi}{\partial y} \\[2mm] \dfrac{\partial \phi}{\partial z} \end{bmatrix} = \begin{bmatrix} \dfrac{\partial f_1}{\partial x} & \dfrac{\partial f_2}{\partial x} & \dfrac{\partial f_3}{\partial x} & \dfrac{\partial f_4}{\partial x} \\[2mm] \dfrac{\partial f_1}{\partial y} & \dfrac{\partial f_2}{\partial y} & \dfrac{\partial f_3}{\partial y} & \dfrac{\partial f_4}{\partial y} \\[2mm] \dfrac{\partial f_1}{\partial z} & \dfrac{\partial f_2}{\partial z} & \dfrac{\partial f_3}{\partial z} & \dfrac{\partial f_4}{\partial z} \end{bmatrix} \begin{bmatrix} \alpha_1 \\[1mm] \alpha_2 \\[1mm] \alpha_3 \\[1mm] \alpha_4 \end{bmatrix}. \tag{5.17}$$

Having performed FEM discretization, the following matrix equation is obtained:

$$[a]\{\alpha\} = \{Q\}, \tag{5.18}$$

where $\{Q\}$ denotes the flux vector, and global matrix $[a]$ is assembled from local finite element matrices $[a]^e_{ji}$.

Therefore, performing certain mathematical manipulations, the finite element matrix is determined by the integral:

$$[a]^e = \int_{\Omega_e} \begin{bmatrix} \dfrac{\partial f_1}{\partial x} & \dfrac{\partial f_1}{\partial y} & \dfrac{\partial f_1}{\partial z} \\[6pt] \dfrac{\partial f_2}{\partial x} & \dfrac{\partial f_2}{\partial y} & \dfrac{\partial f_2}{\partial z} \\[6pt] \dfrac{\partial f_3}{\partial x} & \dfrac{\partial f_3}{\partial y} & \dfrac{\partial f_3}{\partial z} \\[6pt] \dfrac{\partial f_4}{\partial x} & \dfrac{\partial f_4}{\partial y} & \dfrac{\partial f_4}{\partial z} \end{bmatrix}$$

$$\times \begin{bmatrix} \dfrac{\partial f_1}{\partial x} & \dfrac{\partial f_2}{\partial x} & \dfrac{\partial f_3}{\partial x} & \dfrac{\partial f_4}{\partial x} \\[6pt] \dfrac{\partial f_1}{\partial y} & \dfrac{\partial f_2}{\partial y} & \dfrac{\partial f_3}{\partial y} & \dfrac{\partial f_4}{\partial y} \\[6pt] \dfrac{\partial f_1}{\partial z} & \dfrac{\partial f_2}{\partial z} & \dfrac{\partial f_3}{\partial z} & \dfrac{\partial f_4}{\partial z} \end{bmatrix} d\Omega. \quad (5.19)$$

Solving (5.19) in closed form, one obtains the FEM matrix:

$$[a]^e = \frac{1}{D} \begin{bmatrix} a_1^2 + b_1^2 + c_1^2 & a_1 a_2 + b_1 b_2 + c_1 c_2 \\[4pt] a_1 a_2 + b_1 b_2 + c_1 c_2 & a_2^2 + b_2^2 + c_2^2 \\[4pt] a_1 a_3 + b_1 b_3 + c_1 c_3 & a_2 a_3 + b_2 b_3 + c_2 c_3 \\[4pt] a_1 a_4 + b_1 b_4 + c_1 c_4 & a_2 a_4 + b_2 b_4 + c_2 c_4 \end{bmatrix}$$

$$\begin{bmatrix} a_1 a_3 + b_1 b_3 + c_1 c_3 & a_1 a_4 + b_1 b_4 + c_1 c_4 \\[4pt] a_2 a_3 + b_2 b_3 + c_2 c_3 & a_2 a_4 + b_2 b_4 + c_2 c_4 \\[4pt] a_3^2 + b_3^2 + c_3^2 & a_3 a_4 + b_3 b_4 + c_3 c_4 \\[4pt] a_3 a_4 + b_3 b_4 + c_3 c_4 & a_4^2 + b_4^2 + c_4^2 \end{bmatrix}.$$

$$(5.20)$$

Once the potential φ is obtained, the electric field is computed from the potential gradient

$$\vec{E} = -\nabla \varphi, \quad (5.21)$$

as indicated in [10].

5.2.2 Boundary Element Solution

Applying the weighted residual approach, (5.7) is integrated over the calculation domain Ω [21–24]:

$$\int_{\Omega} \nabla^2 \varphi \cdot \psi \, d\Omega = 0, \quad (5.22)$$

where ψ is a given weight function.

Using Green identities, and applying the generalized Gauss theorem, it follows that

$$\int_{\Omega} \psi \nabla^2 \varphi \, d\Omega = \int_{\Gamma} \psi \frac{\partial \varphi}{\partial n} d\Gamma - \int_{\Gamma} \varphi \frac{\partial \psi}{\partial n} d\Gamma + \int_{\Omega} \varphi \nabla^2 \psi \, d\Omega.$$

$$(5.23)$$

Then, ψ is determined as the fundamental solution of the equation

$$\nabla^2 \psi - \delta(\vec{r} - \vec{r}\,') = 0, \quad (5.24)$$

which for a 3D problem becomes

$$\psi = \frac{1}{4\pi R}, \quad (5.25)$$

where δ is the Dirac delta function, $\vec{r}$ denotes the observation points, $\vec{r}\,'$ denotes the source points, and $R = |\vec{r} - \vec{r}\,'|$ is the distance between them.

Now, the domain integral from the left-hand side of (5.23) becomes

$$\int_{\Omega} \varphi \nabla^2 \psi \, d\Omega = -\int_{\Omega} \varphi \delta(\vec{r} - \vec{r}\,') d\Omega = -\varphi_i, \quad (5.26)$$

for any point inside the domain of interest.

Now, combining (5.22)–(5.26) yields

$$\varphi_i = \int_{\Gamma} \psi \frac{\partial \varphi}{\partial n} d\Gamma - \int_{\Gamma} \varphi \frac{\partial \psi}{\partial n} d\Gamma, \quad (5.27)$$

which can be regarded as the Green representation of φ.

When the observation point i is located on the boundary Γ, the boundary integral becomes singular as R approaches zero. In order to deal with this singularity, a small sphere for 3D problems is considered. By sorting out the surface integrals over the small sphere centered at the singular point, the following integral formulation is obtained:

$$c_i \varphi_i = \int_{\Gamma} \Psi \frac{\partial \varphi}{\partial n} d\Gamma - \int_{\Gamma} \varphi \frac{\partial \Psi}{\partial n} d\Gamma, \quad (5.28)$$

where

$$c_i = \begin{cases} 1, & i \in \Omega, \\ \dfrac{1}{2}, & i \in \Gamma \quad \text{(smooth boundary)}. \end{cases} \quad (5.29)$$

Discretizing the boundary Γ into N_e boundary elements, (5.27) becomes

$$c_i u_i = \sum_{j=1}^{N_e} \int_{\Gamma_j} \psi \frac{\partial \varphi}{\partial n} d\Gamma - \sum_{j=1}^{N_e} \int_{\Gamma_j} \varphi \frac{\partial \psi}{\partial n} d\Gamma. \quad (5.30)$$

Each boundary element contains a number (N_{fn}) of subjacent collocation nodes, in which the potential or fluxes are evaluated. Thus, the values of the potential or its normal derivative at any point defined by the local coordinates $\xi = (\xi_1, \xi_2)$ on a given boundary element can be defined in terms of their values at the collocation nodes, and the N_{fn} interpolation functions Φ_k with $k = 1, \ldots, N_{fn}$, as follows:

$$\varphi(\xi) = \sum_{k=1}^{N_{fn}} \Phi_k(\xi)\, \varphi_k, \qquad \frac{\partial \varphi(\xi)}{\partial n} = \sum_{k=1}^{N_{fn}} \Phi_k(\xi)\, \frac{\partial \varphi}{\partial n}\bigg|_k.$$

$$(5.31)$$

Thus, (5.30) can now be rewritten in the following form:

$$c_i u_i = \sum_{j=1}^{N_e} \sum_{k=1}^{N_{fn}} \left(\int_{\Gamma_j} \psi\, \Phi_k d\Gamma_j \right) \frac{\partial \varphi_{jk}}{\partial n}$$

$$- \sum_{j=1}^{N_e} \sum_{k=1}^{N_{fn}} \left(\int_{\Gamma_j} \frac{\partial \psi}{\partial n} \Phi_k d\Gamma_j \right) \varphi_{jk}, \qquad (5.32)$$

while the corresponding matrix notation is

$$[H]\{\phi\} = [G][\partial_n \phi], \qquad (5.33)$$

where

$$H_{il} = \delta_{il} c_i + \int_{\Gamma_j} \frac{\partial \psi_i}{\partial n_j}\bigg|_\xi \Phi_k(\xi)\, d\Gamma_j, \qquad (5.34)$$

$$G_{il} = \int_{\Gamma_j} \psi_i(\xi)\, \Phi_k(\xi)\, d\Gamma_j. \qquad (5.35)$$

Index l, identifying a collocation node within the domain, can be calculated in terms of indices j and k, by means of the nodal connectivity of the mesh, i.e., index $l = 1, \ldots, M$; and M is the total number of collocation nodes in the domain. Basically, index l is used to identify one of the adjacent freedom (collocation) nodes from a global point of view, and is given as a function of the indicator of element, j, and the local collocation node of that element, k. In the case of discontinuous collocation nodes:

$$M = \sum_{j=1}^{N_e} f_j. \qquad (5.36)$$

The boundary differential $d\Gamma_j$ can be expressed in terms of the domain local coordinates ξ via Jacobian of the transformation $|J|$:

$$d\Gamma_j = |J|\, d\xi_1\, d\xi_2. \qquad (5.37)$$

Finally, the application of the prescribed boundary conditions through the boundary discretization and further collocation into N_{fs} degrees of freedom yields an algebraic linear system of equations of the form [18]:

$$[A]\{x\} = \{b\}, \qquad (5.38)$$

where $[A]$ is an $N_{fe} \times N_{fe}$ matrix which contains the coefficients of the single and double layer potential operators, i.e., the coefficients of $[H]$ and $[G]$ matrices given by (5.34) and (5.35), the 1-column array of unknowns x contains the potentials and normal fluxes that were not prescribed as boundary conditions, and the right-hand side term involves boundary conditions.

Generally, the 3D flexible BEM approach provides more accurate results than standard methods as it is based on the fundamental solution of the leading operator for the governing equation.

5.2.3 Computational Examples

The numerical results [18,24] are obtained for both head types shown in Fig. 5.2. Both BEM and FEM meshes are generated by using the general preprocessor and geometry modeler GiD [20]. For person 1, the BEM mesh consists of 4558 constant triangle elements and 4558 nodes, while FEM mesh consists of 80,771 linear tetrahedral elements (20,184 nodes). A view boundary element mesh for different head types is given in Fig. 5.3.

First, FEM results are presented for the two head types. Fig. 5.4 shows the electrostatic field distribution in [V/cm] in mid-face symmetry for person 1.

Fig. 5.5 shows the electrostatic field distribution expressed in [V/cm] in the mid-face symmetry for person 2.

The results clearly demonstrate the decrease of the field away from the screen with the peak occurring at the top of the nose. Behind the head the field becomes relatively low, reaching the value of 100 V/cm. Comparisons of the field strength induced over the face and around nose and eye of persons 1 and 2, respectively, obtained by FEM are presented in Figs. 5.6 and 5.7.

The maximal value of the field at the nose of the person 1 (around 2400 V/cm) is appreciably higher than the value at the same position induced in the case of person 2 (1900 V/cm). The difference occurs due to the

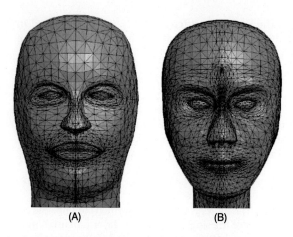

FIG. 5.3 Boundary element mesh: (A) European head type – person 1, (B) Asiatic head type – person 2.

FIG. 5.4 Electrostatic field in the mid-face symmetry for person 1.

FIG. 5.5 Electrostatic field in the mid-face symmetry for person 2.

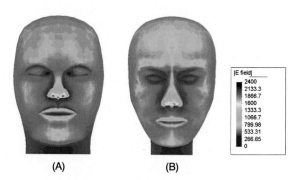

FIG. 5.6 Electrostatic field strength [V/cm] over the face: (A) person 1; (B) person 2.

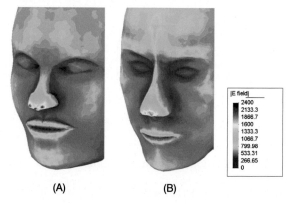

FIG. 5.7 Electrostatic field strength [V/cm] around the nose and eye: (A) person 1; (B) person 2.

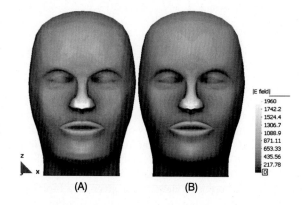

FIG. 5.8 Electrostatic field on the face of person 1: (A) FEM solution, (B) BEM solution.

different geometry of the face, particularly the shape of the nose. For other parts of the face, the differences do not appear to be appreciable. It is worth noting that the field strength around the eyes is relatively low, i.e., around 400–700 V/cm, when compared to the other parts of the face.

The next set of results compares the findings obtained via FEM and BEM.

Fig. 5.8 shows the electrostatic field distribution over the face of person 1 obtained by means of FEM and BEM.

Maximal field values are obtained in the close vicinity of the nose, while significant field values are calculated on the lips, chin and forehead. Note that *hot spots* appear on the nose for the case of FEM ($\approx$1960 V/m) and BEM ($\approx$1760 V/m) model.

Figs. 5.9–5.11 show the x, y, and z components of the electrostatic field induced on the face, both computed via BEM and FEM solution.

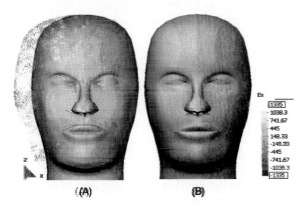

FIG. 5.9 E_x [V/cm] component: (A) FEM solution, (B) BEM solution.

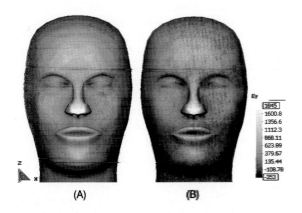

FIG. 5.10 E_y [V/cm] component: (A) FEM solution, (B) BEM solution.

It can be noticed that E_y has the highest value as the screen is located 40 cm away from the face in the y direction. The highest values of E_y are located on the nose, lips, chin and forehead, while the highest values of E_z are located on the bottom of the nose, around nostrils.

For the case of person 2, BEM mesh consists of 6066 constant triangle elements (6066 nodes), while FEM mesh contains 104,082 linear tetrahedral elements (26,659 nodes). Fig. 5.12 shows the electrostatic field E [V/cm] on the face of person 2 obtained by FEM and BEM methods.

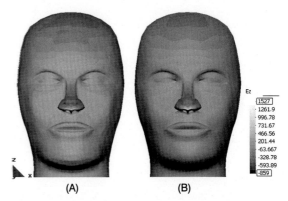

FIG. 5.11 E_z [V/cm] component of the field: (A) FEM solution, (B) BEM solution.

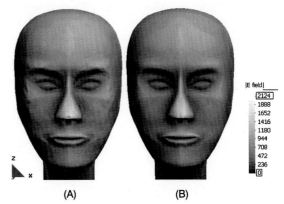

FIG. 5.12 Electrostatic field on the face of person 2: (A) FEM solution, (B) BEM solution.

Higher field levels are found on the nose (maximal), lips, chin, forehead and eyebrows. Eyes and neighboring regions are not so exposed (around 300–400 V/cm) compared with the nose (maximal value around 2100 V/cm).

Comparing the results obtained for persons 1 and 2, appreciable differences in the field values are noticed on certain parts of the face, such as nose, due to the different shape of the face, particularly the shape of the nose.

Fig. 5.13 shows the field computed via BEM versus distance l_s between the nose tip and screen.

The closer the face to the screen, the greater the field on the nose. Also, the field values in the eyes area are less, i.e., the peaky shape of the nose in a certain sense protects the eyes. At distance $l_s = 10$ cm, the field on the nose tip is approximately 3600 V/cm, and 980 V/cm in the eye; while at $l_s = 50$ cm, the field on the nose tip and in the eye is 1230 and 350 V/cm, respectively.

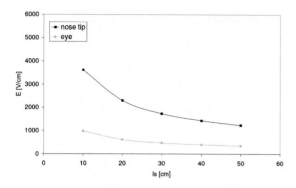

FIG. 5.13 Electrostatic field on the nose tip and eyes versus distance between the display and head for a 17″ display.

Fig. 5.14 shows the field computed via BEM versus the display size d_S with the 40 cm distance between screen and the nose tip. The results are also obtained via BEM. Note that the electrostatic field is almost linearly dependent on the display size.

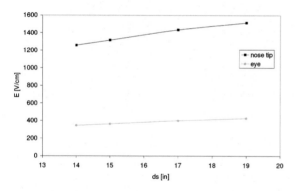

FIG. 5.14 Electrostatic field in the nose tip and eyes versus display size for 40 cm distance between the display and head.

The field increase is slow and, for example, on the nose tip it increases by approximately 19% when increasing the display size from 14″ to 19″ at 40 cm display–face distance.

Generally, BEM solution requires much shorter computational time compared to FEM which is discussed in detail in [23]. The implementation of BEM was proved to be more suitable for the given problem in terms of accuracy, efficiency and stability [18].

5.3 WHOLE BODY EXPOSED TO LF FIELDS

The section deals with the whole human body exposed to extremely low frequency (ELF) electric fields with par-

ticular emphasis on pregnant woman and foetus, featuring the use of the BEM approach.

The tremendous growth in the use of electrical energy for industrial applications is associated with the continuous presence of extremely low frequency (ELF) electric fields in the environment. Generally, high voltage ELF fields are used for power utilities (transmission, distribution and applications) and for strategic global communications with submarines submerged in sea water. At low frequency exposures, i.e., when displacement currents can be neglected, the electric and magnetic fields are assumed to be decoupled. Generally, a human being can be exposed to two kinds of field generated by low frequency (LF) power systems: (1) low voltage/high intensity systems (the principal field is the magnetic one, while the induced currents form close loops in the body); (2) high voltage/low intensity systems (the principal field is the electric one while the induced currents have axial character).

As the magnetic field penetrating into the human body remains mostly unchanged, the evaluation of human exposure to electric fields is much more difficult than to magnetic fields due to the electric field perturbation by the human body, various objects in the environment, and the measuring device. Basically, human exposure to high voltage ELF electric fields results in induced fields and currents in all organs. These induced currents and fields may give rise to thermal and non-thermal effects. While the thermal effects seem to be negligible, certain non-thermal effects often related to the cell level are still possible [12,13,25,26].

The use of numerical methods in the assessment of the current density induced in the human body when exposed to ELF fields has been reported by several researchers [6–11,21,23,27–30].

Chiba et al. [28] developed a Finite Element Method (FEM) based inhomogeneous body model for the calculation of the induced current density inside human beings exposed to 60 Hz electric fields. Some advances in this model were reported in [29,30]. Gandhi and Chen [31] developed a realistic model of the human body and used the finite difference time domain (FDTD) method for the numerical solution of the problem.

On the contrary to the computationally very expensive domain discretization methods such as FEM [28–30] and FDTD [31], an efficient multi-domain human body representation based on a multi-domain implementation of the Boundary Element Method (BEM) has been promoted in [27]. This efficient BEM scheme is more sophisticated approximation than FDTD and at the same time computationally less expensive than

[...] only the domain boundary has to be discretized.

It is worth emphasizing that one of the biggest advantages of the domain decomposition techniques, such as BEM, regarding its capabilities in dealing with piecewise homogeneous material properties, is that the final system of equations is sparse and highly bounded.

The formulation of the problem is based on the quasi-static approximation of the ELF electric field and the related continuity equation for the induced current density. This continuity equation can be reduced to the Laplace equation for the electric scalar potential. The resulting equation is then numerically handled via the BEM based on the domain decomposition concept [20] [21].

Once obtaining the electric scalar potential distribution along the body, as a solution of the corresponding Laplace equation, one can readily calculate the induced current density inside the body. Some illustrative computational examples of human exposure to power line electric field are presented in this section.

5.3.1 Quasi-Static Formulation

It is worth noting that at ELF exposures dielectric properties are assumed to be negligible, i.e., $\sigma \gg \omega\varepsilon$, namely the body is dominantly conducting. The quasi-static approximation can be used as the dimensions of the body are rather small compared to the wavelength of the impressed field, i.e., the body appears to be electrically short.

Upon this assumption, ELF exposures can be modeled and formulated via the Laplace type equation.

The formulation is based on the continuity equation which is readily derived from Maxwell equations. Starting from the second curl Maxwell equation,

$$\nabla \times \vec{H} = \vec{J} + \frac{\partial \vec{D}}{\partial t}, \tag{5.39}$$

and applying the divergence operator,

$$\nabla(\nabla \times \vec{H}) = \nabla \cdot \vec{J} + \nabla(\frac{\partial \vec{D}}{\partial t}), \tag{5.40}$$

as the left side of (5.40) vanishes identically, one obtains

$$\nabla\vec{J} + \frac{\partial}{\partial t}(\nabla\vec{D}) = 0. \tag{5.41}$$

Now, taking into account the third divergence Maxwell equation (differential form of Gauss law),

$$\nabla \cdot \vec{D} = \rho, \tag{5.42}$$

the continuity equation is obtained in the differential form,

$$\nabla\vec{J} + \frac{\partial\rho}{\partial t} = 0, \tag{5.43}$$

where $\vec{J}$ is the current density and ρ represents the volume charge density.

The induced current density can be expressed in terms of the scalar electric potential using the constitutive equation (Ohm's law),

$$\vec{J} = -\sigma\nabla\varphi, \tag{5.44}$$

where σ is the conductivity of the medium.

The volume charge density ρ and electric scalar potential φ are related through the Poisson equation

$$\nabla(\varepsilon\nabla\varphi) = -\rho, \tag{5.45}$$

where ε is the corresponding permittivity of the medium.

Inserting (5.44) and (5.45) into (5.43), we obtain

$$\nabla(\sigma\nabla\varphi) + \frac{\partial}{\partial t}\nabla(\varepsilon\nabla\varphi) = 0, \tag{5.46}$$

which can be written as

$$\nabla\left[(\sigma\nabla\varphi) + \nabla(\varepsilon\frac{\partial\varphi}{\partial t})\right] = 0. \tag{5.47}$$

Finally, for the time-harmonic ELF exposures, it follows that

$$\nabla\left[(\sigma + j\omega\varepsilon)\nabla\varphi\right] = 0, \tag{5.48}$$

where $\omega = 2\pi f$ is the operating frequency.

In the ELF range, all organs behave as good conductors and the continuity equation (5.48), in accordance with the quasi-static approximation, simplifies into Laplace equation of the form

$$\nabla(\sigma\nabla\varphi) = 0. \tag{5.49}$$

On the contrary, the surrounding air is a lossless dielectric medium and the corresponding governing equation is

$$\nabla(\varepsilon\nabla\varphi) = \nabla^2\varphi = 0. \tag{5.50}$$

Thus, the problem has been formulated in terms of the set of piecewise homogeneous Laplace's equations for the scalar potential. Having computed the scalar potential, the induced electric field and the current distribution inside the body can be readily obtained as a potential gradient.

As it has already been shown in [25–27], the field component parallel to the upright cylindrical body axis is the largest within the practical ranges of body height, while the other components can be neglected.

The electric field over flat ground plane is assumed to be vertical and uniform near the ground level, and the human body is located between the parallel plate electrodes, in the middle of the lower one. Therefore, the impressed electric field generated by the power line is assumed to be spatially uniform along the body.

A calculation domain with the corresponding boundary conditions is shown in Fig. 5.15.

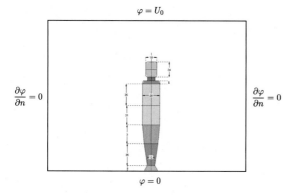

$\varphi = U_0$

$\dfrac{\partial \varphi}{\partial n} = 0$ $\dfrac{\partial \varphi}{\partial n} = 0$

$\varphi = 0$

FIG. 5.15 Calculation domain with the prescribed boundary conditions.

The continuity rules at discontinuities in material properties of conducting and dielectric medium should be implemented in terms of scalar potential. Thus, the continuity condition for the tangential component of the electric field near the two-media interface is given by

$$\hat{n} \times \left(\vec{E}_b - \vec{E}_a \right) = 0, \qquad (5.51)$$

where $\hat{n}$ is the unit normal to the interface and $\vec{E}_a$ and $\vec{E}_b$ represent the fields in the air and body, respectively.

Expressing the electric field in terms of scalar potential, we get

$$\hat{n} \times (\nabla \varphi_b - \nabla \varphi_a) = 0. \qquad (5.52)$$

The condition for the normal component of the induced current density near the body–air surface is given by

$$\hat{n} \cdot \vec{J} = -j\omega \rho_s, \qquad (5.53)$$

where ρ_s denotes the surface charge density.

Now, the current density can be expressed by the scalar potential as

$$\sigma_b \hat{n} \nabla \varphi_b = -j\omega \rho_s, \qquad (5.54)$$

where σ_b is the corresponding tissue conductivity, and φ_b is the scalar potential at the body surface.

The continuity condition for the normal component of the electric flux density at the air–body surface is

$$\hat{n} \cdot \vec{D} = \rho_s, \qquad (5.55)$$

or, expressing the electric flux density in terms of scalar potential,

$$\varepsilon_0 \hat{n} \nabla \varphi_a = \rho_s, \qquad (5.56)$$

where φ_a denotes the potential in the air in the close proximity of the body.

5.3.2 Multi-Domain Model of the Human Body

The multi-domain body model promoted in [27] and shown in Fig. 5.16 is a first step in building a realistic anatomically based body model. Within the framework of this model, the human body consists of various tissues and organs with varying electrical properties in terms of conductivity σ and relative permittivity ε_r. At ELF frequencies near the power frequency $f = 50$ Hz (or 60 Hz in the US), all organs in the body behave as good conductors with related values of conductivity.

The human body model used in this work is similar to that presented in [28] and consists of nine portions, as shown in Fig. 5.16 (dimensions are given in centimeters). The corresponding conductivities of each part of the body are given in Table 5.1.

TABLE 5.1 Conductivities of different body compartments.		
Body part	**Region**	**Conductivity σ [S/m]**
Head	I, II	0.12
Neck	III	0.6
Shoulders	IV	0.04
Thorax	V	0.11
Pelvis and crotch	VI	0.11
Knee	VII	0.52
Ankle	VIII	0.04
Foot	IX	0.11

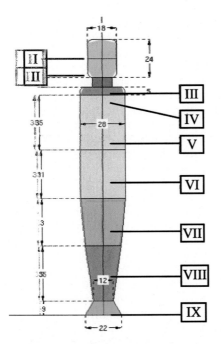

FIG. 5.16 Multi-domain model of the body and conducting properties of different parts at ELF exposures.

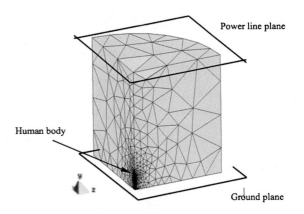

FIG. 5.17 The boundary element mesh.

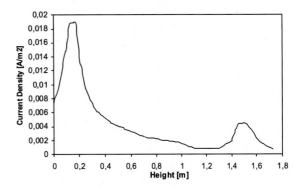

FIG. 5.18 The current density distribution inside the human body.

A calculation domain pertaining to the human body exposed to a power line field with the associated boundary conditions is presented in Fig. 5.15. The lower plate electrode is assumed to be at zero potential while the upper plate electrode is assumed to be at the potential of a high voltage power line.

The computational example deals with the well-grounded human body (multi-domain model) of 175 cm height, exposed to the 60 Hz overhead power line electric field. The height of the power line is 10 m above the ground. The analysis is carried out via the Boundary element Method (BEM) procedure presented in Sect. 5.2.2. This efficient BEM procedure is considered to be more accurate than FDTD and computationally less expensive than FEM, as only the domain boundaries have to be discretized.

A view of the actual boundary element mesh is shown in Fig. 5.17.

The human being is exposed to an external electric field $E = 10$ kV/m. The distribution of the induced current densities inside the body is shown in Fig. 5.18.

It can be clearly observed that the induced current density peaks occur at narrow cross-sections such as ankle and neck.

The values of current densities induced in the ankle obtained using BEM are compared with the results obtained by FEM [28] and with experimental results [28], as well. This comparison is presented in Table 5.2.

TABLE 5.2
Comparison between the BEM, FEM and experimental results for the induced current density at various body portions, expressed in [nA/cm^2].

Body part	BEM	FEM [6]	Experimental [6]
Neck	452	462	466
Pelvis	232	227	225
Ankle	1891	1916	1866

The calculated results via BEM agree rather satisfactorily with FEM and experimental results. The BEM has been demonstrated to be a highly accurate technique according to the comparison with the experimental results. Numerical results obtained by BEM are also in good agreement with FEM results.

5.3.3 Realistic Model of the Human Body – No Arms

This section deals with a realistic geometrical representation of the body with internal organs but no arms, as indicated in Fig. 5.19. The model accounts for some relevant tissues like brain, eyes, heart, liver, kidneys, and intestine. All organs are treated as conductors embedded in a saline fluid with conductivity 0.5 S/m. The values of conductivities used for different organs are shown in Table 5.3.

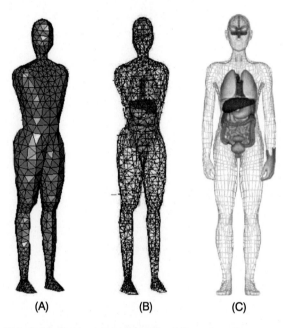

| (A) | (B) | (C) |

FIG. 5.19 Mesh and preprocessing information of the human body: (A) geometry definition, (B) meshed model, (C) internal organs taken into account.

TABLE 5.3
Tissue Conductivities used at ELF exposures.

Tissue type	Conductivity [S/m]
Air	0
Muscle	0.5
Heart	0.11
Brain	0.12
Eye	0.11
Liver	0.13
Kidney	0.16
Intestine	0.16

Average values of physical and geometrical properties have been obtained from available databases and references; see, e.g., [6,25,26,28,32].

The preprocessing and the geometrical introduction represented a major problem which has been solved by adapting a customizable geometry modeler, preprocessor and mesh generator. The meshes generated considering organs have a number of elements close to 40,000.

The analysis has been undertaken for the human body exposed to the electric field generated by an overhead power line. A plain view of the integration domain is shown in Fig. 5.20A.

The person is assumed to be grounded, i.e., standing barefoot on earth. The height of the domain has been fixed to 4 m in order to reduce as much as possible the size of the model. Above that height the electric field is unaffected by the presence of the human body, i.e., the equipotential lines are parallel to the ground level, thus allowing the simplification of the model. The potential has been fixed to zero in the ground and has been set to one at the top of the domain. Thus, due to the linearity between the electric field and current density, the results can be easily scaled to the values of different power lines, typically between 10 and 30 kV. Along the vertical boundary the flux is assumed to be zero.

The calculation domain for the case of realistic body representation is shown in Fig. 5.20A, while related BEM mesh is shown in Fig. 5.20B with the distances given in meters. As the potential at the upper plate has been set to one, the results can be easily scaled to the typical power line values. The reduced computational domain reduced to the parallelepiped (5 m × 4 m) is indicated.

Fig. 5.21 shows the scalar potential and related electric field distribution in the air around the human body.

The front and lateral view to the equipotential lines around the body (the grounded human model exposed to a planar field) is depicted in Fig. 5.22.

Fig. 5.23 shows the related current density distribution calculated along the longitudinal axis of the realistic human body model without arms.

The presence of peaks in current density values again, as in the body of revolution model, corresponds to the position of the ankle and neck. The results are comparable with those obtained by different methods and presented elsewhere, e.g., in [28,29].

The current density obtained by the realistic model is compared with the results calculated with the simplified cylindrical representation of the human body as shown in Fig. 5.24.

It is evident that oversimplified cylindrical body model fails to capture the current density peaks in the

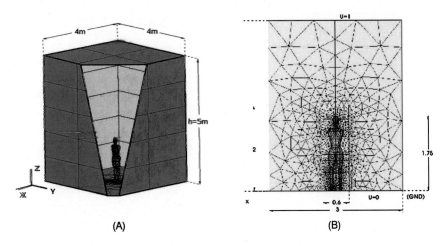

FIG. 5.20 Calculation domain: (A) a plan view of the integration domain, (B) BEM mesh.

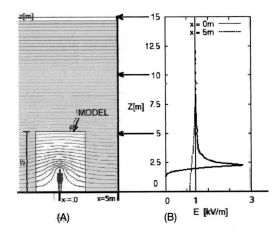

FIG. 5.21 Electric field in the air near the human body.

regions with narrow cross-section. Yet, the cylindrical approximation is at least useful for determining the order of average value of current density in the torso.

Note that the results obtained via the HNA model exceed the ICNIRP basic restrictions [13] for public exposure (2 mA/m²) in the knee (8.6 mA/m²) and neck (9.8 mA/m²) and for occupational exposure (10 mA/m²) in the ankle (32 mA/m²).

5.3.4 Realistic Model of the Human Body – Arms Included

This section deals with an improved realistic body model including the arms. The input geometrical data, material properties and the exposure conditions are the same as in Sect. 5.3.3. The meshing of the improved realistic body model with the arms included and posi-

tioned up is shown in Fig. 5.25, while the corresponding scalar potential distribution around the body is presented in Figs. 5.26 and 5.27. The current density induced inside the human body is determined by solving the corresponding Laplace equation using the Boundary Element Method with domain decomposition. The number of elements used for the discretization is nearly 20,000.

A comparison of the longitudinal current density values from pelvis to head obtained via different body models is shown in Fig. 5.28.

It is clearly demonstrated that an oversimplified cylindrical representation of the human body suffers from inability to capture the effect of high current density values in regions with reduced cross-section. Again, there is a significant increase of the current density in narrow cross-sections, such as neck. The arms extended upwards cause a screening of the electric field from the top, thus reducing the peak of current density in the neck.

Table 5.4 shows the peak value of the current density induced in the neck for some typical values of the electric field under power lines near the ground level.

Therefore, as predicted by the HNA model, the results obtained by the improved model with the arms up exceed the ICNIRP basic restrictions [13] for public and occupational exposure.

The current density induced inside the human body is determined by solving the corresponding Laplace equation using the Boundary Element Method with domain decomposition.

Finally, the last computational example deals with human being positioned inside a transformer substa-

FIG. 5.22 Equipotentials around the body (near field) $x–y$ plane view.

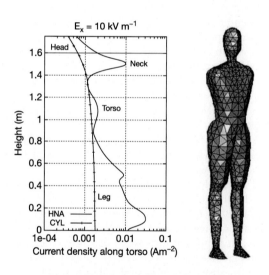

FIG. 5.23 Current density distribution.

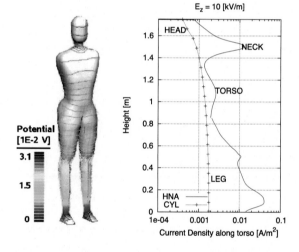

FIG. 5.24 Induced current density along the body.

TABLE 5.4
Peak values of the current density in the neck versus electric fields.

E [kV/m]	J [mA/m^2]
1	2
5	10
10	19

tion room [33], as depicted in Fig. 5.29. The boundary conditions are indicated, as well.

Fig. 5.30A shows the conceptual model of a realistic 1.75 m tall human body inside a transformer substation room (2.5 m × 1.6 m × 2 m) touching a control panel (1 m × 2 m) at the potential $\varphi_0 = 400$ V with actual BEM mesh. Two scenarios for dry-air between worker's hand and panel are considered: $d = 0.016$ m and $d = 0.116$ m. The floor is kept grounded, and all other surfaces of the room are considered with Neumann adiabatic type conditions [33]. Fig. 5.30B shows the current density induced along the body using homogeneous (HO) and heterogeneous (H) models for two different scenarios. In this case the values of internal current density do not exceed ICNIRP basic restrictions.

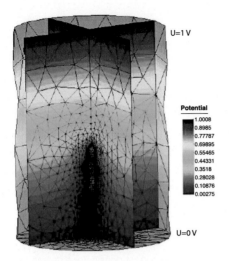

FIG. 5.25 Meshing for the realistic model of the body with arms up.

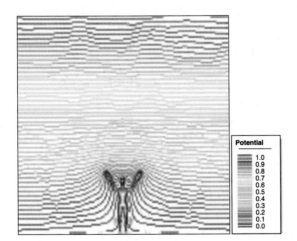

FIG. 5.26 Scalar potential distribution in the vicinity of the human body.

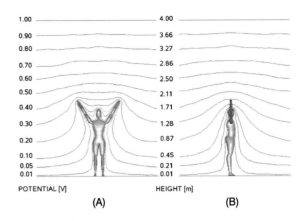

FIG. 5.27 Scalar potential around the body: (A) front view, (B) lateral view.

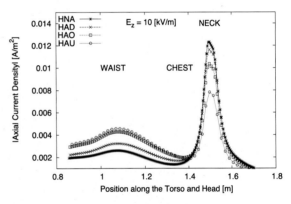

FIG. 5.28 Induced current density distribution for the various body models.

5.3.5 Pregnant Woman/Fetus Exposed to ELF Electric Field

This section deals with a rather demanding issue of woman/foetus exposure to EFL fields. Various physical and geometrical properties of corresponding tissues are taken from medical data available in the relevant literature. For the case of LF exposures, a realistic, anatomically based, model of a pregnant woman/foetus uses a quasi-static approximation based on the Laplace equation formulation and a solution via three-dimensional multi-domain Boundary Element Method (BEM). The model accounts for variations in geometry, body mass,

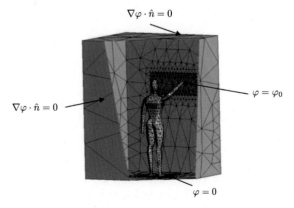

FIG. 5.29 Human inside a substation touching a control panel.

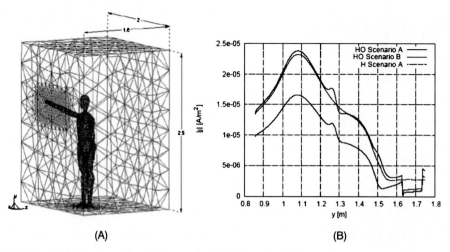

FIG. 5.30 The conceptual model and the results: (A) BEM mesh, (B) internal current density for different scenarios.

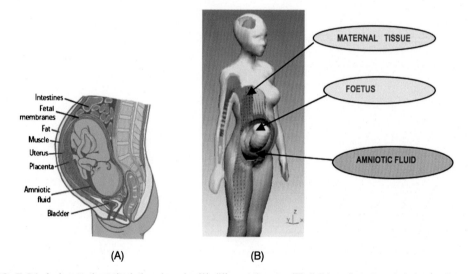

FIG. 5.31 A view to the calculation domain: (A) different tissues, (B) division of maternal abdomen into equivalent subdomains.

fat and overall chemical composition in the female body.

At LF exposures, human tissues and organs are good conductors with tissue conductivity σ varying from 0.2 to 0.5 S/m, while the relative permittivity is assumed to be around 100.

Through the fetal period (from 8th to 40th gestational week), the growth, development and maturation of the already formed structures occur. Pregnancy stages studied in this model are the 8th, 13th, 26th and 38th gestational weeks [9,18,21].

For the maternal abdomen, the decomposition into subdomains is based on the different properties of the tissues, as depicted in Fig. 5.31A.

The amniotic fluid (AF) has the highest conductivity which also varies through the periods of gestation. Kidney, muscle bone cortical, bladder, spleen, and skin have conductivity very close to 0.1 S/m, while the ovary and cartilage conductivity is around 0.2 S/m. Therefore, all these tissues can be grouped into one subdomain – maternal tissue. Furthermore, the uterus conductivity is 0.23 S/m (similar to the conductivity of the maternal tissue) while the placenta is assumed to have the same conductivity as the blood and considered as part of the maternal-tissue subdomain. Consequently, the maternal abdomen is divided into 3 subdomains: maternal tissue, amniotic fluid, contained within the uterus and foetus, as shown in Fig. 5.31B.

The foetal length change is greatest in the second trimester, while foetal weight variation is greatest in the final weeks of development. Therefore, the foetus is free to move inside the maternal abdomen, mostly until the 24th week. Since then, the movement is more constrained. Fig. 5.32 shows a general 3D view of a 1.7 m tall mother to be at the 26th week of pregnancy (foetus in the cephalic position).

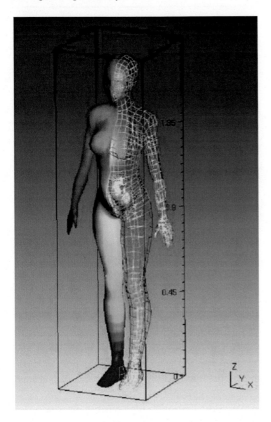

FIG. 5.32 A 3D view of the model at the 26th week of pregnancy (cephalic presentation).

Both cephalic and breech position of the foetus are taken into account within this model.

At low frequencies, adopting the quasi-static approximation, the body dimensions become electrically short, i.e., negligible when compared to the external field wavelength.

Furthermore, constant values of σ and ε are assumed within a subdomain, and the static approximation for the ELF exposure model can be formulated via the Laplace equation (5.48) where the solution is carried out via multi-domain BEM.

More mathematical details on the actual BEM procedure can be found elsewhere; see [24].

As already underlined, the sparse and highly banded resulting system of equations is one of the main advantages of the domain decomposition technique, in addition to its capabilities in dealing with piecewise homogeneous material properties.

The electric field over a flat ground plane is assumed to be vertical and uniform near the ground level, the human body is located between the parallel circular plate electrodes, in the middle of the lower one. A calculation domain with the associated boundary conditions is shown in Fig. 5.33.

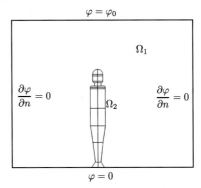

FIG. 5.33 The calculation domain with the associated boundary conditions.

The lower plate electrode represents the ground zero potential while the upper plate electrode is at the potential of a high voltage power line. Three different conductivity scenarios, shown in Table 5.5, are used, as discussed in [9].

Applying the multi-domain Boundary Element Method (BEM) [9] to the solution of the Laplace equation and determine internal currents, potentials and electric fields are determined.

Fig. 5.34 shows a lateral view of the sliced pregnant woman model at the 8th, 13th, 26th and 38th gestational week. The direction of the electric field in the maternal tissues is assigned with black arrows and the iso-lines represent the related scalar potential.

Fig. 5.35 shows a view to the sliced model of the pregnant woman/foetus with the electric field/scalar potential lines for the foetus in cephalic and breach presentation.

TABLE 5.5
Conductivity scenarios.

Scenario	[S/m]	Week 8	Week 13	Week 26	Week 38
1	σ_f	0.23	0.23	0.23	0.23
	σ_{AF}	1.28	1.28	1.27	1.10
	σ_m	0.20	0.20	0.20	0.20
2	σ_f	0.996	0.996	0.574	0.574
	σ_{AF}	1.70	1.70	1.64	1.64
	σ_m	0.52	0.52	0.52	0.52
3	σ_f	0.732	0.732	0.396	0.396
	σ_{AF}	1.70	1.70	1.64	1.64
	σ_m	0.17	0.17	0.17	0.17

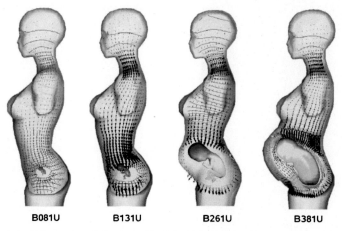

B081U   B131U   B261U   B381U

FIG. 5.34 Lateral view of a pregnant woman at the 8th, 13th, 26th and 38th gestational week (breech presentation).

Observing the results presented in Figs. 5.34 and 5.35, it is visible that the uterus, due to its higher conductivity compared to the maternal tissue, focuses the field lines.

The maximal value of current density is induced at the 8th gestational week and decreases progressively as the foetus develops due to the following reasons: First, both the foetus and AF conductivity decrease with age; Second, as the foetus grows the extremities are drawn in towards the center of the chest and the head is tucked down to the chest. Thus, the foetus surface is getting smoother and the cross-sectional area becomes more regular.

Fig. 5.36 presents the mean, maximum and minimum values of current density calculated in the foetus at different gestational weeks.

The induced current density decreases with age due to the increasing uniformity of the foetus model, i.e., the smoother the geometry the less fluctuations in the results are expected. The maximum value of current density in the foetus occurs during the 8th gestation week for all conductivity scenarios.

Another body part with high current density is in the mother's brain and neck. The maximum value of internal current density in the brain decreases from 0.43 to 0.41 mA/m^2 from week 8 to week 38. As expected, this value is not sensitive to the different conductivity scenarios related to pregnancy.

Furthermore, in all conductivity scenarios the current density in breech configuration is higher than in the cephalic. This effect becomes less evident in the final pregnancy stage (38 week). In particular, for a given

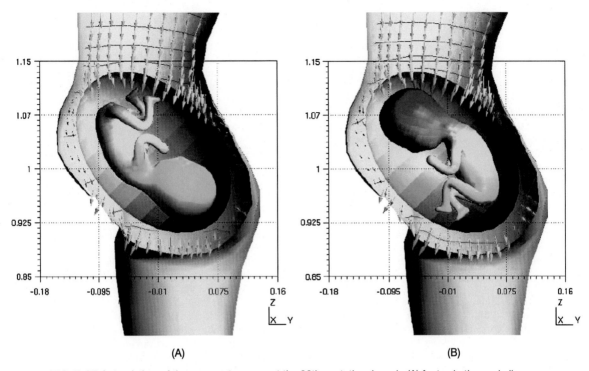

FIG. 5.35 Lateral view of the pregnant woman at the 26th gestational week: (A) foetus in the cephalic presentation, (B) foetus in the breach presentation.

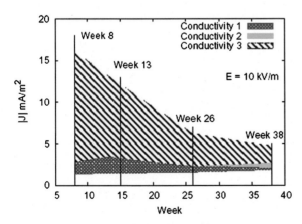

FIG. 5.36 Mean current density in the foetus during the pregnancy for different conductivity scenarios.

exposure, the maximum value of the current density induced in the foetus is during the 8th gestational week. In particular, the maximum current density obtained in the foetus for an incident external field $E = 10$ kV/m is 7.4 mA/m^2 while the maximum value recommended for public exposure by ICNIRP [12] is 2 mA/m^2.

REFERENCES

1. Riadh W.Y. Habash, Electromagnetic Fields and Radiation: Human Bioeffects and Safety, CRC Press, Boca Raton, Florida, 2001.
2. J.W. Hand, Modelling the interaction of electromagnetic fields (10 MHz – 10 GHz) with the human body: methods and applications, Physics in Medicine & Biology 53 (16) (2008) 243–286.
3. Dragan Poljak, Electromagnetic fields: environmental exposure, in: Encyclopedia of Environmental Health, Elsevier, 2011, pp. 259–268.
4. K. Nagasawa, H. Kawai, M. Takahashi, K. Saito, K. Ito, T. Ueda, M. Saito, H. Ito, H. Osada, Y. Koyanagi, et al., Experimental evaluation of the EM exposure in the simple abdomen solid phantom, in: Proceedings of ISAP, 2005, pp. 881–884.
5. Niels Finderup Nielsen, J. Michelsen, J.A. Michelsen, T. Schneider, Numerical calculation of electrostatic field surrounding a human head in visual display environments, Journal of Electrostatics 36 (3) (1996) 209–223.
6. Om P. Gandhi, Some numerical methods for dosimetry: extremely low frequencies to microwave frequencies, Radio Science 30 (1) (1995) 161–177.
7. Mara Cristina Gonzalez, Andres Peratta, Dragan Poljak, Boundary element modeling of the realistic human body exposed to extremely-low-frequency (ELF) electric fields:

computational and geometrical aspects, IEEE Transactions on Electromagnetic Compatibility 49 (1) (2007) 153–162.

8. Ilkka Laakso, Akimasa Hirata, Reducing the staircasing error in computational dosimetry of low-frequency electromagnetic fields, Physics in Medicine & Biology 57 (4) (2012) N25–N34.

9. Cristina Gonzalez, A. Peratta, D. Poljak, Electromagnetic modelling of foetus and pregnant woman exposed to extremely low frequency electromagnetic fields, in: Boundary Elements and Other Mesh Reduction Methods XXX, vol. 47, 2008, pp. 85–94.

10. Cristina Peratta, Andrés Peratta, Dragan Poljak, BEM modelling of high voltage ELF electric field applied to a 3D pregnant woman model, Journal of Communications and Software Systems 6 (1) (2010) 31–38.

11. Dragan Poljak, Mario Cvetković, Andres Peratta, Cristina Peratta, Hrvoje Dodig, Akimasa Hirata, On some integral approaches in electromagnetic dosimetry, in: The Joint Annual Meeting of the Bioelectromagnetics Society and the European BioElectromagnetics Association, 2016, pp. 289–295.

12. International Commission on Non-Ionizing Radiation Protection (ICNIRP), Guidelines for limiting exposure to time-varying electric, magnetic and electromagnetic fields (up to 300 GHz), Health Physics 74 (4) (1998) 494–522.

13. International Commission on Non-Ionizing Radiation Protection (ICNIRP), Guidelines for limiting exposure to time-varying electric and magnetic fields (1 Hz to 100 kHz), Health Physics 99 (6) (2010) 818–836.

14. International Radiation Protection Association, International Non-Ionising Radiation Committee, International Radiation Protection Association, Visual Display Units: Radiation Protection Guidance, 70, International Labour Organization, Geneva, 1994.

15. E.R. Adair, R. Ashley, C.K. Chou, R. Curtis, L.S. Erdreich, G.D. Lapin, K.H. Mild, R. Petersen, J.P. Reilly, A.R. Sheppard, et al., Biological and health effects of electric and magnetic fields from video display terminals – a technical information statement, IEEE Engineering in Medicine and Biology Magazine (ISSN 0739-5175) 16 (3) (1997) 87–92.

16. Australian Radiation Protection, Nuclear Safety Agency: Radiation Emission From Video Display Terminals, Australian Radiation Protection and Nuclear Safety Agency, 2003.

17. Niels Finderup Nielsen, Thomas Schneider, Particle deposition onto a human head: influence of electrostatic and wind fields, Bioelectromagnetics: Journal of the Bioelectromagnetics Society, the Society for Physical Regulation in Biology and Medicine, the European Bioelectromagnetics Association 19 (4) (1998) 246–258.

18. Dragan Poljak, Damir Cavka, Hrvoje Dodig, Cristina Peratta, Andres Peratta, On the use of the boundary element analysis in bioelectromagnetics, Engineering Analysis with Boundary Elements 49 (2014) 2–14.

19. NASA, Man-systems integration standards: anthropometry and biomechanics, 1995.

20. International center for numerical methods in engineering: GID, http://gid.cimne.upc.es.

21. Dragan Poljak, On the use of boundary integral methods in bioelectromagnetics, in: Numerical Methods and Advanced Simulation in Biomechanics and Biological Processes, Elsevier, 2018, pp. 119–143.

22. Dragan Poljak, Khalil El Khamlichi Drissi, Computational Methods in Electromagnetic Compatibility: Antenna Theory Approach Versus Transmission Line Models, John Wiley & Sons, New York, 2018.

23. Damir Čavka, Dragan Poljak, Andres Peratta, Comparison between finite and boundary element methods for analysis of electrostatic field around human head generated by video display units, Journal of Communications and Software Systems 7 (1) (2011) 22–28.

24. Dragan Poljak, Advanced Modeling in Computational Electromagnetic Compatibility, Wiley-Interscience, New Jersey (NJ), 2007.

25. Ronold W.P. King, Sheldon S. Sandler, Electric fields and currents induced in organs of the human body when exposed to ELF and VLF electromagnetic fields, Radio Science 31 (5) (1996) 1153–1167.

26. Ronold W.P. King, Fields and currents in the organs of the human body when exposed to power lines and VLF transmitters, IEEE Transactions on Biomedical Engineering 45 (4) (1998) 520–530.

27. Dragan Poljak, Andres Peratta, Carlos Brebbia, A 3D BEM modelling of human exposure to extremely low frequency (ELF) electric fields, in: 27th World Conference on Boundary Elements; 7th International Seminar on Computational Methods in Electrical Engineering and Electromagnetics, 2005, pp. 441–451.

28. Atsuo Chiba, Katsuo Isaka, Yoshihide Yokoi, Masayoshi Nagata, Mioru Kitagawa, Tsuneo Matsuo, Application of finite element method to analysis of induced current densities inside human model exposed to 60-Hz electric field, IEEE Transactions on Power Apparatus and Systems 103 (7) (1984) 1895–1902.

29. Atsuo Chiba, Katsuo Isaka, Yukio Onogi, Analysis of current densities induced inside human model exposed to AC electric field, Electronics and Communications in Japan (Part III: Fundamental Electronic Science) 77 (7) (1994) 58–68.

30. Atsuo Chiba, Katsuo Isaka, Analysis of current densities induced inside a human model by the two-step process method combining the surface-charge integral equation and the finite-element method, Electronics and Communications in Japan (Part II: Electronics) 79 (4) (1996) 102–111.

31. Om P. Gandhi, Jin-Yuan Chen, Numerical dosimetry at power-line frequencies using anatomically based models, Bioelectromagnetics 13 (S1) (1992) 43–60.

32. Nicolas Siauve, Riccardo Scorretti, Noel Burais, Laurent Nicolas, Alain Nicolas, Electromagnetic fields and human body: a new challenge for the electromagnetic field computation, COMPEL-The International Journal for Compu-

tation and Mathematics in Electrical and Electronic Engineering 22 (3) (2003) 457–469.

33. Dragan Poljak, C. Peratta, A. Peratta, A. Sarolic, V. Doric, Dosimetry methods for human exposure to non-ionising radiation, in: Proceedings of the Eighth Symposium of the Croatian Radiation Protection Association, Krk, 2011, pp. 501–506.

CHAPTER 6

Realistic Models for Human Exposure to High Frequency (HF) Radiation

The number of high frequency (HF) electromagnetic interference (EMI) sources in the human environment is steadily increasing. Of partial interest are base station antennas for GSM, UMTS, LTE frequency range and cell phones.

6.1 INTERNAL ELECTROMAGNETIC FIELD DOSIMETRY METHODS

Compared to low frequency (LF) fields that may elicit excitation of sensory, nerve or muscle cells, high frequency (HF) fields, due to the resonance effect, are strongly absorbed by the body. At HF, this is related to the dimensions of the human body and organs being comparable to the wavelength of the external field. Thus, the thermal effects are considered dominant at HF.

The main goal of HF dosimetry is to quantify the thermal effects, i.e., to assess the level and distribution of the electromagnetic energy absorbed by the body. The main dosimetric quantity for HF fields is the specific absorption rate (SAR) defined as the rate of energy W absorbed by, or dissipated in, the unit body mass m:

$$\text{SAR} = \frac{dP}{dm} = \frac{d}{dm}\frac{dW}{dt} = C\frac{dT}{dt}, \qquad (6.1)$$

where C is the specific heat capacity of tissue, T is the temperature, and t denotes time.

Also, SAR is proportional to the square of the internal electric field:

$$\text{SAR} = \frac{dP}{dm} = \frac{dP}{\rho dV} = \frac{\sigma}{2\rho}|E|^2 = \frac{\sigma}{\rho}|E_{\text{rms}}|^2, \qquad (6.2)$$

where E and E_{rms} are the peak and root-mean-square values of the electric field, respectively, ρ is the tissue density, and σ is the tissue conductivity.

The distribution of SAR inside the human body generally depends on the incident field parameters, but also on the parameters of the exposed body. However, to determine SAR, it is first necessary to find the corresponding electric field distribution inside the human body.

6.1.1 Surface Integral Equation Formulation

The homogeneous lossy dielectric model of the human brain exposed to HF radiation is based on the surface integral equation (SIE) formulation [1,2]. If considered as a classical scattering problem, it can be readily derived from the surface equivalence theorem and the appropriate boundary conditions for the electric or magnetic field.

To apply the equivalence theorem, the human brain is first replaced by an arbitrarily shaped dielectric body S with homogeneous properties (ε_1, μ_1), placed in free-space, as shown in Fig. 6.1A.

The homogeneous brain is considered as a lossy material with complex permittivity and permeability (ε_2, μ_2). Due to the fact that biological tissues do not posses magnetic properties, the value for the permeability of the brain is taken to be μ_0, i.e., the free-space permeability, while the complex permittivity of the brain is given by

$$\varepsilon_2 = \varepsilon_0 \varepsilon_r - j\frac{\sigma}{\omega}, \qquad (6.3)$$

where ε_0 is the permittivity of the free-space, ε_r is the relative permittivity, σ is the electrical conductivity of the brain, and $\omega = 2\pi f$ is the operating frequency.

The HF electromagnetic field $(\vec{E}^{\text{inc}}, \vec{H}^{\text{inc}})$ is incident on the lossy homogeneous object representing the brain. Due to the presence of the scattering object, i.e., brain, a scattered field denoted by $(\vec{E}^{\text{sca}}, \vec{H}^{\text{sca}})$ is also present. The fields exterior and interior to the surface S of the object are $(\vec{E}_1, \vec{H}_1)$ and $(\vec{E}_2, \vec{H}_2)$, respectively. Also, on the surface S, a unit vector $\hat{n}$ is placed, pointing from region 2 into region 1, as seen in Fig. 6.1.

Applying the equivalence theorem for regions 1 and 2, two equivalent problems can be formulated, in terms of the equivalent electric and magnetic current densities $\vec{J}$ and $\vec{M}$ placed on the surface S [3–6]. These equivalent problems are shown in Figs. 6.1B–C, for the external and internal region, respectively.

In the case of an external equivalent problem, shown in Fig. 6.1B, the field inside is assumed to be zero $(\vec{E}_2 = 0, \vec{H}_2 = 0)$, so that material properties of this region could be chosen arbitrarily. Selecting the same properties as that of an exterior region leads to the ho-

Human Interaction with Electromagnetic Fields. https://doi.org/10.1016/B978-0-12-816443-3.00014-5

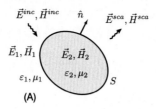

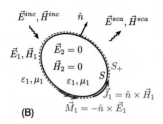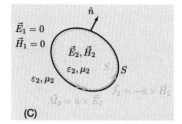

FIG. 6.1 Human brain as a lossy homogeneous dielectric (ε_2, μ_2) placed in the incident HF field $(\vec{E}^{inc}, \vec{H}^{inc})$: (A) original problem, (B) equivalent problem for region 1, (C) equivalent problem for region 2.

mogeneous domain of (ε_1, μ_1), allowing one to use the Green's function for free-space. To satisfy the boundary conditions on the surface S, equivalent surface currents $\vec{J}_1$ and $\vec{M}_1$ are introduced at this surface. Following the same procedure for the interior equivalent problem, another homogeneous domain with properties (ε_2, μ_2) is obtained. Again, the equivalent surface currents $\vec{J}_2 = -\vec{J}_1$ and $\vec{M}_2 = -\vec{M}_1$ are introduced to the surface S, as indicated in Fig. 6.1C.

Since the obtained equivalent problems represent equivalent current densities radiating in a homogeneous medium, the following expressions for the scattered fields due to these sources can be used:

$$\vec{E}_i^{sca}(\vec{J}, \vec{M}) = -j\omega \vec{A}_i - \nabla \varphi_i - \frac{1}{\varepsilon_i} \nabla \times \vec{F}_i, \quad (6.4)$$

$$\vec{H}_i^{sca}(\vec{J}, \vec{M}) = -j\omega \vec{F}_i - \nabla \psi_i + \frac{1}{\mu_i} \nabla \times \vec{A}_i, \quad (6.5)$$

where the index $i = 1, 2$ denotes the medium where equivalent surface currents radiate, and φ, $\vec{F}$, ψ and $\vec{A}$ are scalar and vector, electric and magnetic potentials, respectively.

These potentials can be written in terms of surface integrals over the sources, i.e.,

$$\vec{A}_i(\vec{r}) = \mu_i \int_S \vec{J}(\vec{r}')G_i(\vec{r}, \vec{r}')\,dS', \quad (6.6)$$

$$\vec{F}_i(\vec{r}) = \varepsilon_i \int_S \vec{M}(\vec{r}')G_i(\vec{r}, \vec{r}')\,dS', \quad (6.7)$$

$$\varphi_i(\vec{r}) = \frac{1}{\varepsilon_i} \int_S \rho(\vec{r}')G_n(\vec{r}, \vec{r}')\,dS', \quad (6.8)$$

$$\psi_i(\vec{r}) = \frac{1}{\mu_i} \int_S \rho_m(\vec{r}')G_n(\vec{r}, \vec{r}')\,dS', \quad (6.9)$$

where ρ and ρ_m are the electric and magnetic charge densities, respectively, and $G_i(\vec{r}, \vec{r}')$ is the Green's function for the homogeneous medium i given by

$$G_i(\vec{r}, \vec{r}') = \frac{e^{-jk_i R}}{4\pi R}, \quad R = |\vec{r} - \vec{r}'|. \quad (6.10)$$

In (6.10), R is the distance from the observation point $\vec{r}$ to the source point $\vec{r}'$, and $k_i = \omega\sqrt{\mu_i \varepsilon_i}$ is a wave number of medium i.

Using (6.6)–(6.9), the scattered field from (6.4) and (6.5) can be expressed in terms of the equivalent surface currents. Applying the boundary conditions at the surface S, which is the interface of the two equivalent problems, a set of four equations for the electric and magnetic fields is obtained. Choosing two expressions for the electric field,

$$-\hat{n} \times \vec{E}_i^{sca}(\vec{J}, \vec{M}) = \begin{cases} \hat{n} \times \vec{E}^{inc}, & i = 1, \\ 0, & i = 2, \end{cases} \quad (6.11)$$

the electric field integral equation (EFIE) formulation in the frequency domain for the lossy homogeneous human brain is obtained.

In (6.11), $\vec{E}^{inc}$ is the known incident field, while $\vec{J}$ and $\vec{M}$ represent the unknown surface currents.

Using the continuity equation, the electric and magnetic charges in (6.8) and (6.9) can be replaced using the divergence of the electric and magnetic currents, respectively, and after inserting (6.6)–(6.9) in (6.4), the set of coupled integral equations is obtained:

$$
j\omega\mu_i \int_S \vec{J}(\vec{r}\,')G_i(\vec{r},\vec{r}\,')\,\mathrm{d}S'
$$

$$
-\frac{j}{\omega\varepsilon_i}\nabla\int_S \nabla'_S\cdot\vec{J}(\vec{r}\,')G_i(\vec{r},\vec{r}\,')\,\mathrm{d}S'
\qquad (6.12)
$$

$$
+\nabla\times\int_S \vec{M}(\vec{r}\,')G_i(\vec{r},\vec{r}\,')\,\mathrm{d}S' =
\begin{cases}
\vec{E}^{\mathrm{inc}}, & i=1,\\
0, & i=2.
\end{cases}
$$

Performing certain mathematical manipulations on the second and third integral of (6.12), the nabla operator can be transferred to the Green's function, leading to

$$
j\omega\mu_i \int_S \vec{J}(\vec{r}\,')G_i(\vec{r},\vec{r}\,')\,\mathrm{d}S'
$$

$$
-\frac{j}{\omega\varepsilon_i}\int_S \nabla'_S\cdot\vec{J}(\vec{r}\,')\nabla G_i(\vec{r},\vec{r}\,')\,\mathrm{d}S'
$$

$$
+\int_S \vec{M}(\vec{r}\,')\times\nabla'G_i(\vec{r},\vec{r}\,')\,\mathrm{d}S' =
\begin{cases}
\vec{E}^{\mathrm{inc}}, & i=1,\\
0, & i=2,
\end{cases}
$$

$$(6.13)$$

where the property of Green's function gradient, $\nabla G_i(\vec{r}, \vec{r}\,') = -\nabla'G_i(\vec{r},\vec{r}\,')$, has been used.

6.1.1.1 Numerical Solution Using Method of Moments

The corresponding numerical solution is carried out using the Method of Moments (MoM). The equivalent electric and magnetic currents $\vec{J}$ and $\vec{M}$ in (6.13) are first expanded by a linear combination of basis functions $\vec{f}_n$ and $\vec{g}_n$, respectively, [7] as

$$
\vec{J}(\vec{r}) = \sum_{n=1}^N J_n\vec{f}_n(\vec{r}), \quad \vec{M}(\vec{r}) = \sum_{n=1}^N M_n\vec{g}_n(\vec{r}), \quad (6.14)
$$

where J_n and M_n are unknown coefficients, and N is the number of elements used to discretize the surface S.

The surface of the brain S is discretized with triangles facilitating the use of Rao–Wilton–Glisson (RWG) basis functions [8], defined on a pair of triangles sharing an edge. Namely, the function is given by

$$
\vec{f}_n^\pm(\vec{r}) =
\begin{cases}
\dfrac{l_n}{2A_n^\pm}\vec{\rho}_n^\pm, & \vec{r}\in T_n^\pm,\\[2mm]
0, & \vec{r}\notin T_n^\pm,
\end{cases}
\qquad (6.15)
$$

where l_n is the length of the triangles shared edge, the surface areas of those triangles are given by A_n^+ and A_n^-, respectively, while $\vec{\rho}_n^+$ and $\vec{\rho}_n^-$ are the vectors from each triangle free vertex, as shown on Fig. 6.2.

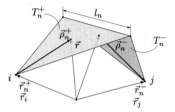

FIG. 6.2 RWG basis function $\vec{f}_n(\vec{r})$ defined on a pair of triangles.

The equivalent electric current $\vec{J}$ is approximated using the RWG function $\vec{f}_n$, and the equivalent magnetic current $\vec{M}$ is approximated by a pointwise orthogonal function $\vec{g}_n = \hat{n}\times\vec{f}_n$. The unknown equivalent currents $\vec{J}(\vec{r}\,')$ and $\vec{M}(\vec{r}\,')$ in (6.13) are substituted with (6.14). Eq. (6.13) is then multiplied by the set of a test functions $\vec{f}_m$, where $\vec{f}_m = \vec{f}_n$, and integrated over the surface S. After some mathematical manipulations, it follows that

$$
j\omega\mu_i \sum_{n=1}^N J_n \int_S \vec{f}_m(\vec{r})\cdot\int_{S'}\vec{f}_n(\vec{r}\,')G_i\,\mathrm{d}S'\,\mathrm{d}S
$$

$$
+\frac{j}{\omega\varepsilon_i}\sum_{n=1}^N J_n \int_S \nabla_S\cdot\vec{f}_m(\vec{r})\int_{S'}\nabla'_S\cdot\vec{f}_n(\vec{r}\,')G_i\,\mathrm{d}S'\,\mathrm{d}S
$$

$$
\pm\sum_{n=1}^N M_n \int_S \vec{f}_m(\vec{r})\cdot[\hat{n}\times\vec{g}_n(\vec{r}\,')]\,\mathrm{d}S
$$

$$
+\sum_{n=1}^N M_n \int_S \vec{f}_m(\vec{r})\cdot\int_{S'}\vec{g}_n(\vec{r}\,')\times\nabla'G_i\,\mathrm{d}S'\,\mathrm{d}S
$$

$$
=
\begin{cases}
\displaystyle\int_S \vec{f}_m(\vec{r})\cdot\vec{E}^{\mathrm{inc}}\,\mathrm{d}S, & i=1,\\[3mm]
0, & i=2,
\end{cases}
\qquad (6.16)
$$

where subscript i is now the index of the medium. The last two terms from the left-hand side of (6.16) represent the residual term and Cauchy principal value, respectively, of the last integral from (6.13).

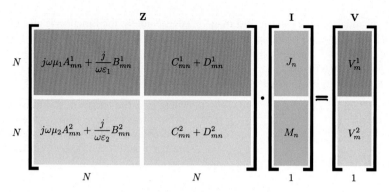

FIG. 6.3 MoM matrix system.

Taking the sum out, (6.16) is written as the following system of linear equations:

$$\sum_{n=1}^{N}\left(j\omega\mu_i A_{mn,i}+\frac{j}{\omega\varepsilon_i}B_{mn,i}\right)J_n$$
$$+\sum_{n=1}^{N}\left(C_{mn,i}+D_{mn,i}\right)M_n$$
$$=\begin{cases} V_m, & i=1, \\ 0, & i=2, \end{cases} \tag{6.17}$$

where $A_{mn,i}$, $B_{mn,i}$, $C_{mn,i}$ and $D_{mn,i}$ are the surface integrals calculated for each m-n combination of basis and testing functions, respectively. The procedure for solving these integrals is given in the Appendix.

Eq. (6.17) can be written in the matrix form

$$[Z]\cdot\{I\}=\{V\}, \tag{6.18}$$

where [Z] represents the $2N\times 2N$ system matrix, while {V} is the source vector of dimension $2N$, as illustrated on Fig. 6.3.

Vector {I} containing the unknown coefficients J_n and M_n is the solution of the matrix equation (6.18). From these coefficients, the equivalent electric and magnetic currents $\vec{J}$ and $\vec{M}$ placed on the surface S of the dielectric object, i.e., homogeneous biological body, can be determined using (6.14). From these currents, the electric and magnetic field, respectively, can be easily determined.

6.1.2 Tensor Volume Integral Equation

The whole body exposure to HF radiation can be evaluated using tensor volume integral equation (VIE) [9]. The electric field induced inside the biological body is the sum of the incident field E^{inc} and the scattered field E^{sca}:

$$\vec{E}(\vec{r})=\vec{E}^{\text{inc}}(\vec{r})+\vec{E}^{\text{sca}}(\vec{r}). \tag{6.19}$$

To obtain VIE formulation, one can start from the differential form of the first two Maxwell's equations applied to the scattered fields $\vec{E}^{\text{sca}}$ and $\vec{H}^{\text{sca}}$:

$$\nabla\times\vec{E}^{\text{sca}}=-j\omega\mu\vec{H}^{\text{sca}},$$
$$\nabla\times\vec{H}^{\text{sca}}=j\omega\varepsilon\vec{E}^{\text{sca}}+\vec{J}, \tag{6.20}$$

where

$$\vec{J}=\sigma\vec{E} \tag{6.21}$$

is the polarization current related to the electric field within the scatterer, i.e., current density induced inside the human body.

Elimination of $\vec{E}^{\text{sca}}$ and $\vec{H}^{\text{sca}}$ leads to:

$$\nabla\times\nabla\times\vec{E}^{\text{sca}}-k_0^2\vec{E}^{\text{sca}}=-j\omega\mu_0\vec{J},$$
$$\nabla\times\nabla\times\vec{H}^{\text{sca}}-k_0^2\vec{H}^{\text{sca}}=\nabla\times\vec{J}, \tag{6.22}$$

where k_0 is the free-space wave number.

The boundary condition for the normal component of $\vec{J}$ near the body-air surface is given by

$$\hat{n}\cdot\vec{J}=-j\omega\rho_s, \tag{6.23}$$

where ρ_s denotes the surface charge density.

The boundary condition for the normal component of the electric flux density at the air(or free-space)–body surface is

$$\hat{n}\cdot\vec{D}=\rho_s. \tag{6.24}$$

Combining (6.23) and (6.24), one can obtain the free-space current density

$$\vec{J}_1 = \sigma \vec{E} + j\omega(\varepsilon - \varepsilon_0)\vec{E}, \tag{6.25}$$

where index 1 denotes the free-space (air) region.

The induced current in the body may be accounted for by replacing the body with the above-mentioned equivalent free-space current density

$$\vec{J}_{eq}(\vec{r}) = \left[\sigma(\vec{r}) + j\omega(\varepsilon(\vec{r}) - \varepsilon_0)\right]\vec{E}(\vec{r}) = \tau(\vec{r})\vec{E}(\vec{r}). \tag{6.26}$$

The solutions of equations (6.22) are:

$$\vec{E}^{\text{sca}} = -j\omega\left[1 + \frac{1}{k_0^2}\nabla\nabla\cdot\right]\vec{A},$$
$$\vec{H}^{\text{sca}} = \frac{1}{\mu_0}\nabla \times \vec{A}, \tag{6.27}$$

where $\vec{A}$ is the magnetic vector potential expressed using the integral over the sources,

$$\vec{A}(\vec{r}) = \mu_0 \int_v G_0(\vec{r},\vec{r}')\vec{J}_{eq}(\vec{r}')\,dV', \tag{6.28}$$

and $G_0(\vec{r},\vec{r}')$ is the free-space scalar Green's function,

$$G_0(\vec{r},\vec{r}') = \frac{e^{-jk_0 R}}{4\pi R}, \tag{6.29}$$

where $R = |\vec{r} - \vec{r}'|$ is the distance between the observation point $\vec{r}$ and the source point $\vec{r}'$.

If $\vec{J}_{eq}$ is assumed to be an infinitesimal, elementary source at $\vec{r}'$ oriented in the x-direction so that

$$\vec{J}_{eq}(\vec{r}') = \delta(\vec{r} - \vec{r}')\hat{e}_x, \tag{6.30}$$

the corresponding vector potential (6.28) can be written as

$$\vec{A}(\vec{r}) = \mu_0 G_0(\vec{r},\vec{r}')\hat{e}_x. \tag{6.31}$$

Hence, if $\vec{G}_{0x}(\vec{r} - \vec{r}')$ represents the electric field produced by the elementary source, then $\vec{G}_{0x}(\vec{r} - \vec{r}')$ has to satisfy

$$\nabla \times \nabla \times \vec{G}_{0x}(\vec{r} - \vec{r}') - k_0^2\vec{G}_{0x}(\vec{r} - \vec{r}')$$
$$= -j\omega\mu_0\delta(\vec{r} - \vec{r}')\hat{e}_x, \tag{6.32}$$

whose solution is

$$\vec{G}_{0x}(\vec{r} - \vec{r}') = -j\omega\left[1 + \frac{1}{k_0^2}\nabla\nabla\cdot\right]G_0(\vec{r} - \vec{r}')\hat{e}_x, \tag{6.33}$$

where $\vec{G}_{0x}(\vec{r} - \vec{r}')$ represents the free-space vector Green's function in case of a source oriented in the x-direction.

In a similar way, the corresponding infinitesimal, elementary sources $\vec{G}_{0y}$ and $\vec{G}_{0z}$, pointing in the y- and z-direction, respectively, can be introduced.

Therefore, the following function storing the three vector Green's functions is introduced:

$$\underline{G}_0(\vec{r} - \vec{r}') = \vec{G}_{0x}(\vec{r} - \vec{r}')\hat{e}_x + \vec{G}_{0y}(\vec{r} - \vec{r}')\hat{e}_y$$
$$+ \vec{G}_{0z}(\vec{r} - \vec{r}')\hat{e}_z. \tag{6.34}$$

According to [10], $\underline{G}_0$ is the free-space dyadic Green's function that represents the solution to the following dyadic differential equation:

$$\nabla \times \nabla \times \underline{G}_0(\vec{r} - \vec{r}') - k_0^2\underline{G}_0(\vec{r} - \vec{r}') = -j\omega\mu_0\delta(\vec{r} - \vec{r}')\tilde{I}, \tag{6.35}$$

where $\tilde{I}$ denotes the unit dyad defined by

$$\tilde{I} = \hat{e}_x\hat{e}_x + \hat{e}_y\hat{e}_y + \hat{e}_z\hat{e}_z. \tag{6.36}$$

From Eqs. (6.22) and (6.35), the solution for the electric field at a field point $\vec{r}$ due to infinitesimal source at $\vec{r}'$ is given by

$$\vec{E}^{\text{sca}}(\vec{r}) = -j\omega\mu_0 \int_v \underline{G}_0(\vec{r},\vec{r}') \cdot \vec{J}_{eq}(\vec{r}')\,dV'. \tag{6.37}$$

Eq. (6.33) clearly shows that $\vec{G}_0(\vec{r},\vec{r}')$ has a singularity of order R^3, therefore, when $R \to 0$, i.e., $\vec{r} \to \vec{r}'$, a small volume around the field point has to be excluded and the integral in (6.37) has to be evaluated in the limiting case.

Thus, (6.37) can be expressed as [11]:

$$\vec{E}^{\text{sca}}(\vec{r}) = \text{PV} \int_v \vec{J}_{eq}(\vec{r}) \cdot \underline{G}(\vec{r},\vec{r}')\,dV' - \frac{\vec{J}_{eq}(\vec{r})}{j3\omega\varepsilon_0}, \tag{6.38}$$

where PV stands for the Cauchy principal value of integral (6.37) and the last term denotes the correction term.

Combining Eqs. (6.19), (6.26) and (6.38) leads to the desired volume integral equation for the electric field inside the body:

$$\left[1 + \frac{\tau(\vec{r})}{3j\omega\varepsilon_0}\right]\vec{E}(\vec{r}) - \text{PV} \int_v \tau(\vec{r}')\vec{E}(\vec{r}) \cdot \underline{G}(\vec{r},\vec{r}')\,dv'$$
$$= \vec{E}^{\text{inc}}(\vec{r}), \tag{6.39}$$

where the incident electric field $\vec{E}^{\text{inc}}$ and properties of biological body denoted by

$$\tau(\vec{r}) = \sigma(\vec{r}) + j\omega\left[\varepsilon(\vec{r}) - \varepsilon_0\right] \qquad (6.40)$$

are known quantities, while the total electric field $\vec{E}$ inside the biological body is unknown to be determined by MoM.

6.1.2.1 Numerical Solution Using Method of Moments

The inner product from (6.39) can be expressed as

$$\vec{E}(\vec{r}) \cdot \underline{G}(\vec{r}, \vec{r}') = \begin{bmatrix} \vec{G}_{xx}(\vec{r}, \vec{r}') & \vec{G}_{xy}(\vec{r}, \vec{r}') & \vec{G}_{xz}(\vec{r}, \vec{r}') \\ \vec{G}_{yx}(\vec{r}, \vec{r}') & \vec{G}_{yy}(\vec{r}, \vec{r}') & \vec{G}_{yz}(\vec{r}, \vec{r}') \\ \vec{G}_{zx}(\vec{r}, \vec{r}') & \vec{G}_{zy}(\vec{r}, \vec{r}') & \vec{G}_{zz}(\vec{r}, \vec{r}') \end{bmatrix}$$
$$\cdot \begin{bmatrix} E_x(\vec{r}') \\ E_y(\vec{r}') \\ E_z(\vec{r}') \end{bmatrix}, \qquad (6.41)$$

where each element can be written in the following form:

$$G_{x_p x_q}(\vec{r}, \vec{r}') = -j\omega\mu_0\left[\delta_{pq} + \frac{1}{k_0^2}\frac{\partial^2}{\partial x_p \partial x_q}\right]G_0(\vec{r}, \vec{r}'),$$
$$p, q = 1, 2, 3, \qquad (6.42)$$

where $x_1 = x$, $x_2 = y$, $x_3 = z$.

Applying the method of moments (MoM), (6.39) can be transformed into a matrix equation. The body volume is first discretized using N subvolumes or cells, denoted by V_m ($m = 1, 2, \ldots, N$). The electric field $\vec{E}(\vec{r})$ and $\tau(\vec{r})$ are assumed constant within each cell [11]. Furthermore, if $\vec{r}_m$ is the center of the mth cell, by enforcing each scalar component of Eq. (6.39) (point matching) at the cell centers, it follows that

$$\left[1 + \frac{\tau(\vec{r})}{3j\omega\varepsilon_0}\right]E_{x_p}(r_m)$$
$$- \sum_{q=1}^{3}\left[\sum_{n=1}^{N}\tau(\vec{r})\,\text{PV}\int_{V_m}G_{x_p x_q}(\vec{r}_m, \vec{r}')dv'\right]E_{x_q}(\vec{r}_n)$$
$$= E_{x_p}^{\text{inc}}(\vec{r}_m). \qquad (6.43)$$

If one lets $G_{x_p x_q}^{mn}$ be an $N \times N$ matrix with elements

$$G_{x_p x_q}^{mn} = \tau(\vec{r}_n)\,\text{PV}\int_{V_n}G_{x_p x_q}(\vec{r}_m, \vec{r}')dV'$$
$$- \delta_{pq}\delta_{mn}\left[1 + \frac{\tau(\vec{r})}{3j\omega\varepsilon_0}\right], \qquad (6.44)$$

where $m, n = 1, 2, \ldots, N$, $p, q = 1, 2, 3$, and also lets $[E_{x_p}]$ and $[E_{x_p}^{\text{inc}}]$ be column matrices with elements

$$E_{x_p} = \begin{bmatrix} E_{x_p}(\vec{r}_1) \\ \vdots \\ E_{x_p}(\vec{r}_N) \end{bmatrix}, \qquad E_{x_p}^{\text{inc}} = \begin{bmatrix} E_{x_p}^{\text{inc}}(\vec{r}_1) \\ \vdots \\ E_{x_p}^{\text{inc}}(\vec{r}_N) \end{bmatrix}, \qquad (6.45)$$

then from Eqs. (6.39) and (6.43) one obtains $3N$ simultaneous equations for E_x, E_y and E_z at the centers of N cells, written in the matrix form as

$$\begin{bmatrix} [G_{xx}] & [G_{xy}] & [G_{xz}] \\ \hline [G_{yx}] & [G_{yy}] & [G_{yz}] \\ \hline [G_{zx}] & [G_{zy}] & [G_{zz}] \end{bmatrix}\begin{bmatrix} [E_x] \\ [E_y] \\ [E_z] \end{bmatrix}$$
$$= -\begin{bmatrix} [E_x^{\text{inc}}] \\ [E_y^{\text{inc}}] \\ [E_z^{\text{inc}}] \end{bmatrix}, \qquad (6.46)$$

or more compactly as

$$[G][E] = -\left[E^{\text{inc}}\right], \qquad (6.47)$$

where $[G]$ is a $3N \times 3N$ matrix, while $[E]$ and $\left[E^{\text{inc}}\right]$ are $3N$ column vectors.

6.1.3 Hybrid Finite Element/Boundary Element Approach

The human eye and head exposed to plane wave, and the corresponding internal field distribution, can be dealt with using a hybrid boundary element method/finite element method (BEM/FEM) approach. As the eye and head exposed to plane wave represent an exterior field problem, it can be treated as an unbounded scattering problem.

Using the Stratton–Chu integral expression, the time harmonic electric field in the domain exterior to the head is expressed by the following boundary integral equation [12,13]:

$$\alpha\vec{E}'_{\text{ext}} = \vec{E}'_{\text{inc}} + \oint_{\partial V}\hat{n} \times (\nabla \times \vec{E}_{\text{ext}})G\,dS$$
$$+ \oint_{\partial V}\left[(\hat{n} \times \vec{E}_{\text{ext}}) \times \nabla G + (\hat{n} \cdot \vec{E}_{\text{ext}})\nabla G\right]dS, \qquad (6.48)$$

where $\hat{n}$ is an outer normal to surface ∂V bounding the volume V and α is the solid angle subtended at the observation point. The total and incident electric fields are denoted by $\vec{E}_{ext}$ and $\vec{E}_{inc}$, respectively. The free-space Green's function given by

$$G = G(\vec{r}, \vec{r}\,') = \frac{e^{-jkR}}{4\pi R}, \qquad R = |\vec{r} - \vec{r}\,'|, \qquad (6.49)$$

where R is the distance from the observation point $\vec{r}$ to the source point $\vec{r}\,'$, and k denotes the wave number.

After some work, (6.48) can be rewritten in terms of tangential components of electric and magnetic fields [13]:

$$\alpha \vec{E}'_{ext} = \vec{E}'_{inc} - j\omega\mu \oint_{\partial V} \hat{n} \times \vec{H}_{ext} G \, dS$$

$$+ \oint_{\partial V} \left[(\hat{n} \times \vec{E}_{ext}) \times \nabla G \right.$$

$$\left. - \frac{1}{\sigma + j\omega\mu} \nabla_s \cdot (\hat{n} \times \vec{H}_{ext}) \nabla G \right] dS. \qquad (6.50)$$

Eq. (6.50) is written in the form appropriate for coupling to the governing differential equations of the interior inhomogeneous domain, given by the vector Helmholtz equation [13]:

$$\nabla \times \left(\frac{j}{\omega\mu} \nabla \times \vec{E}_{int} \right) - (\sigma + j\omega\varepsilon) \vec{E}_{int} = 0. \qquad (6.51)$$

The electric and magnetic fields are approximated using the edge elements [14] preserving the tangential continuity on the boundary surface as

$$\vec{E} = \sum_{i=1}^{n} \delta_i \vec{w}_i e_i, \qquad \vec{H} = \sum_{i=1}^{n} \delta_i \vec{w}_i h_i. \qquad (6.52)$$

The unknown coefficients e_i and h_i, associated with each edge of the model, are determined from the global system of equations, while the coefficient $\delta_i = \pm 1$ depends on the direction of local edge coinciding with the chosen global edge direction or not.

Vector base function w_k is given by [15]

$$\vec{w}_k = N_i \nabla N_j - N_j \nabla N_i, \qquad (6.53)$$

where N_i and N_j represent the first order barycentric shape functions.

Applying the weighted residual approach to (6.51), followed by the dot product of (6.51) with test function w_i, and utilizing the Galerkin–Bubnov procedure, one can write

$$\int_V \delta_i \vec{w}_i \cdot \left[\nabla \times \left(\frac{j}{\omega\mu} \nabla \times \vec{E}_{int} \right) - (\sigma + j\omega\varepsilon) \vec{E}_{int} \right] dV = 0. \qquad (6.54)$$

After application of some standard vector identities, followed by the use of the divergence theorem, the weak form can be obtained as

$$\int_V \left[\frac{j}{\omega\mu} \nabla \times \delta_i \vec{w}_i \cdot \vec{E}_{int} - (\sigma + j\omega\varepsilon) \delta_i \vec{w}_i \cdot \vec{E}_{int} \right] dV$$

$$= \oint_{\partial V} d\vec{S} \cdot \delta_i \vec{w}_i \times \vec{H}_{int}. \qquad (6.55)$$

The coupling between FEM and BEM can now be employed via the electric and magnetic fields tangential components' continuity across the surface ∂V, leading to $\hat{n} \times \vec{E}_{int} = \hat{n} \times \vec{E}_{ext}$ and $\hat{n} \times \vec{H}_{int} = \hat{n} \times \vec{H}_{ext}$. Thus, the following can be written:

$$\oint_{\partial V} d\vec{S}' \cdot \delta_i \vec{w}_i \times \vec{E}'_{int} = \oint_{\partial V} d\vec{S}' \cdot \delta_i \vec{w}_i \times \vec{E}'_{ext}. \qquad (6.56)$$

After substituting $\vec{E}_{ext}$ and $\vec{H}_{ext}$ in (6.50) with $\vec{E}_{int}$ and $\vec{H}_{int}$, respectively, and after inserting (6.50) into (6.56), the following double surface integral is obtained:

$$\oint_{\partial V} d\vec{S}' \cdot \delta_i \vec{w}_i \times \alpha \vec{E}'_{ext} = \oint_{\partial V} d\vec{S}' \cdot \delta_i \vec{w}_i \times \vec{E}'_{inc}$$

$$- j\omega\mu \oint_{\partial V} d\vec{S}' \cdot \delta_i \vec{w}_i \times \oint_{\partial V} \hat{n} \times \vec{H}_{int} G \, dS$$

$$+ \oint_{\partial V} d\vec{S}' \cdot \delta_i \vec{w}_i \times \oint_{\partial V} (\hat{n} \times \vec{E}_{int}) \times \nabla G \, dS$$

$$- \frac{1}{\omega + j\omega\mu} \oint_{\partial V} d\vec{S}' \cdot \delta_i \vec{w}_i \times \oint_{\partial V} \nabla_s \cdot (\hat{n} \times \vec{H}_{ext}) \nabla G \, dS. \qquad (6.57)$$

Inserting (6.52) into (6.57) leads to the system of equations associated to the edges at the boundary surface of the problem:

$$[E_{bem}] \{e_{bem}\} = \{e_{inc}\} + [H_{bem}] \{h_{bem}\}, \qquad (6.58a)$$

$$[E_{fem}] \{e_{fem}\} = [H_{fem}] \{h_{fem}\}, \qquad (6.58b)$$

where e_{bem} and h_{bem} are unknown coefficients related to the problem boundary surface, e_{inc} are known coefficients determined from the incident field, and matrices

$[E_{\text{bem}}]$ and $[H_{\text{bem}}]$ stem from the boundary integral equation (6.57), while matrices $[E_{\text{fem}}]$ and $[H_{\text{fem}}]$ arise from FEM equation (6.55).

Eqs. (6.58a) and (6.58b) can be combined into a single matrix equation as shown in Fig. 6.4.

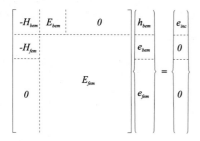

FIG. 6.4 Coupled BEM and FEM matrices.

The resulting submatrices E_{fem} and H_{fem} are sparse and symmetric while submatrices E_{bem} and H_{bem} are fully populated matrices. Solving (6.58) yields the vector of edge-element coefficients $\{h_{\text{bem}}, e_{\text{bem}}, e_{\text{fem}}\}$, used to compute the electric and magnetic fields.

6.1.4 The Human Eye Exposure

This section deals with the human eye exposed to HF plane wave. As the eye exposed to plane wave represents an unbounded electromagnetic scattering problem, the hybrid BEM/FEM approach using edge elements is applied to assess the related internal fields. The results are obtained using two different eye models, the so-called extracted and compound models; the former is considered when the organ is positioned in free-space, and the latter is incorporated into realistic head model obtained from the magnetic resonance imaging scans.

6.1.4.1 Model of the Human Eye

The human eye is a complex organ consisting of many fine tissues, each performing an important function. To account for these small but important subtleties, a detailed model of the eye is crucial. The cross-section of the human eye depicting its parts is shown in Fig. 6.5.

The magnetic resonance imaging (MRI) technique can be used to accurately capture details of the human anatomy, however, the fine geometrical details of the human eye are still unattainable due to insufficient resolution of MRI. Some recent studies, using an ultra-high resolution 7T MRI techniques, reported being able to capture the anatomical data with resolutions of 0.6 and 0.7 mm [16] and $0.5 \times 0.5 \times 0.6$ mm^3 [17]. Moreover, the most accurate MRI scans performed using the 9.4T MRI [18] achieved voxel volumes of about $0.13 \times 0.13 \times 0.8$ mm^3. However, to capture the fine de-

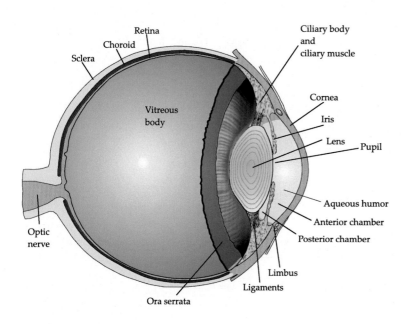

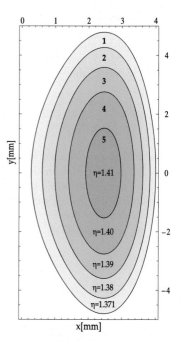

FIG. 6.5 Sagittal cross-section of the human eye depicting various tissues and the crystalline lens modeled using five layers.

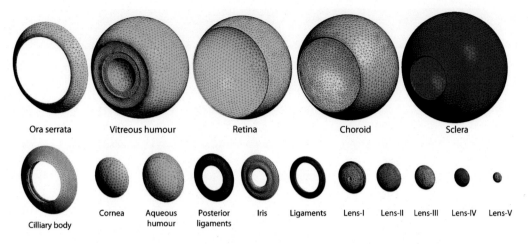

Ora serrata   Vitreous humour   Retina   Choroid   Sclera

Cilliary body

Cornea   Aqueous humour   Posterior ligaments   Iris   Ligaments   Lens-I   Lens-II   Lens-III   Lens-IV   Lens-V

FIG. 6.6 Modeled tissues from the extracted and the compound eye models.

tails of the eye tissues such as choroid, retina, iris, etc., a much finer resolution is necessary. It is estimated that to achieve the resolution with at least several layers of MRI voxels in the retina, one would require the spatial MRI resolution of about $30 \times 30 \times 30$ μm^3 or less. Any lower resolution would result in boundary surfaces between tissues becoming stair-cased [19].

Thus, the detailed geometrical model of the human eye has been developed from the available MRI scans in addition to data from various medical measurements. This model is dubbed the extracted, or the single eye, model, as it is considered to be isolated from the rest of the head tissues. Modeled tissues from the extracted eye model are shown in Fig. 6.6. The details of generating this eye model are reported elsewhere [13].

The eye model consists of 16 tissues, whose parameters for two particular frequencies are given in Table 6.1. The frequency dependent dielectric parameters (electrical conductivity σ and relative permittivity ε) are modeled using the 4-Cole–Cole method [20]. Table 6.1 includes also the tissue mass density ρ.

The eye crystalline lens is modeled as five homogeneous layers with different relative permittivity, according the Gradient Refraction Index (GRIN) model [21,22]. It is depicted in Fig. 6.5.

The interior domain of the eye is discretized using 415,429 tetrahedral (FEM) elements, while the boundary surface of the eye model is discretized using 7,986 triangular (BEM) elements.

6.1.4.2 Human Eye Exposed to Plane Wave

The first results are related to a human eye exposed to the incident plane wave of different frequencies. The power density of incident wave is 10 W/m^2. The numerical results for the electric field induced in the eye, obtained using the hybrid BEM/FEM procedure, are shown in Fig. 6.7.

The corresponding SAR results are shown in Fig. 6.8.

The maximum values of SAR at different frequencies, averaged over the whole eye, are presented in Table 6.2. It is worth noting that the results shown in Fig. 6.8 are in a good agreement with the results reported in [23].

The results show that as frequency increases SAR distribution becomes more localized, promoting the formation of *hot spots* between 1 and 4 GHz. At higher frequencies, the region where electromagnetic energy is concentrated is the eye surface.

Furthermore, in the frequency range from 1 to 2 GHz, the absorbed electromagnetic energy is concentrated in the lens area of an eye, whereas at frequencies between 2 and 6 GHz, the highest concentration of energy is found in the vitreous body.

Also, as obvious from Table 6.2, the whole eye averaged SAR values stay below the ICNIRP exposure limits for localized SAR in the head for general public population (2 W/kg) [24].

Regarding the computational aspects, it is worth emphasizing that, in addition to treating electromagnetic scattering from the human eye as an unbounded problem, the hybrid formulation enables one to successfully deal with tissue inhomogeneities. Furthermore, due to the singularities arising from the SIE formulation, an adaptive contour integration techniques is used [13]. This technique is accurate, fast, and immune to the proximity of the observation point.

TABLE 6.1
Tissue parameters used in eye models.

Tissue	900 MHz		1800 MHz		ρ [g/m^3]
	σ [S/m]	ε [–]	σ [S/m]	ε [–]	
Brainstem	0.622	38.577	0.915	37.011	1043
Cerebellum	1.308	48.858	1.709	46.114	1039
Head skin	0.899	40.936	1.185	38.872	1050
Liquor	1.667	68.875	2.032	68.573	1035
Skull	0.364	20.584	0.588	19.343	1900
Mandible	0.364	20.584	0.588	19.343	1900
Gray Matter	0.985	52.282	1.391	50.079	1039
Anterior chamber	1.667	68.875	2.032	68.573	1003
Choroid	0.729	44.561	1.066	43.343	1000
Ciliary body	0.978	54.811	1.341	53.549	1040
Cornea	1.438	54.835	1.858	52.768	1076
Iris	0.978	54.811	1.341	53.549	1040
Ligaments	0.760	45.634	1.201	44.252	1000
Ora serrata	0.882	45.711	1.232	43.850	1000
Posterior chamber	1.667	68.875	2.032	68.573	1000
Retina	1.206	55.017	1.602	53.568	1039
Sclera	1.206	55.017	1.602	53.568	1076
Vitreous body	1.667	68.875	2.032	68.573	1009
Lens-I	0.824	46.399	1.147	45.353	1100
Lens-II	0.824	47.011	1.147	45.925	1100
Lens-III	0.824	47.694	1.147	46.221	1100
Lens-IV	0.824	48.383	1.147	46.883	1100
Lens-V	0.824	49.076	1.147	47.554	1100

TABLE 6.2
Whole eye averaged specific absorption rate (SAR) at frequencies of 1, 2, 4, and 6 GHz.

Frequency [GHz]	SAR [W/kg]
1	0.335
2	0.619
4	1.262
6	1.069

6.1.4.3 Compound Versus Extracted Eye Models

The computational models employed for dosimetric assessment can be classified as realistic models of the human body (or particular organs of interest) based on the magnetic resonance imaging (MRI), e.g., [25], or the simplified models, computationally much less demanding but failing to provide accurate results in most of the exposure scenarios [26].

There are many readily available detailed models of the complete human body nowadays; see, e.g., [27,28]. However, a detailed human body model puts a significant burden in the computational model preparation at the same time, putting strain on the available computational resources. In addition to this, there are cases when only a particular organ or body parts are of a research interest, such as in the initial dosimetric assessment.

The human eye modeling is considered to be a rather challenging topic [29]. The decision between using a simplified single eye model and a more detailed and

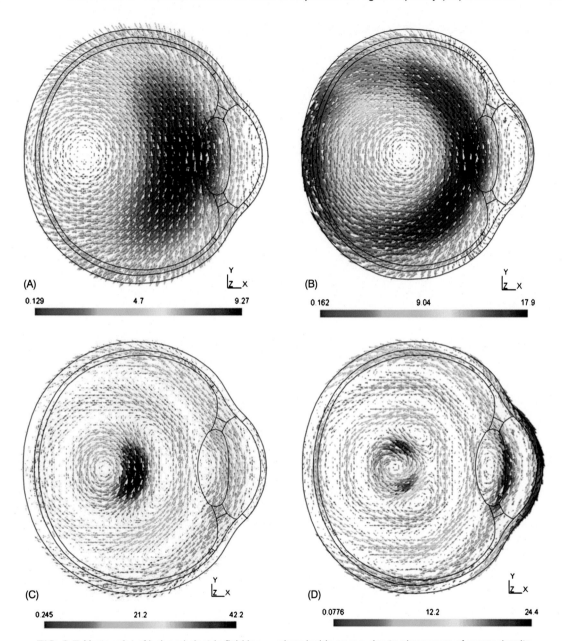

FIG. 6.7 Vector plot of induced electric field in $x–y$ plane inside an eye due to plane wave of power density 10 W/m^2 at a frequency of: (A) 1 GHz, (B) 2 GHz, (C) 4 GHz, and (D) 6 GHz.

complete body model is neither simple nor straightforward.

Thus, this section is about the use of a single organ model, also named the extracted model, and that same model incorporated into a detailed human head model (introduced latter in Sect. 6.1.6), termed the compound

model, used in the numerical assessment of the induced electric field due to high frequency EM radiation [19].

The numerical results for the electric field induced in the extracted and compound models of the eye are respectively given in Figs. 6.9–6.12. The incident plane wave of 1 GHz and 1800 MHz is horizontally and ver-

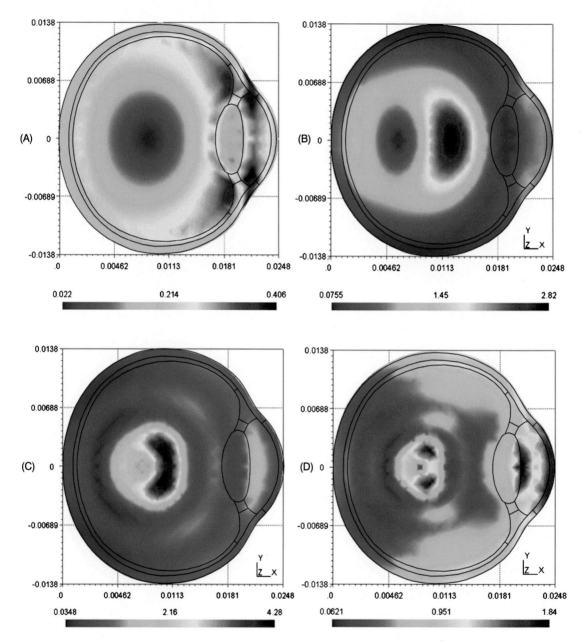

FIG. 6.8 Computed SAR in the eye induced by plane wave of power density 10 W/m^2 at a frequency of: (A) 1 GHz, (B) 2 GHz, (C) 4 GHz, and (D) 6 GHz.

tically polarized, and directed toward the eye anterior surface, perpendicular to the coronal eye cross-section. The amplitude of the incident EM wave is 1 V/m.

Fig. 6.9 shows the results for the induced field on the surface of the extracted and compound eye models, due to incident wave of 1 GHz. From Fig. 6.9 it is

obvious that the distribution of electric field on the surface of the compound eye model (cornea and sclera) is not showing the symmetrical nature as is the case for the extracted eye model. In addition, the obtained peak values in the compound model are higher than the corresponding values in the extracted model, in addition

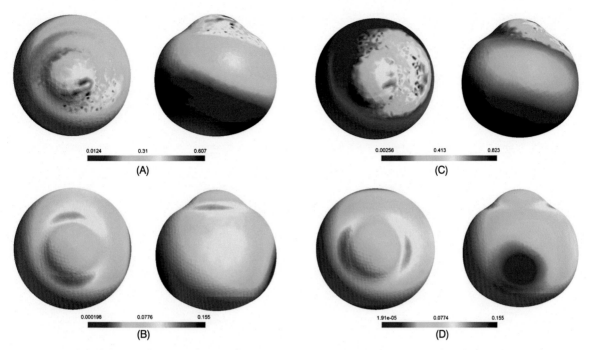

FIG. 6.9 Induced electric field due to 1 GHz EM wave on the surface of the eye (anterior and top view). (Top row) Compound eye model: (A) horizontal polarization and (C) vertical polarization. (Bottom row) Extracted eye model: (B) horizontal polarization and (D) vertical polarization. From [19].

to being different for both polarization, again, contrary to the extracted model where the same values were obtained. These results suggest a significant influence of the neighboring head tissues on the induced electric field values.

A similar result was obtained for the 1800 MHz case, as shown on Fig. 6.10, although the discrepancy in the maximum values between the two models is much less.

It should also be noted that the extracted eye model gave the highest values in the posterior parts denoting sclera, while the compound model obtained a similar trend at both frequencies, i.e., the highest obtained values were in the cornea.

More details on the induced field distribution inside the eye can be seen on Figs. 6.11 and 6.12, where the results are given on the transverse cross-sections and along the eye visual axis, respectively.

The results from an earlier study [30], obtained for vertically polarized wave at 1 GHz, showed similar distribution of the induced electric field along the pupillary axis in both compound and extracted eye models, as depicted again on Fig. 6.12A, suggesting the usefulness of the extracted eye model in the initial EM exposure assessment. However, the results for horizon-

tal polarization showed more significant discrepancy. Moreover, as is obvious from Fig. 6.12B, it is even more pronounced at 1800 MHz, for both polarizations.

One interesting fact that can be seen from Fig. 6.12 is that, due to the symmetrical nature of the employed eye model, the extracted model basically gives the same results for both polarizations, as shown by almost ideally overlapping graphs, suggesting that this model does not discriminate between the two polarizations of the incident plane wave.

The corresponding SAR results are shown on Fig. 6.13. Again, an asymmetric distribution of SAR in the compound eye model, contrary to the symmetric distribution obtained in the extracted eye model, will result in a somewhat different distribution of SAR along the eye pupillary axis, as shown on Fig. 6.13.

6.1.5 The Brain Exposure

Before dealing with a more detailed models of the human head, it is important to emphasize that it is reasonable to start the dosimetric assessment using a more simple model such as a homogeneous brain model [31]. This section deals with an efficient dosimetry procedure for the assessment of the human brain exposed to

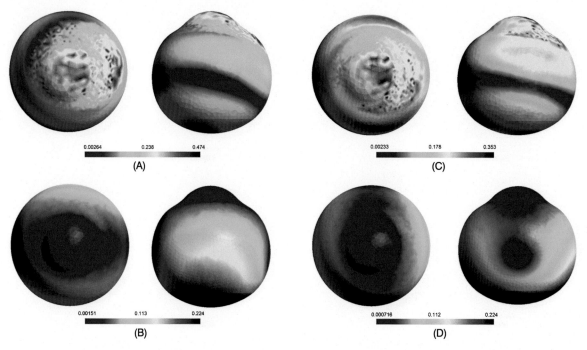

FIG. 6.10 Induced electric field due to 1800 MHz EM wave on the surface of the eye (anterior and top view). (Top row) Compound eye model: (A) horizontal polarization and (C) vertical polarization. (Bottom row) Extracted eye model: (B) horizontal polarization and (D) vertical polarization. From [19].

plane wave radiation. The formulation is based on the use of SIE approach, given in Sect. 6.1.1. The numerical solution to the corresponding set of coupled integral equations are carried out via MoM scheme, outlined in Sect. 6.1.1.1.

6.1.5.1 Model of the Human Brain

As a template, the brain model available from the Google Sketchup, as shown on Fig. 6.14A is used in this work. This model is then imported into open source program called MeshLab [32], where meshing using the triangular elements has been performed, as shown on Fig. 6.14B. Discretizing the brain surface in this manner enables one to use the Rao–Wilton–Glisson (RWG) basis functions [8], specially developed for triangular patches. The final geometrical model of the brain used in dosimetry assessment is shown on Fig. 6.14C. The dimensions of an average adult human brain are used (length 167 mm, width 140 mm, height 93 mm, volume of 1400 cm^3) [33].

The surface of the brain is first discretized using the $N_{tri} = 696$ triangular elements and $N_{edg} = 1044$ edge elements, according to [7]. The entire interior of the brain is then discretized by the $N_{tet} = 1871$ tetrahedral

elements using the existing triangles as faces to these tetrahedra.

The presented model neglects the complex foldings of the brain cortex by flattening in the mesh model. Although the SIE approach could be applied in the analysis of a more detailed brain model, e.g., developed from magnetic resonance images, this would place a significant computational burden due to a large number of elements required for the discretization of the brain surface.

6.1.5.2 Human Brain Exposed to Plane Wave

The numerical results for a homogeneous brain model exposed to plane wave radiation are presented next.

The frequency dependent parameters of the human brain are taken from [20] as average values between white and gray matter. The brain parameters for two particular frequencies are shown in Table 6.3.

The incident plane wave with power density of $P = 5$ mW/cm^2 is incident perpendicular to the right side of the brain (positive x coordinate). The frequency of the incoming EM wave is 900 MHz and 1800 MHz. Both horizontal (y coordinate) and vertical (z coordinate) polarizations of the plane wave are considered.

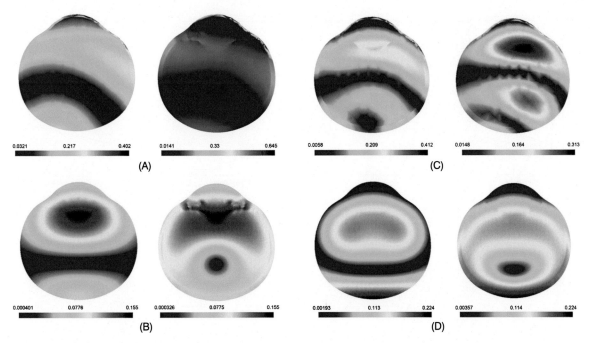

FIG. 6.11 Induced electric field in the transverse cross-section of the compound eye model (top row), and the extracted eye model (bottom row). Incident EM wave of: (A)–(B) 1 GHz horizontal (left) and vertical (right) polarization, (C)–(D) 1800 MHz horizontal (left) and vertical (right) polarization. From [19].

TABLE 6.3
Homogeneous human brain parameters.

Human brain parameters			
900 MHz		**1800 MHz**	
ε_r	σ [S/m]	ε_r	σ [S/m]
45.8	0.76	43.5	1.15

The electric field inside the human brain can be determined using

$$\vec{E}_2(\vec{r}) = -j\omega\mu_2 \int_S \vec{J}(\vec{r}')G_2(\vec{r},\vec{r}')\,dS'$$
$$-\frac{j}{\omega\varepsilon_2}\int_S \nabla_S' \cdot \vec{J}(\vec{r}')G_2(\vec{r},\vec{r}')\,dS'$$
$$-\int_S \vec{M}(\vec{r}') \times \nabla G_2(\vec{r},\vec{r}')\,dS', \qquad (6.59)$$

where ε_2 is the permittivity of the brain and μ_2 is the brain permeability, assumed to be that of the free-space.

Figs. 6.15–6.18 show the results for the distribution of electric and magnetic fields, respectively, on the surface of the brain, due to incident plane wave obtained using the SIE formulation.

Furthermore, from the electric field values in the brain interior, SAR can be determined using the following relation:

$$\text{SAR} = \frac{\sigma}{2\rho}|\vec{E}|^2, \qquad (6.60)$$

where σ and ρ are the electric conductivity and brain tissue density, respectively.

The obtained maximum and average SAR values for the four cases of incidence are shown in Table 6.4.

The calculated results show that the maximum SAR value in all four cases does not exceed the limit set by ICNIRP [24] as a basic restriction for localized SAR (in the head and trunk), for the occupational exposure (10 W/kg). On the other hand, the limit for the general public exposure has been exceeded in case of 1800 MHz (both polarizations).

Regarding the whole brain average SAR value, it can be observed from Table 6.4 that for a 1800 MHz, horizontally polarized wave, the calculated value exceeds 0.4 W/kg, the basic restriction for the professional pop-

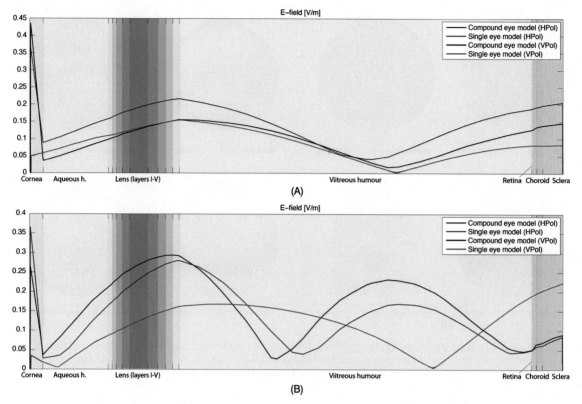

FIG. 6.12 Comparison of the induced electric field along the pupillary axis of the compound and the extracted eye models, respectively, for both polarizations: (A) 1 GHz, (B) 1800 MHz. From [19].

TABLE 6.4
Comparison of maximum and average SAR values in the human brain considering the four exposure scenarios.

	H		V	
	900 MHz	1800 MHz	900 MHz	1800 MHz
SAR$_{max}$ [W/kg]	0.85	4.39	0.86	2.68
SAR$_{avg}$ [W/kg]	0.17	0.41	0.16	0.35

ulation. However, the power density ($P = 5$ mW/cm^2) used in the calculation is higher than the reference level proposed for this frequency ($f/40$) by ICNIRP [24].

Figs. 6.19 and 6.20 show the effects of incident plane wave polarization on the obtained SAR in the homogeneous brain model due to 900 MHz and 1800 MHz wave, respectively.

Fig. 6.21 shows maximum and average SAR values for both polarizations, in case when the frequency of the incident plane wave is varied between 100 MHz and 2.45 GHz.

From Fig. 6.21 it is evident that above 900 MHz, homogeneous brain maximum SAR is higher for the vertical polarization, while SAR averaged over whole brain is similar for both polarizations [34].

The study on the influence of the wave polarization on whole-body SAR [35] for frequencies above 2 GHz showed higher values for the horizontally polarized plane wave compared to the vertically polarized one. The same study attributed this result to the component of the surface area perpendicular to the incident electric field.

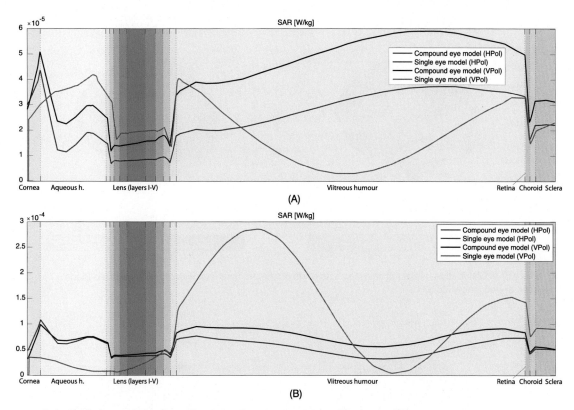

FIG. 6.13 Comparison of specific absorption rate along the pupillary axis of the compound and the extracted eye models, respectively, for both polarizations: (A) 1 GHz, (B) 1800 MHz.

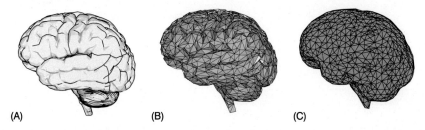

FIG. 6.14 Development of a homogeneous model of a human brain: (A) detailed three-dimensional model from Google Sketchup, (B) discretization of brain surface using the triangular elements done in MeshLab, (C) final geometrical model imported into MATLAB.

Similar conclusion could be drawn from Fig. 6.21, however, in this case, due to greater brain dimension in transversal plane, this result is obtained for the vertically polarized wave.

6.1.5.3 Child Brain Exposed to Plane Wave

The currently prescribed safety limits for electromagnetic radiation exposure are given by the International Commission on Non-Ionizing Radiation Protection (ICNIRP) [24]. Although based on a large body of a verifiable, published scientific work, some controversies still remain [36]. One of controversies is due to the claim that the amount of EM energy absorbed in the head of a child when using mobile device is higher than in adult individuals [37–40].

Moreover, there is a concern that existing methods for determining SAR in the child head, based on the use of adult head model, are inadequate because of a

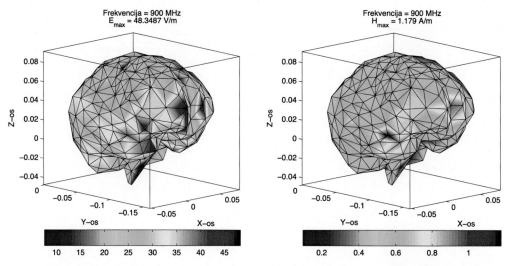

FIG. 6.15 Distribution of electric and magnetic fields on the brain surface. Horizontally polarized plane wave of frequency 900 MHz.

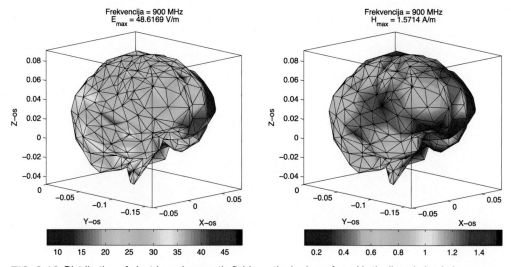

FIG. 6.16 Distribution of electric and magnetic fields on the brain surface. Vertically polarized plane wave of frequency 900 MHz.

smaller size of a child head [41]. On the other hand, if simply scaled adult models are used, anatomical inaccuracies may result. Anatomically correct models of a child's head based on MRI data should be used when available, although many studies using scaled models are still consistent with anatomically correct ones [36].

Interested readers may find many studies on the differences of absorbed EM energy in adult and child models [36,41–47], mainly based on the use of finite difference methods.

This section compares the absorbed electromagnetic energy in the adult and two scaled child brains obtained using the model based on the surface integral equation formulation.

The models of a 10- and 5-year old child brain, respectively, are obtained by linearly scaling an adult brain, as this is the most easily implemented approach. According to [41], the brain models of a 10-year old (10-yo) and 5-year old (5-yo) child, respectively, are obtained by scaling factors of 0.805 and 0.693, respec-

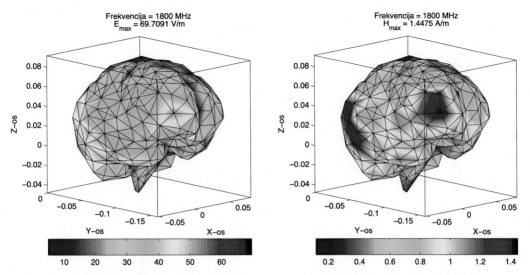

FIG. 6.17 Distribution of electric and magnetic fields on the brain surface. Horizontally polarized plane wave of frequency 1800 MHz.

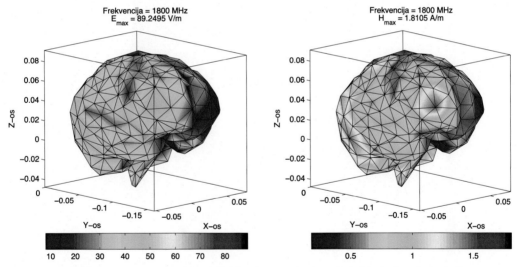

FIG. 6.18 Distribution of electric and magnetic fields on the brain surface. Vertically polarized plane wave of frequency 1800 MHz.

tively, in the horizontal (XY) plane, while the vertical (Z) axis is scaled by factors of 0.782 and 0.635, respectively. The applied scaling factors and the dimensions of the models are given in Table 6.5, while Fig. 6.22 depicts a comparison between the three models.

Although the child's brain is not a scaled version of the adult brain, because the surrounding tissues such as the skin or the skull develop at a different rate [46], it can provide some insights on the sensitivity of the re-

sults due to variability of the brain size. The question of assessing and managing these and other uncertainties is now recognized as the main challenge of the dosimetry [46].

The parameters of biological tissues significantly affect the absorption of EM energy within these tissues.

It is well known that biological tissues contain a high proportion of water (TBW – Total Body Water content), which is reduced during the lifetime. This fact has led

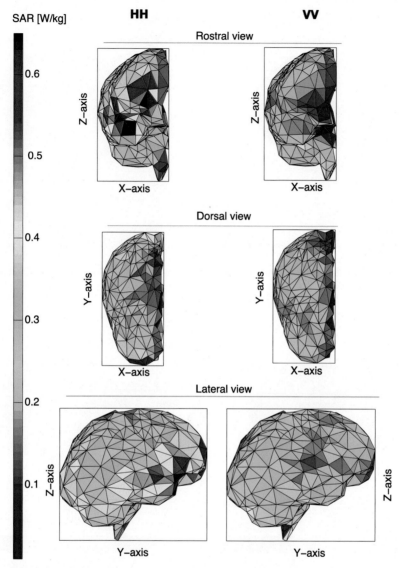

FIG. 6.19 Distribution of SAR for the case of horizontally and vertically polarized plane wave of frequency 900 MHz, power density $P = 5$ mW/cm^2.

TABLE 6.5
Dimensions of 10- and 5-year old child brain obtained by scaling the adult brain. The scaling factors according to [41].

	Adult	10-yo	scaling factor	5-yo	scaling factor
Width (X) [cm]	13.18	10.61	0.805	9.13	0.693
Length (Y) [cm]	16.11	12.97	0.805	11.17	0.693
Height (Z) [cm]	13.90	10.87	0.782	8.82	0.635

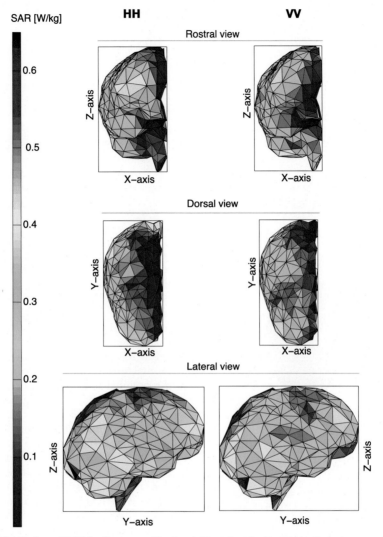

FIG. 6.20 Distribution of SAR for the case of horizontally and vertically polarized plane wave of frequency 1800 MHz, power density $P = 5$ mW/cm^2.

to the expressions that can evaluate the permittivity and conductivity of biological tissues depending on the age [48], i.e., on TBW content.

The following expression can be used to derive the complex permittivity of a child brain:

$$\varepsilon = \varepsilon_{rW}^{\frac{\alpha-\alpha_A}{1-\alpha_A}} \, \varepsilon_{rA}^{\frac{1-\alpha}{1-\alpha_A}} \left(1 - j\frac{1}{\omega\tau}\right), \qquad (6.61)$$

where ε_{rW} is the relative permittivity of water, ε_{rA} is the relative permittivity of adult tissue, while α_A and α are the tissue hydration rates of adult and child, respec-

tively, given by $\alpha = \rho \cdot$TBW, where ρ is the tissue density. The proposed fitting function for TBW is given by [48]:

$$\text{TBW} = 784 - 241e^{-\left(\frac{\ln(\text{Age}/55)}{6.9589}\right)^2}. \qquad (6.62)$$

More details can be found in [48].

The parameter values of two child brains taking into account age dependence via TBW are given in Table 6.6.

The results for the absorbed electromagnetic (EM) energy in brain models of adult and child are obtained in case of two wave polarizations (vertical = TM,

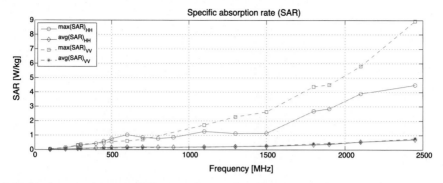

FIG. 6.21 Maximum and average SAR, respectively, versus frequency of incident plane wave, in case of horizontal (HH) and vertical (VV) polarization.

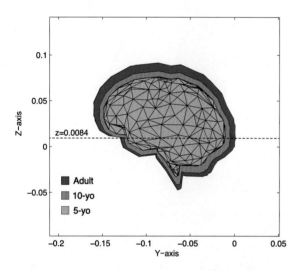

FIG. 6.22 Comparison of adult and 10- and 5-year old brain models. The transversal plane $z = 0.00836$ cm, where SAR is sampled, is indicated by a dashed line.

TABLE 6.6
Age related brain parameters based on total body water content (TBW), according to [48]. The adult human brain permittivity and conductivity [20], as found as average between the white and gray matter.

		ε_r	σ [S/m]
900 MHz	Adult	45.805	0.765
	10-yo	46.750	0.760
	5-yo	47.650	0.795
1800 MHz	Adult	43.545	1.1531
	10-yo	44.443	1.1456
	5-yo	45.302	1.1984

horizontal = TE) of 900 and 1800 MHz [1]. The power density of impinging plane wave is $P = 5$ mW/cm^2, and wave is directed perpendicular to the right side of the brain (positive x coordinate).

Fig. 6.23 shows the electric field distribution on the surface of three models, obtained for a 1800 MHz, horizontally polarized incident wave.

In all three models, the highest field values are obtained at the side where the EM wave is incident. At horizontal polarization, the highest values are obtained in the occipital and frontal lobe regions in all three models. In these regions, the highest values for SAR are expected, as well as the related temperature rise.

The distribution of SAR along the transversal plane of brain, along the x- and y-axis is shown on Fig. 6.24, for a 1800 MHz, vertically polarized wave.

A comparison of the maximum values for the electric field, peak and average SAR, maximum obtained temperature, as well as temperature increase in the adult and 10- (10-yo) and 5 year-old (5-yo) brain models is given in Table 6.7.

Table 6.7 shows that maximum and average SAR values are obtained in smaller brain models at 900 MHz with a horizontally polarized wave. In case of vertical polarization at 900 MHz, the highest average and maximum SAR values are obtained in the 10-year old and adult brain models, respectively.

At 1800 MHz vertical polarization, the maximum and average SAR values decrease in smaller models, while at horizontal polarization, the highest maximum and average SAR is obtained in the adult brain.

A similar analysis of the head exposed to mobile phone near field showed maximum values of SAR averaged over 1 g of tissue in 5- and 8-year old models, which were two times higher compared to the adult [46]. On the other hand, the review from [36] does not support these conclusions and leaves the question open.

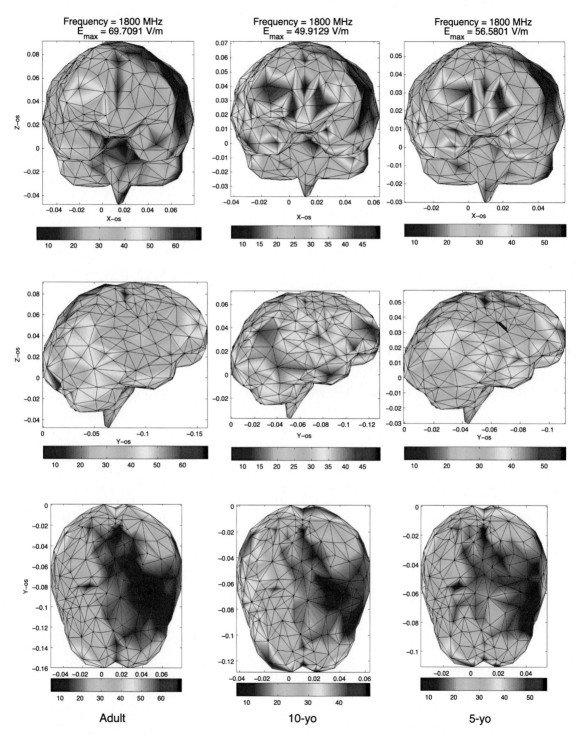

FIG. 6.23 Comparison of the induced electric field on the surface of adult and 10- and 5-year old brain models for a 1800 MHz, horizontally polarized incident wave. Top row shows front view, middle row presents right side view, and bottom row illustrates top view.

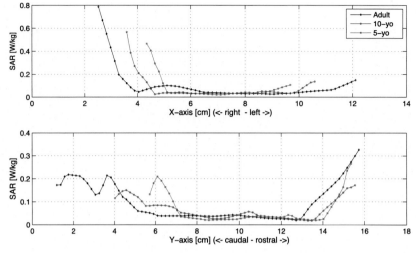

FIG. 6.24 SAR distribution in the transversal plane $z = 0.00836$ cm in three brain models, along x-axis (top), and y-axis (bottom); a 1800 MHz, vertically polarized wave is considered.

TABLE 6.7
Maximum electric field values, maximum and average SAR, in the adult, 10-yo, and 5-yo brain. Frequency of the incident field is 900 and 1800 MHz, both polarizations.

		E [V/m]	SAR_{max} [W/kg]	SAR_{avg} [W/kg]
900 MHz	Adult	48.349	0.856	0.174
	10-yo	51.835	0.976	0.213
TE	5-yo	61.314	1.428	0.248
900 MHz	Adult	48.617	0.866	0.158
	10-yo	46.475	0.784	0.173
TM	5-yo	44.666	0.758	0.163
1800 MHz	Adult	69.709	2.678	0.348
	10-yo	49.913	1.364	0.295
TE	5-yo	56.580	1.833	0.303
1800 MHz	Adult	89.249	4.390	0.411
	10-yo	74.374	3.029	0.302
TM	5-yo	67.529	2.612	0.282

From the results shown in Table 6.7, it is evident that incident wave polarization plays a role. However, it is difficult to draw a firm conclusion, as different trends were obtained at 900 and 1800 MHz.

Although the anatomical accuracy of these models is low compared to MRI based child models, they may be found useful for rapid dosimetric assessment [1]. How-

ever, as the surrounding tissues such as of skull and scalp may influence the distribution of the induced electric field and the related SAR, it is important to take into account the rest of the head tissues, as will be done next.

6.1.6 The Human Head Exposure

The results of the dosimetric assessment of the human head exposed to high frequency (HF) electromagnetic radiation are presented next [49]. The results are obtained using the hybrid finite element/boundary element method (FEM/BEM) approach [13]. The hybrid method, combining the ability of FEM to deal with arbitrary material properties, and to accurately model curved geometries, and the BEM suitability for the open boundary problems, has been proved to be highly desirable for the accurate characterization of the biological interactions of EMFs [12] compared to the already existing and well-established approaches [7,9].

The model of the human head including the various head tissues is shown on Fig. 6.25. The model was constructed from magnetic resonance imaging (MRI) of a 24-year old male [50]. The current implementation of the model consists of 8 tissues, whose frequency-dependent dielectric parameters, modeled using the 4-Cole–Cole method [20] are given in Table 6.8. The boundary surface of the head model is discretized using 6,838 triangular elements, while the interior domain of the head is discretized using 1,034,641 tetrahedral elements.

The computational examples are given for the human head exposed to an incident plane wave of 900 and

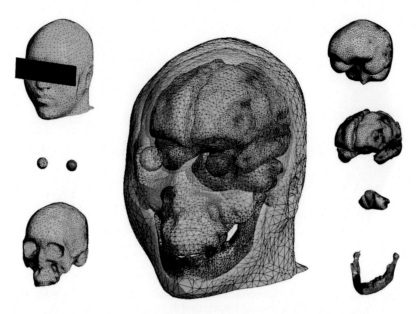

FIG. 6.25 Model of the human head with insets showing various tissues.

TABLE 6.8
Tissue dielectric parameters at 900 and 1800 MHz.

Tissue	900 MHz		1800 MHz	
	σ [S/m]	ε [−]	σ [S/m]	ε [−]
Brainstem	0.591	38.886	0.915	37.011
Cerebellum	1.263	49.444	1.709	46.114
Eye (vitreous)	1.636	68.902	2.032	68.573
Head skin	0.867	41.405	1.185	38.872
Skull and mandible	0.339	20.788	0.588	19.343
Gray matter	0.942	52.725	1.391	50.079
Muscle tissue	0.943	55.032	1.341	53.549

1800 MHz [49]. Both horizontal and vertical polarizations are considered. The incident field $\vec{E}_{inc} = 1$ V/m, directed toward the nose, is perpendicular to the head coronal cross-section. Numerical results for the horizontal and vertical polarization are denoted by HP and VP, respectively.

The illustrative results for the electric field induced on the surface of the head model are shown on Fig. 6.26 for a 900 MHz horizontally polarized EM wave. The highest value of the electric field is obtained at the tip of the nose.

Figs. 6.27–6.29 show the results for the induced field on the sagittal, coronal, and transverse cross-sections of the head, respectively, due to 900 and 1800 MHz, vertically and horizontally polarized EM wave, respectively.

As evident from Figs. 6.27 and 6.29, the highest values of the induced field are in regions around the nose and eyes, suggesting possible formation of hot-spots. According to international guidelines [51], next to the human brain, the temperature elevation in the human eye due to absorbed EM energy is listed as one of the most important concerns.

Finally, considering the influence of polarization of the incident electromagnetic wave on the induced field, the results show somewhat higher values for the vertical polarization. The study in [35] attributed this fact to

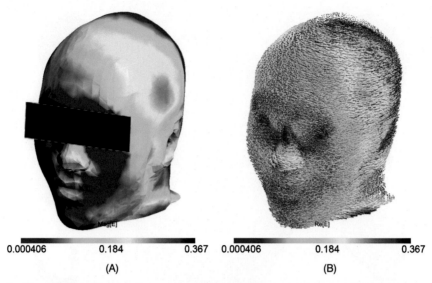

| 0.000406 | 0.184 | 0.367 | 0.000406 | 0.184 | 0.367 |

(A)   (B)

FIG. 6.26 Electric field induced on the surface of the human head model due to 900 MHz horizontally polarized plane wave (A); vectors of the induced electric field (B).

TABLE 6.9
Tissue parameters used in the three-compartment model.

Tissue	900 MHz		1800 MHz		ρ [g/m³]
	σ [S/m]	ε [–]	σ [S/m]	ε [–]	
Scalp	0.899	40.936	1.185	38.872	1050
Brain	0.985	52.282	1.391	50.079	1039
Skull	0.364	20.584	0.588	19.343	1900

the component of the surface area perpendicular to the incident electric field. The results obtained in this work suggest the same.

6.1.6.1 Brain Dosimetry Comparison

The exposure of a human brain to high frequency EM radiation is considered again, using three different models, as shown on Fig. 6.30.

The numerical results are obtained using a detailed head model introduced in Sect. 6.1.6, comprising 7 tissues with additional 16 ocular tissues from the extracted eye model given in Sect. 6.1.4. The head tissue parameters are also modeled using the 4-Cole–Cole method [20] and are given in Table 6.8. The boundary surface of the complete head model is discretized using 5,934 triangular elements, while the region inside the head is discretized using 4,073,250 tetrahedral elements. The numerical results are sought out using the hybrid FEM/BEM formulation.

The second model is a very simple representation of the human brain, consisting of a single, homogeneous structure, introduced in Sect. 6.1.5.1. Frequency dependent parameters of the homogeneous model are taken from [7]. In addition, linear and isotropic behavior is assumed for the electrical properties of tissues. The numerical solution of the SIE formulation based approach is found using an efficient MoM technique.

The final model is a so-called three-shell, or three-compartment, head model, being a sort of a compromise between the detailed and homogeneous models. This realistic head model is routinely used in experimental magnetoencephalography (MEG) [52], consisting of homogeneous compartments of brain, skull, and scalp, as depicted on Fig. 6.30B. The model is freely available and can be easily implemented in the BEM analysis [52]. The parameters of the three-shell model are given in Table 6.9.

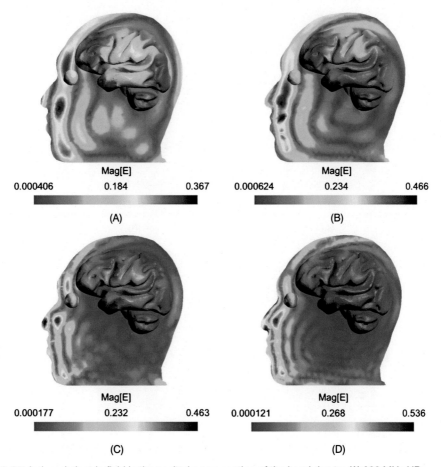

FIG. 6.27 Induced electric field in the sagittal cross-section of the head due to: (A) 900 MHz HP, (B) 900 MHz VP, (C) 1800 MHz HP, and (D) 1800 MHz VP.

The hybrid FEM/BEM method used for detailed model is also featured in the analysis pertaining to the three-compartment head model.

Four different scenarios are considered, taking into account two polarizations (vertical and horizontal) and two frequencies of the incident plane wave (900 and 1800 MHz). In each case the amplitude of the incident EM wave directed toward the anterior part of the brain/head models is 1 V/m.

The induced electric field on the brain surfaces of three models is shown on Figs. 6.31–6.34.

As evident from colormaps, similar electric field distribution was obtained, however, lower maximum values were obtained in all four cases in the homogeneous brain model compared to the other two brain models.

One drawback of the current implementation of the homogeneous model is a relatively low numbers of tri-angular elements used for brain surface discretization, namely, only $T = 696$ triangular elements and $N = 1044$ edge elements were used. The reason for such low numbers is that employed SIE formulation results in a fully populated matrices, thus preventing the current implementation from running on larger system matrices. As shown in [19], the low numbers of elements lead to some numerical artifacts evident as peaks on superficial areas of the model, shown on Figs. 6.35–6.38.

On the other hand, the two detailed models, showed similar distributions for induced surface fields, as evident on Figs. 6.31–6.34. Moreover, although the surface of a three-compartment model is rather smooth, due to lacking cortex details featured in the conformed model, similar maximum values were obtained.

More details on the distribution of the electric field along the sagittal axis of the three brain models, ob-

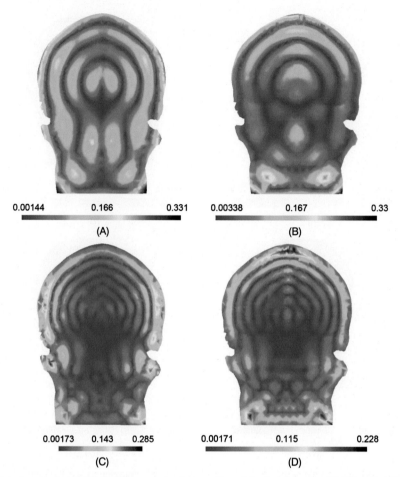

FIG. 6.28 Induced electric field in the coronal cross-section of the head due to: (A) 900 MHz HP, (B) 900 MHz VP, (C) 1800 MHz HP, and (D) 1800 MHz VP.

tained approximately at the medial prefrontal cortex, are seen on Figs. 6.35–6.38. Overall, similar trends were obtained in all models, however, only the detailed head model shows the wavelike behavior in the interior domain.

The results from both 1800 MHz cases showed that the three-compartment model obtained very uniform distribution of the field in the central part of the brain. Compared to this, the homogeneous model showed a field distribution in the central part of the brain, which was similar to the detailed head model.

6.1.7 The Whole Body Exposure

The electric field induced inside the human body exposed to the plane wave incident field can be evaluated using the tensor volume integral equation (6.39) [9].

The computational example is related to the whole body exposure to a base station antenna system mounted on the building roof at the height $h = 15$ m. The assessment is carried out for two frequency carriers, from the GSM and UMTS downlink bands. Thus, two frequencies of the incident wave are considered: 936 MHz of GSM, and 2140 MHz of UMTS.

A double layered three-dimensional (3D) model of an adult torso with a height of 1.7 m and a shape as shown in Fig. 6.39 is considered. Because of the symmetry, only one half of the body, constructed from 62 cubic cells, has to be examined.

The human being is assumed to be located approximately 24 m from the radiation source and illuminated by the antenna radiated electric field. The values of the electric field, obtained by narrowband measurements,

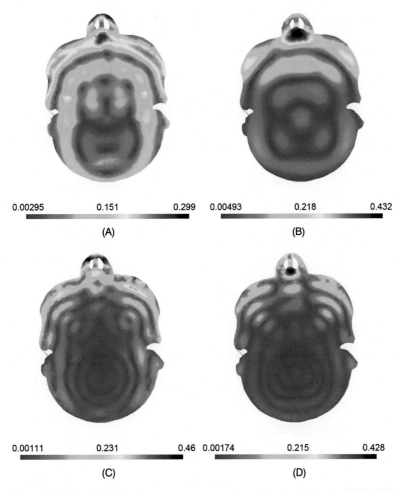

FIG. 6.29 Induced electric field in the transverse cross-section of the head due to: (A) 900 MHz HP, (B) 900 MHz VP, (C) 1800 MHz HP, and (D) 1800 MHz VP.

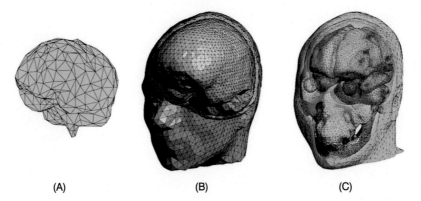

FIG. 6.30 Models for the brain comparison: (A) homogeneous brain model, (B) three-compartment head model, and (C) detailed head model. Overlay on the two latter models shows various head tissues surrounding the brain. From [19].

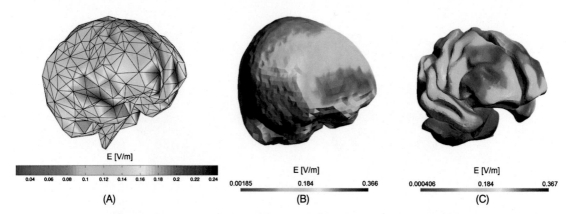

FIG. 6.31 Induced electric field on the surface of the brain due to 900 MHz horizontally polarized EM wave: (A) homogeneous model, (B) three-compartment model, and (C) detailed head model. From [19].

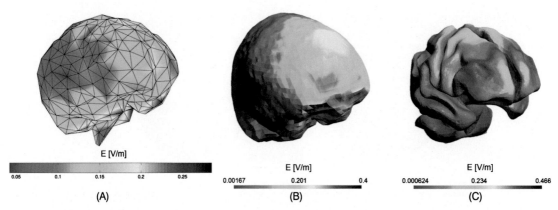

FIG. 6.32 Induced electric field on the surface of the brain due to 900 MHz vertically polarized EM wave: (A) homogeneous model, (B) three-compartment model, and (C) detailed head model. From [19].

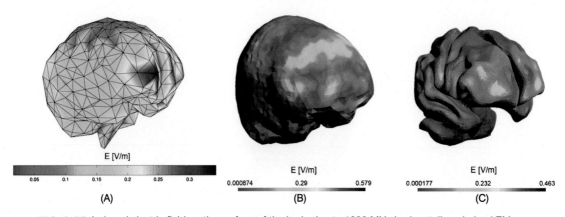

FIG. 6.33 Induced electric field on the surface of the brain due to 1800 MHz horizontally polarized EM wave: (A) homogeneous model, (B) three-compartment model, and (C) detailed head model. From [19].

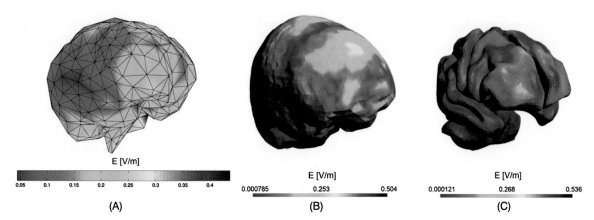

FIG. 6.34 Induced electric field on the surface of the brain due to 1800 MHz vertically polarized EM wave: (A) homogeneous model, (B) three-compartment model, and (C) detailed head model. From [19].

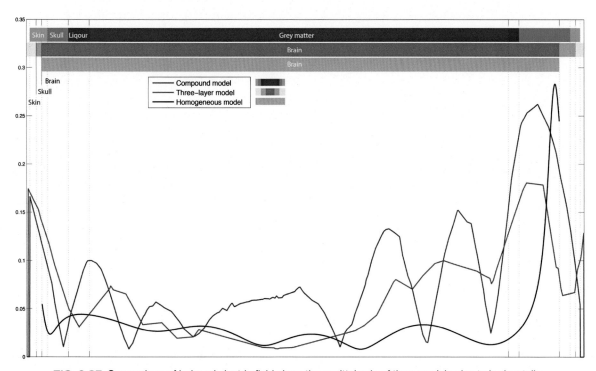

FIG. 6.35 Comparison of induced electric field along the sagittal axis of three models, due to horizontally polarized 900 MHz EM wave. From [19].

are used as an input data to the numerical model and are given in Table 6.10.

The electrical properties of the human body are given in Table 6.11. The parameter τ, determined from (6.40), at GSM and UMTS frequencies corresponds to $\tau = 1.4 + j2.8901$ and $\tau = 1.54 + j1.61978$, respectively.

The results for the specific absorption rate in the human body model are shown on Fig. 6.40.

A comparison between numerical and measured results is given in Table 6.12.

At GSM frequency, the maximum value of SAR is 0.064 µW/kg, occurring in the 18th cell, while the cal-

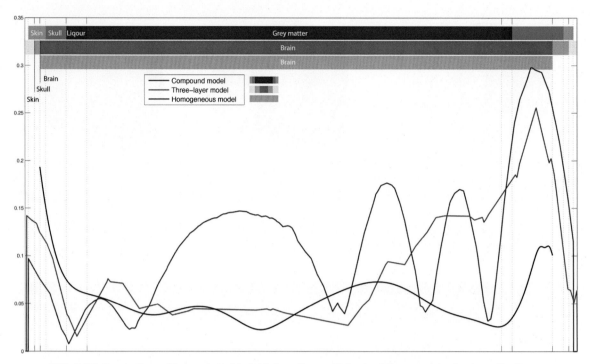

FIG. 6.36 Comparison of induced electric field along the sagittal axis of three models, due to vertically polarized 900 MHz EM wave. From [19].

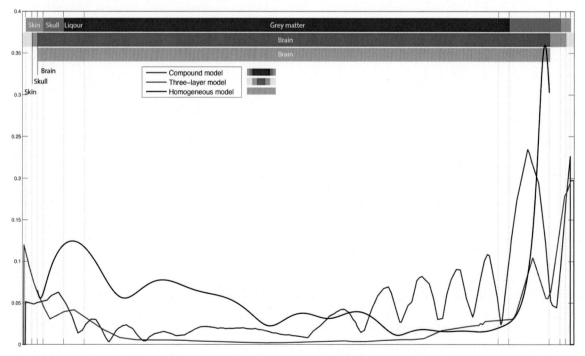

FIG. 6.37 Comparison of induced electric field along the sagittal axis of three models, due to horizontally polarized 1800 MHz EM wave. From [19].

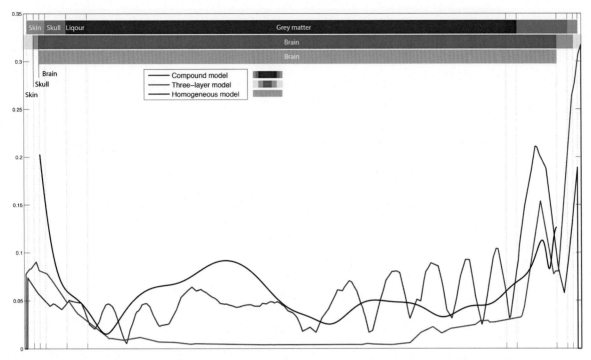

FIG. 6.38 Comparison of induced electric field along the sagittal axis of three models, due to vertically polarized 1800 MHz EM wave. From [19].

TABLE 6.10
The electric field obtained by narrowband measurements and the related calculated whole-body SAR.

Frequency MHz	E^{inc} (measured) [V/m]	SAR_{avg} (measured) [μW/kg]
936	0.141	0.0251
2140	14.873	267.5

TABLE 6.11
Electrical properties of the human body at two frequencies.

Frequency MHz	Permittivity (ε) [–]	Conductivity (σ) [S/m]
936	55	1.4
2140	53.11	1.54

culated average value is 0.0212 μW/kg. According to the ICNIRP Guidelines, the limit value for the whole-body average SAR for the general public exposure should not exceed 0.08 W/kg. As the whole-body average SAR provided by measurements [53] takes value of 0.0251 μW/kg, the results obtained via the numerical and the measurement techniques are shown to be in a very good agreement, as seen from Table 6.12.

At UMTS frequency, the maximum value of SAR is 0.31 mW/kg, in the region of the waist, whereas the calculated value of the whole-body average SAR is 82.4701 μW/kg, which stays far below the ICNIRP exposure limit. Local SAR in any cube is also below the exposure limits proposed by relevant international guidelines. Therefore the corresponding heating effects are negligible. A comparison of the numerical and measurement results is shown in Table 6.12. Taking into account the differences in the methodology of the numerical procedure and measurement technique, the obtained results are rather plausible.

6.2 THERMAL DOSIMETRY PROCEDURES

It is a well established fact that the dominant biological effect of high frequency EM radiation is essentially thermal in nature [54–56]. According to international

TABLE 6.12
Comparison of MoM results with the results obtained by measurements [53].

Frequency MHz	SAR$_{avg}$, measured [μW/kg]	SAR$_{avg}$, MoM [μW/kg]	SAR$_{max}$, MoM [μW/kg]
936	0.0251	0.0212	0.064
2140	267.5	82.47	314.38

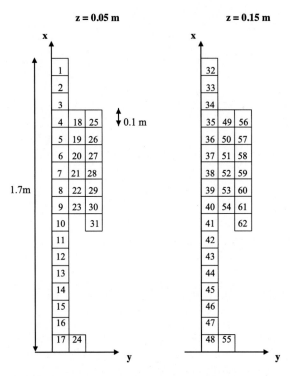

FIG. 6.39 Geometry and dimensions of one half of the human body model constructed from 62 cells.

standards/guidelines [51], the temperature elevation in organs such as human brain and eye is listed as one of the biggest concerns. The limit in HF region is thus prescribed in terms of specific absorption rate (SAR) as a surrogate of temperature elevation. According to [51], this limit is 2 W/kg for general population and 10 W/kg for professional exposure. As direct experimental measurements of thermal response on healthy human subjects is not possible, many computational studies aim to relate SAR and temperature elevation in the brain [57–60]. Moreover, the indirect methods such as MRI lack sufficient resolution necessary to record minute temperature variations, while animal studies, on the other hand, are questionable due to species difference. Hence, computational modeling emerges as a power-

ful alternative for a study of body thermal response, especially when it is important to take into account individual variability of induced physical quantities [24].

6.2.1 Bioheat Transfer Equation

From a thermodynamics point of view, the two most important factors for sustaining a biological system are metabolism and the blood flow [61]. These two factors, in addition to the complex network of blood vessels, appreciably complicate the mathematical modeling of heat transfer in biological tissues, unless a distributed heat source or sink is assumed.

Thus, due to the importance of the overall heat transfer via bloodstream, the main goal is to include the effects of blood flow. The most commonly used model taking this into account is the so called Pennes' or bioheat transfer equation [62], whose derivation is given in the following.

6.2.1.1 Heat Conduction Equation

Heat conduction through a medium is described by Fourier law:

$$\vec{q} = -\lambda \nabla T, \tag{6.63}$$

where $\vec{q}$ is the heat flow density, T denotes a temperature, and λ [J/kg/°C] is the heat conductivity of the medium.

The heat through surface element dS in differential dt is:

$$Q = \int_t \int_S \vec{q}\, d\vec{S}\, dt - \int_t \int_S \lambda \nabla T d\vec{S}\, dt. \tag{6.64}$$

Now, Gauss divergence theorem yields

$$Q = -\int_t \int_S \lambda \nabla T\, d\vec{S}\, dt = -\int_t \int_S \nabla \cdot (\lambda \nabla T)\, dV\, dt. \tag{6.65}$$

6.2.1.2 Heat Conduction Equation

The differential of total heat delivered into a volume V, due to internal and external heat sources in time differ-

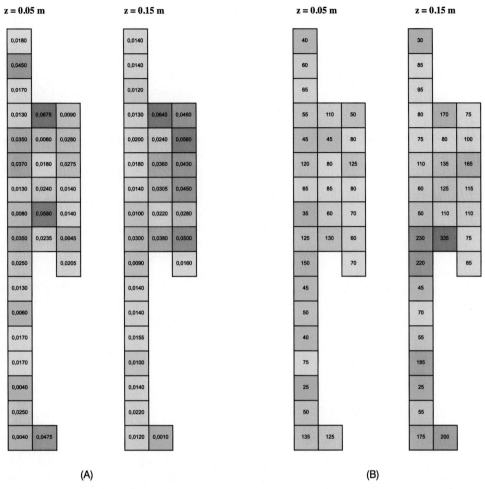

FIG. 6.40 Specific absorption rate [μW/kg] in the human body model determined at $z = 0.05$ m and $z = 0.15$ m due to incident field at: (A) 936 MHz and (B) 2140 MHz.

ential dt, is given by

$$dQ_{tot} = dQ_{int} + dQ_{ext}. \qquad (6.66)$$

The differential of internal heat sources can be written as

$$dQ_{int} = Q_i \, dV dt. \qquad (6.67)$$

The differential of external heat sources is

$$dQ_{ext} = \nabla \cdot (\lambda \nabla T) \, dV dt. \qquad (6.68)$$

Finally, the differential of total heat is

$$dQ_{tot} = \rho C \frac{\partial T}{\partial t} dV dt. \qquad (6.69)$$

Combining previous equations, one obtains

$$\rho C \frac{\partial T}{\partial t} dV dt = Q_i \, dV dt + \nabla \cdot (\lambda \nabla T) \, dV dt. \qquad (6.70)$$

Taking spatial and temporal integration yields

$$\rho C \int_t \int_S \frac{\partial T}{\partial t} dV dt = \int_t \int_S Q_i \, dV dt \\ + \int_t \int_S \nabla \cdot (\lambda \nabla T) \, dV dt. \qquad (6.71)$$

Rearranging the integral expression

$$\int_t \int_S \left[\rho C \frac{\partial T}{\partial t} - Q_i - \nabla \cdot (\lambda \nabla T) \right] dV dt = 0, \quad (6.72)$$

the function under spatial-temporal integral can be referred to as Fourier heat conduction equation

$$\rho C \frac{\partial T}{\partial t} = Q_i + \nabla \cdot (\lambda \nabla T). \quad (6.73)$$

For stationary phenomena, (6.73) simplifies to Poisson equation

$$\nabla \cdot (\lambda \nabla T) = Q_i. \quad (6.74)$$

For source-free areas, the latter equation becomes

$$\nabla \cdot (\lambda \nabla T) = 0. \quad (6.75)$$

Expression (6.75) is referred to as Laplace equation.

6.2.1.3 Convective Heat Transfer

Convection refers to heat transfer between a surface and a fluid. Convective heat transfer between a fluid at temperature T_e flowing at velocity v and a surface at temperature T_s is depicted in Fig. 6.41.

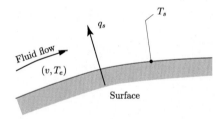

FIG. 6.41 Convective heat transfer.

In an external forced flow, the rate of heat transfer is approximately proportional to the difference between the surface temperature T_s, and the temperature of the free stream fluid T_e.

Therefore, the heat flux density q can be expressed as

$$q_s = -\lambda \frac{\partial T}{\partial n} = h_c (T_s - T_a), \quad (6.76)$$

which represents Neumann boundary condition for the air–body interface.

The factor of proportionality h_c is called the convective heat transfer coefficient [W/m²K].

6.2.1.4 Pennes' Equation

Volume blood perfusion of tissue Q_b can be expressed as

$$Q_b = W_b C_{pb} (T_a - T), \quad (6.77)$$

where W_b is the volumetric blood perfusion rate and T_a is the arterial blood temperature.

Finally, the space-time bioheat transfer equation can be written in the form [62]

$$\nabla \cdot (\lambda \nabla T) + w \rho_b c_b (T_a - T) + Q_m + Q_{ext} = \rho C \frac{\partial T}{\partial t}. \quad (6.78)$$

According to (6.78), the temperature rise in the given volume of tissue is based on the energy balance between the conductive heat transfer, heat generated due to metabolic processes Q_m, heat loss (generation) due to blood perfusion, and influence of external heat sources Q_{ext}. Various modes of heat transfer in case of a human brain are illustrated on Fig. 6.42.

The volumetric perfusion rate is given by w, ρ_b and c_b are the density and specific heat capacity of blood, respectively, λ is the thermal conductivity of the tissue, while T_a is the temperature of the arterial blood.

The bioheat equation (6.78) is accompanied by the corresponding boundary conditions. Generally, two types of boundary conditions, Neumann and Dirichlet, or their combination, can be prescribed on the boundary.

The thermal steady-state phenomena can be analyzed via stationary form of the bioheat equation (6.78):

$$\nabla \cdot (\lambda \nabla T) + W_b C_{pb} (T_a - T) + Q_m + Q_{ext} = 0, \quad (6.79)$$

where Q_{ext} represents the electromagnetic power deposited in biological body, i.e., the amount of heat generated per unit time and per unit volume due to absorption of EM energy. This term is obtained from the electromagnetic modeling, by the following relation:

$$Q_{ext} = \frac{\sigma}{2} |\vec{E}|^2, \quad (6.80)$$

where E is the peak value of the electric field induced inside the body, and σ is the conductivity of the particular tissue.

The deposited power density is directly related to the specific absorption rate (SAR) as follows:

$$Q_{ext} = \rho \cdot SAR \quad (6.81)$$

where ρ is tissue density and SAR is defined as

$$SAR = \frac{\sigma}{2\rho} |\vec{E}|^2, \quad (6.82)$$

representing a standard measure in thermal dosimetry.

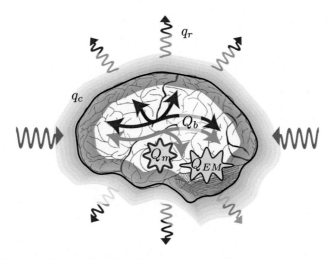

FIG. 6.42 Illustration of various modes of heat transfer in case of human brain: the heat generated due to metabolic processes Q_m, heat loss/generation due to blood perfusion Q_b, absorbed electromagnetic energy Q_{EM}, thermal radiation q_r, heat conduction q_c.

6.2.2 Finite Element Solution

The finite element formulation of (6.79) is based on the weighted residual approach. The approximate solution of (6.79) is expressed in terms of linear combination of known basis functions N_i and the unknown coefficients α_i,

$$T(x, y, z) = \sum_{i=1}^{m} N_i(x, y, z)\alpha_i, \qquad (6.83)$$

where i is the node index, m is the number of nodes per finite element, and N_i is the basis function given by

$$N_i(x, y, z) = \frac{1}{D}(V_i + a_i x + b_i y + c_i z), \quad i = 1, 2, 3, 4, \qquad (6.84)$$

where expressions for the coefficients a_i, b_i, c_i, V_i and D can be found in [63].

Multiplying (6.79) by weighting functions W_j, followed by integration over problem domain Ω, yields

$$\int_{\Omega} \left[\nabla \cdot (\lambda \nabla T) + W_b (T_a - T) + Q_m + Q_{ext} \right] W_j \, d\Omega = 0. \qquad (6.85)$$

Applying the same procedure to the natural boundary condition (6.116) gives

$$-\lambda \int_{\partial\Omega} \frac{\partial T}{\partial \hat{n}} W_j \, dS = \int_{\partial\Omega} h_s T W_j \, dS - \int_{\partial\Omega} h_s T_{amb} W_j \, dS. \qquad (6.86)$$

Integrating by parts the first term of (6.85), and using the Gauss divergence theorem, results in

$$\int_{\Omega} \nabla \cdot \left[(\lambda \nabla T) W_j \right] d\Omega = \lambda \int_{\partial\Omega} \frac{\partial T}{\partial \hat{n}} W_j \, dS. \qquad (6.87)$$

Inserting (6.87) into (6.85), after some mathematical manipulations, the integral formulation of Pennes' bioheat equation (6.79) is obtained [64], namely

$$\int_{\Omega} \lambda \nabla T \cdot \nabla W_j \, d\Omega + \int_{\Omega} W_b T W_j \, d\Omega + \int_{\partial\Omega} h_s T W_j \, dS$$
$$= \int_{\Omega} (W_b T_a + Q_m + Q_{ext}) W_j \, d\Omega + \int_{\partial\Omega} h_s T_{amb} W_j \, dS. \qquad (6.88)$$

Implementing the Galerkin–Bubnov procedure, i.e., choosing $N_i = W_i$, followed by the standard finite element discretization of (6.88), the weak formulation for the finite element domain Ω_e can be written in the matrix form as

$$[K]^e \{T\}^e = \{M\}^e + \{P\}^e, \qquad (6.89)$$

where $[K]^e$ is the finite element matrix

$$[K]^e_{ji} = \int_{\Omega_e} \lambda^e \nabla W_i \cdot \nabla W_j \, d\Omega_e + \int_{\Omega_e} W^e_b W_i W_j \, d\Omega_e,$$

(6.90)

$\{M\}^e$ is the flux vector on the boundary $\partial\Omega_e$ of the finite element

$$\{M\}^e_j = \int_{\partial\Omega_e} \lambda^e \frac{\partial T}{\partial \hat{n}} W_j \, dS_e,$$

(6.91)

and $\{P\}^e$ is the finite element source vector

$$\{P\}^e_j = \int_{\Omega_e} (W^e_b T_a + Q^e_m + Q^e_{ext}) W_j \, d\Omega_e.$$

(6.92)

The integrals (6.90)–(6.92), can be evaluated using the expressions found in Appendix D.

Solving (6.90)–(6.92) for each of N elements, the global matrix is assembled from the contribution of the local finite element matrices, while the global flux and source vectors are assembled from the local flux and local source vectors, respectively:

$$[K]\{T\} = \{M\} + \{P\}.$$

(6.93)

The solution to the matrix system (6.93) yields vector $\{T\}$ whose elements represent the temperature values at finite element nodes.

6.2.3 Boundary Element Solution

The steady state variant of Pennes' bioheat equation (6.79) is solved using a three-dimensional direct boundary element method (BEM) [65]. This approach features domain decomposition and use of a dual reciprocity method (DRM) for the integration of the source terms, according to [66].

The problem of steady state heat transfer is solved by finding the solution of the diffusion equation in the homogeneous medium with the corresponding boundary conditions and variable source terms:

$$\nabla \cdot (\lambda \nabla \phi) = \rho(\phi) \quad \text{in} \quad \Omega,$$

(6.94)

$$q = \bar{q} \quad \text{on} \quad \Gamma_1,$$

(6.95)

where the terms due to metabolic activity, electromagnetic power deposition, and due to the volumetric perfusion rate were all accounted by the generalized source term given by

$$\rho(\phi) = W_b C_{pb} \phi + Q_m + Q_{EM},$$

(6.96)

and the scalar field ϕ represents the variation of temperature given by

$$\phi = T - T_a.$$

(6.97)

Applying a weighted residual technique and integrating by parts, the following integral relation for an isolated subdomain is obtained:

$$c(\xi)\phi(\xi) + \int_{\Gamma_s} \phi \frac{\partial \phi^*}{\partial n} d\Gamma - \int_{\Gamma_s} \frac{\partial \phi}{\partial n} \phi^* d\Gamma = \int_{\Omega_s} \phi^* \rho \, d\Omega_s,$$

(6.98)

where

$$\phi^* = \frac{1}{4\pi R}$$

(6.99)

is the 3D fundamental solution of the Laplace equation

$$\nabla^2 \phi^* + \delta(\xi, y) = 0,$$

(6.100)

where $R(\xi, y)$ is the distance between the field point ξ and source point y, $\partial\phi^*/\partial n$ is the derivative in the normal direction to the boundary. The geometrically dependent free term, $c(\xi)$, is due to the Cauchy type singularity of the integral on the left-hand side of Eq. (6.98), e.g., $c = 1/2$ for a field node placed on a smooth boundary.

Discretizing expression (6.98) with N_e elements leads to the following equation:

$$c_i \phi_i + \sum_{j=1}^{N_e} \int_{\Gamma_{s,j}} \phi \frac{\partial \phi^*}{\partial n} d\Gamma - \sum_{j=1}^{N_e} \int_{\Gamma_{s,j}} \frac{\partial \phi}{\partial n} \phi^* d\Gamma = \int_{\Omega_s} \rho \phi^* d\Omega_s,$$

(6.101)

where i is the source point and $\Gamma_{k,j}$ is the jth boundary element of Ω_k.

The isoparametric elements with quadratic interpolation functions over triangular and quadrilateral elements have been used.

The number of degrees of freedom associated to a given subdomain s due to the pure boundary element is

$$N_{ds} = \sum_{k=1}^{N_e} n_k,$$

(6.102)

where n_k is the number of freedom nodes per boundary element ($n_k = 6$ for triangular elements, and $n_k = 9$ for quadrilateral elements).

Using quadratic elements, the potential or its normal derivative, at any point on the jth boundary element can be expressed in terms of their corresponding values at the n_k collocation nodes by means of the interpolation functions ψ_a:

$$\phi(\xi) = \sum_{a=1}^{n_k} \psi_a(\xi)\phi_a, \qquad (6.103)$$

$$\frac{\partial\phi(\xi)}{\partial n} = \sum_{a=1}^{n_k} \psi_a(\xi)\phi_a. \qquad (6.104)$$

The dimensionless coordinate ξ spans from the computational domain to the physical patch or boundary element.

The description of the elements used, and their corresponding shape and interpolation functions, can be found in [67].

To this point, the BEM approach for the Laplace part of the operator has been considered. The rest of the equation is integrated with the DRM approach in which the right-hand side of Eq. (6.101) is approximated, allowing to express the domain integral in terms of the information available at the boundary, and maybe in terms of a number L_s of some optional internal collocation nodes. This enables the domain integral to be expressed in terms of a boundary integral.

A series of particular solutions $\hat{u}_{ij}$ is used to relate the ith source node with $N_{ds} + L_s$ field points, by approximating the source term by:

$$\rho_i \approx \sum_{j=1}^{N_{ds}+L_s} \alpha_j f_{ij}, \qquad (6.105)$$

where α_j is unknown coefficients and f_{ij} is arbitrary approximating functions. The particular solution then follows from

$$\nabla^2 u_p^{ij} = f_{ij}, \qquad (6.106)$$

where f_{ij} is the radial basis functions with first order augmentation polynomials of the form

$$f_{ij} = r_{ij} + 1 + x + y \qquad (6.107)$$

and r_{ij} is the distance between source (i) and field (j) points [68,69].

Combining previous equations yields

$$\nabla^2\phi(x) = \sum_{j=1}^{N_{ds}+L_s} \alpha_j\left(\nabla^2\phi_p^{ij}(x,y)\right). \qquad (6.108)$$

Multiplying Eq. (6.108) with the fundamental solution of the Laplace operator and integrating leads to the following integral equation:

$$\sum_{j=1}^{N_{ds}+L_s} \alpha_j \int_\Omega \phi^* \nabla^2\phi_p^{ij}\, d\Omega = \sum_{j=1}^{N_{ds}+L_s} \alpha_j \int_\Omega \phi^* f_{ij}\, d\Omega. \qquad (6.109)$$

Applying Green's identity once again on Eq. (6.101) results in:

$$\begin{aligned} c_i\phi_i + \sum_{j=1}^{N_e}\int_{\Gamma_{k,j}} \phi\frac{\partial\phi^*}{\partial n}\,d\Gamma - \sum_{j=1}^{N_e}\int_{\Gamma_{k,j}} \frac{\partial\phi}{\partial n}\phi^*\,d\Gamma \\ = \sum_{l=1}^{N_{sd}+L_s} \alpha_l\left(c_i\phi_p^{il} + \sum_{j=1}^{N_e}\int_{\Gamma_{k,j}} \phi_p^{il}\frac{\partial\phi_i^*}{\partial n}\,d\Gamma \right. \\ \left. - \sum_{j=1}^{N_e}\int_{\Gamma_{k,j}} \frac{\partial\phi_p^{il}}{\partial n}\phi_i^*\,d\Gamma\right), \end{aligned} \qquad (6.110)$$

where k indexes the different subdomains, index j stands for the element adjacent to a given subdomain, and l lists the adjacent collocation nodes.

Furthermore, performing some straightforward mathematical manipulations, the following system of equations for each subdomain is obtained, expressed in matrix notation as:

$$[H]\{\phi\} - [G]\left\{\frac{\partial\phi}{\partial n}\right\} = \left([H]\{\phi\} - [G]\left\{\frac{\partial\phi}{\partial n}\right\}\right)[F]^{-1}\{\rho\}, \qquad (6.111)$$

where $[H]$ and $[G]$ are the BEM matrices defined by

$$[H] = h_{ij}^a = \int_{\Gamma_{k,j}} \psi_a\left(\frac{\partial\phi^*}{\partial n}\right)_j\,d\Gamma \qquad (6.112)$$

and

$$[G] = g_{ij}^a = \int_{\Gamma_{k,j}} \psi_a\,\phi^*\,d\Gamma, \qquad (6.113)$$

where $a = 1, \ldots, N_{ds} + L_s$ stands for the collocation nodes inside the jth field element ($j = 1, \ldots, N_e$), and $i = 1, \ldots, N_{ds} + L_s$ stands for the source point, while ϕ is the vector of potentials at the collocation nodes, and $\partial\phi/\partial n$ is the vector of normal derivatives of potentials at the collocation nodes.

The next step is to assemble the individual systems of equations provided by each subdomain.

The needed matching conditions are provided by the continuity of potential and normal current density, imposed at the interface between two adjacent subdomains, as shown on Fig. 6.43.

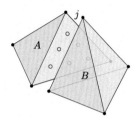

FIG. 6.43 Assembling subdomains A and B system of equations by imposing continuity of potentials and fluxes at the common interface j.

The matching between subdomains A and B, can be established via nodes $a = 1, \ldots, 6$, associated with the common element j according to

$$\phi_{jA}^{\alpha} = \phi_{jB}^{\alpha} \qquad (6.114)$$

and

$$\left(-\tau_A \frac{\partial \phi}{\partial n} \Big|^{\alpha}_j \right)_A = \left(\tau_A \frac{\partial \phi}{\partial n} \Big|^{\alpha}_j \right)_B . \qquad (6.115)$$

The assembled system of equations is similar to (6.111), except that the matrices [H] and [G] are now banded.

Finally, by imposing the global boundary conditions of the problem, a square system of m equations with m unknowns is formed, where $m = \sum_{s=1}^{N} N_{ds}$.

Each subdomain contributes with N_{ds} equations, and each collocation node carries two magnitudes, potential and flux. In case of large decomposition with tetrahedral elements, the relation between the total number of collocation nodes in the whole domain N_{fn} and the total number of collocation nodes at the boundary N_{fb} is $N_{fn} = 12m + N_{fb}/2$. From here it is clear that the total number of unknowns corresponds to the number of equations, thus the system becomes determined.

Using the domain decomposition technique, in addition to its capacity in dealing with piecewise homogeneous material properties, the resulting system of equations is sparse and highly banded, contrary to the fully populated dense system matrices resulting by the use of pure boundary element technique.

6.2.4 Computational Examples
6.2.4.1 Thermal Dosimetry for the Homogeneous Human Brain Model

The governing equation for heat transfer in a homogeneous human brain (6.79) is accompanied with two types of boundary conditions, i.e., natural and forced, prescribed on the brain surface, as shown on Fig. 6.44.

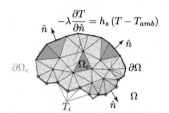

FIG. 6.44 Illustration of the finite element mesh with two types of boundary conditions on the brain surface.

If the heat transfer by radiation and forced convection is neglected, the heat exchange between the brain surface and environment can be satisfactorily described by

$$-\lambda \frac{\partial T}{\partial \hat{n}} = h_s \left(T - T_{\text{amb}} \right), \qquad (6.116)$$

where h_s is the convection coefficient between the surface and surroundings, while T and T_{amb} denote the surface and ambient temperature, respectively.

As heat transfer depends both on the distance from the heat source and the properties of the tissues, it is important to take this into account if a homogeneous brain model is considered. The actual human brain is separated from the head surface by the meninges, cerebrospinal fluid, skull, scalp muscles and a thin layer of fat, hence, to account for the heat exchange through all of these tissues, the effective thermal convection coefficient h_{eff} [70] between the brain and the surroundings is employed [2].

Finding the effective thermal coefficient h_{eff} usually involves determining the direct heat transfer coefficient beforehand, which is a rather demanding process. The interested reader can find more details in [70].

The widely adopted value for the human head heat transfer coefficient is $h = 40$ W/m$^{2\circ}$C, and the corresponding effective coefficient, typical for the human brain, is $h_{\text{eff}} = 12$ W/m$^{2\circ}$C [71], utilized in the homogeneous thermal model of the brain. An analysis of the impact of the effective heat transfer coefficient value on the temperature distribution is carried out in [2].

When forced boundary conditions are considered on the brain surface, $T = T_i$, the pertinent value is 35.4°C,

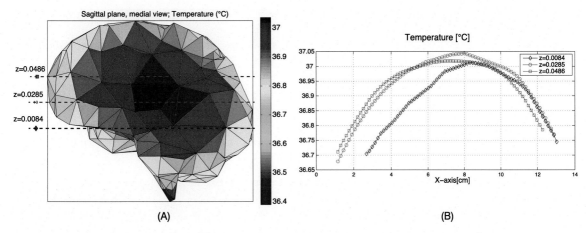

FIG. 6.45 Steady state temperature distribution, natural boundary conditions, $h_{eff} = 12$ W/m^2°C, $T_{amb} = 20$°C: (A) sagittal cross-section, (B) temperature distribution along the brain x-axis.

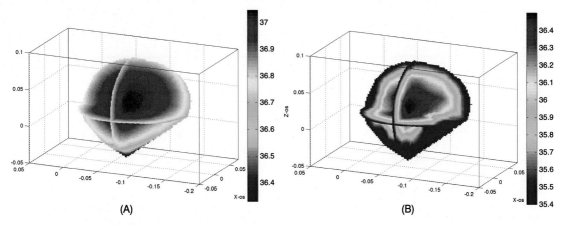

FIG. 6.46 Interpolated steady state temperature distribution in the three cross-sections obtained using: (A) Neumann and (B) Dirichlet boundary conditions.

as this is the average subdural brain temperature under the intact skull, according to [72].

The steady state temperature distribution on the sagittal cross-section is shown on Fig. 6.45.

From Fig. 6.45A, it is evident that the temperature at the center of the brain is almost uniform, while the temperature gradients are noticeable in the superficial layers near the brain surface.

A comparison of the interpolated results for the temperature on transverse, sagittal and coronal cross-sections of the brain, when natural and forced boundary conditions are respectively used, is given in Fig. 6.46.

Temperature profiles along the transversal x-axis of the brain are shown on Fig. 6.45B. The maximum temperature obtained in the brain center is around 37.05°C,

while near the surface the temperature drops to 36.7°C, corresponding to the measured steady-state temperature distribution [73] and to the results obtained from the models using more different tissues [72,74].

The current findings [72,73,75] indicate that there is a generally uniform temperature distribution within the brain, except in the superficial few millimeter thick layer, where the notable temperature gradients exist. Various studies [72,74,76] report a higher brain temperature compared to the core temperature, i.e., $T_a = 37$°C [77]. This 0.3–0.4°C difference is attributed to heat generated due to the metabolic processes [70,78]. The maximum temperature of 37.05°C is obtained using a homogeneous model, while the corresponding temperature difference is 0.35°C.

TABLE 6.13
Effects of the effective surface heat transfer coefficient h_{eff}.

h_{eff} W/m$^{2\circ}$C	T_{min} °C	T_{max} °C	T_{avg} °C	$T_{max} - T_{min}$ °C
2	37.0422	37.1669	37.1266	0.1248
4	36.8943	37.1423	37.0619	0.2480
12	36.3206	37.0459	36.8089	0.7254
20	35.7735	36.9526	36.5641	1.1792
50	33.9329	36.6279	35.7143	2.6949

TABLE 6.14
Effects of the ambient air temperature T_{amb}.

T_{amb} °C	T_{min} °C	T_{max} °C	T_{avg} °C	$T_{max} - T_{min}$ °C
10	35.8138	36.9611	36.5861	1.1473
15	36.0672	37.0035	36.6975	0.9363
20	36.3206	37.0459	36.8089	0.7254
25	36.5740	37.0884	36.9202	0.5144
30	36.8273	37.1308	37.0316	0.3034

The influence of the ambient temperature and heat transfer coefficient h, respectively, on the maximum, minimum and average calculated value of the steady-state temperature is shown in Tables 6.13 and 6.14.

Table 6.13 shows that an increased value of the effective heat transfer coefficient will result in the increased heat exchange between the environment and brain surface. This will result not only in an overall temperature decrease, but also in higher temperature gradients. In case of $h_{eff} = 50$ W/m$^{2\circ}$C, this gradient corresponds to 2.69°C, dropping to only 0.72°C when $h_{eff} = 12$ W/m$^{2\circ}$C is used.

Table 6.14 shows that only the superficial layers of the brain will be affected by the change in the ambient temperature T_{amb}, as evidenced from the obtained minimum temperature. The central regions of the brain are only slightly affected by the change of this parameter, as is evident from a small variation of the maximum temperature [2].

In case of human brain exposed to HF electromagnetic radiation, the absorbed energy will result in a corresponding temperature increase which will affect the steady state temperature field. These effects can be assessed using the thermal dosimetry model based on the finite element solution to the Pennes' equation (6.78).

The human brain thermal parameters are [79] as follows: heat conductivity $\lambda = 0.513$ W/m°C, volumetric perfusion rate of blood $W_b = 33297$ kg/m^3, specific heat capacity of blood $c_b = 1$ J/kg°C, heat generated due to metabolism $Q_m = 6385$ W/m^3, and arterial blood temperature $T_{art} = 37$°C.

Considering the exposure of the human brain to incident plane wave of 900 and 1800 MHz, carried out in Sect. 6.1.5.2, the results for the related temperature rise are shown on Figs. 6.47 and 6.48.

Figs. 6.47 and 6.48 show that at horizontal polarization (HH), higher temperature rise (due to higher SAR) is found in the occipital and frontal lobes (near longitudinal fissure), while at vertical polarization, the highest temperature rise is found in the cerebellum and the brain-stem region (lower brain parts) [34].

The maximum temperature increase for the four exposure scenarios is shown in Table 6.15.

The maximum temperature rise of $\Delta T = 2.09 \cdot 10^{-2}$°C is obtained in case of a 1800 MHz horizontally polarized wave. This temperature increase is rather negligible compared to the values proven to cause adverse health effects.

Fig. 6.49 shows the maximum and average temperature increase for both polarizations, in case when the frequency of the incident plane wave is varied between 100 MHz and 2.45 GHz.

Fig. 6.49 shows similar increasing trends both for the maximum and average temperature, in case of both polarizations of the incident plane wave [34].

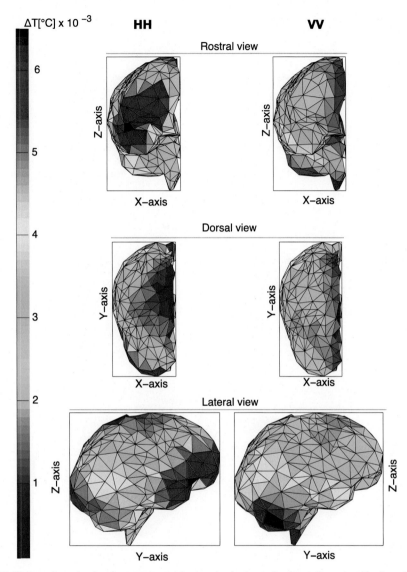

FIG. 6.47 Temperature rise in a homogeneous human brain due to horizontally and vertically polarized plane wave of frequency 900 MHz and power density $P = 5$ mW/cm^2.

Finally, the thermal dosimetry model of a homogeneous human brain can be used to assess the differences between the adult and child models. The temperature rise for the above-mentioned four exposure scenarios is calculated using scaled versions of the human brain, taking into account the age dependent electrical parameters of the brain, as explained in Sect. 6.1.5.3.

The temperature rise (compared to steady-state temperature) is computed in case of the adult and 10- and 5-year old brain. The results in the brain transversal plane, along x- and y-axis, are shown on Fig. 6.50.

The results shown on Fig. 6.50 are computed along the line denoted on Fig. 6.22.

A comparison of the maximum obtained temperature, as well as temperature increase in the adult, 10-year old (10-yo), and 5 year-old (5-yo) brain models, is given in Table 6.16.

Table 6.16 shows that the maximum temperature and the corresponding temperature rise decreases with

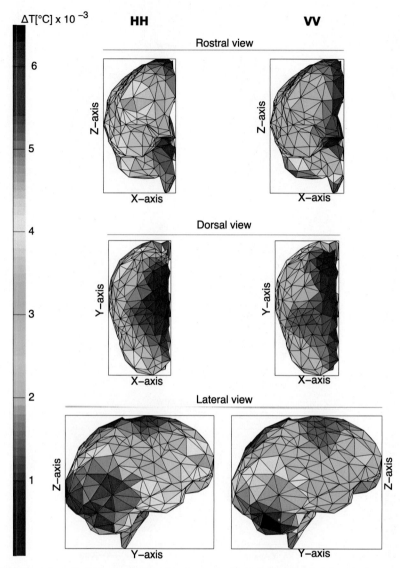

FIG. 6.48 Temperature rise in homogeneous human brain due to horizontally and vertically polarized plane wave of frequency 1800 MHz and power density $P = 5$ mW/cm^2.

TABLE 6.15
Comparison of maximum temperature increase in the human brain.

	Horizontal polarization		Vertical polarization	
	900 MHz	**1800 MHz**	**900 MHz**	**1800 MHz**
ΔT [°C]	$7.110 \cdot 10^{-3}$	$2.088 \cdot 10^{-2}$	$6.167 \cdot 10^{-3}$	$1.827 \cdot 10^{-2}$

the decrease in model size. In all four cases, the temperature increase is found to be negligible. However, since the human brain is not thermally isolated from the remaining head parts, the power absorbed in the

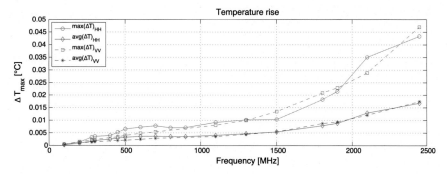

FIG. 6.49 Maximum and average temperature increase versus frequency of incident plane wave, in case of horizontal (HH) and vertical (VV) polarization.

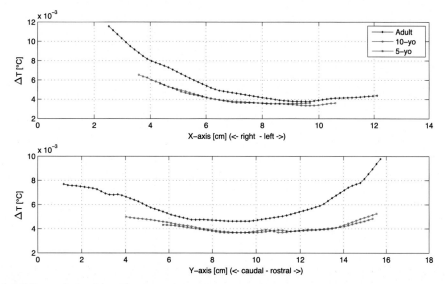

FIG. 6.50 Temperature rise ΔT in the transversal plane $z = 0.00836$ cm of three brain models, exposed to 1800 MHz, vertically polarized wave.

surrounding tissues may cause additional temperature elevation, as shown in [80] for the human eye exposure.

6.2.4.2 Temperature Increase in the Human Eye

The thermal analysis of the human eye exposed to high frequency electromagnetic field is based on the finite element solution to the stationary Pennes' bioheat equation [9,10]. The results for the specific absorption rate obtained using the hybrid boundary element/finite element method approach (Sect. 6.1.4.2) are utilized in the thermal model.

The incident plane wave is irradiating the eye from the corneal part. The human eye is discretized using 36,027 tetrahedral elements.

The eye model consists of 12 tissues, whose thermal parameters are given in Table 6.17 [81]. The density ρ of various eye tissues can be found in Table 6.1.

The natural boundary conditions are imposed on the boundary surface of the model

$$\lambda \frac{\partial T}{\partial \vec{n}} = -h_s (T - T_{amb}) \qquad (6.117)$$

where the convection coefficient between the eye surface and the surroundings is $h_s = 20$ W/m²°C. The ambient air temperature, denoted by T_{amb} is set to 22°C, while the body core temperature is 37°C [13,82].

Fig. 6.51 shows the temperature increase in the eye due to the induced SAR presented in Fig. 6.8.

TABLE 6.16
Maximum temperature and temperature increase in the model of adult, 10-yo, and 5-yo brain. Frequency of the incident field is 900 and 1800 MHz.

		T_{max} [°C]	ΔT [°C]
900 MHz	Adult	37.047	$7.110 \cdot 10^{-3}$
TE	10-yo	36.957	$8.048 \cdot 10^{-3}$
	5-yo	36.874	$8.261 \cdot 10^{-3}$
900 MHz	Adult	37.047	$6.167 \cdot 10^{-3}$
TM	10-yo	36.956	$5.539 \cdot 10^{-3}$
	5-yo	36.872	$4.757 \cdot 10^{-3}$
1800 MHz	Adult	37.050	$1.827 \cdot 10^{-2}$
TE	10-yo	36.958	$1.075 \cdot 10^{-2}$
	5-yo	36.873	$1.038 \cdot 10^{-2}$
1800 MHz	Adult	37.050	$2.088 \cdot 10^{-2}$
TM	10-yo	36.958	$1.217 \cdot 10^{-2}$
	5-yo	36.873	$8.857 \cdot 10^{-3}$

The maximum temperature increase obtained is 0.1°C at 4 GHz while the *hot spot* region is within the vitreous body as the absorbed energy is focused in vitreous region at 4 GHz, and there is also a lack of the blood perfusion within vitreous body.

As seen from Fig. 6.51D, at 6 GHz, the *hot spot* region is moved to anterior parts of eye such as anterior chamber and cornea, as the electromagnetic energy is concentrated at the eye surface.

6.2.4.3 Temperature Rise in Compound and Extracted Eye Models

The thermal eye model from the previous section is now incorporated in the human head model, hence, a comparison between the extracted and compound eye models is carried out in the following. The details on the implementation of the two eye models can be found in Sect. 6.1.4.3.

In addition to ocular tissues, thermal parameters of additional head tissues have to be considered, as given in Table 6.18.

The boundary surface of the extracted eye model is discretized using 7,986 triangles, while the eye interior regions are discretized using 415,429 tetrahedra. The surface of the detailed head model is discretized using 5,934 triangular elements, while the interior of the head model itself is discretized with 4,073,250 tetrahedra.

In case of the compound eye model, two values of convection coefficient h_c are considered, 20 W/m² on the interface between an eye and surrounding air, and 40 W/m² on the interface between an eye and surrounding head tissue [81,83]. The forced convection and heat loss due to radiation can be neglected.

The steady-state temperature distribution in two eye models, when no EM wave is present, is shown on Fig. 6.52.

TABLE 6.17
Tissue parameters used in human eye model.

Tissue	c_b [J/kg/°C]	λ [W/m/°C]	W_b [W/m³/°C]	Q_m [W/m³]
Anterior chamber	3900	0.580	0	0
Choroid	3840	0.530	0	0
Ciliary body	3430	0.498	2700	690
Cornea	4200	0.580	0	0
Iris	3430	0.498	2700	690
Ligaments	3300	0.420	2700	690
Ora serrata	2500	0.250	520	180
Posterior chamber	3997	0.578	0	0
Retina	3680	0.565	35000	10000
Sclera	4200	0.580	13500	0
Vitreous body	3997	0.594	0	0
Lens	3000	0.400	0	0

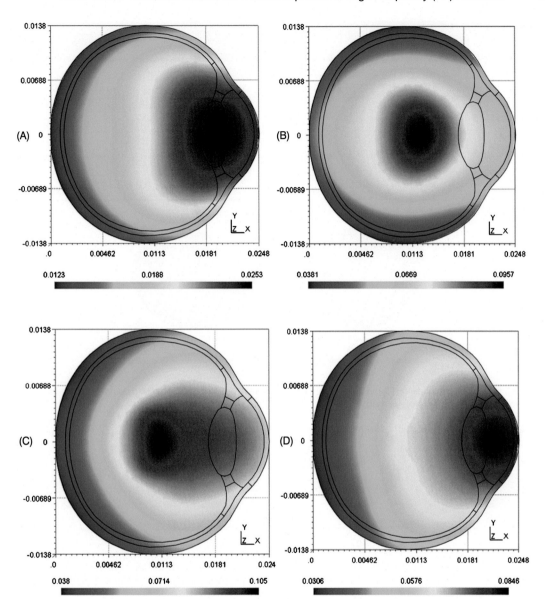

FIG. 6.51 Temperature increase in the eye exposed to plane wave of power density of 10 W/m^2 at a frequency of: (A) 1 GHz, (B) 2 GHz, (C) 4 GHz, and (D) 6 GHz.

Fig. 6.52 shows the steady-state temperature distribution on the horizontal axial slice of two eye models. A very similar temperature gradient from the cornea to sclera can be observed, although different maximum values are obtained. However, the maximum temperature obtained in the scleral region of the compound eye is close to the body temperature as the eye is surrounded with other head tissues. Compared to this, the extracted

eye model is surrounded with ambient air set to 22°C, resulting in much lower maximum temperature [81].

The temperature increase due to eye exposed to incident plane wave is considered next. Fig. 6.53 shows the temperature increase due to 1800 MHz EM wave, obtained on the horizontal axial slice of the two eye models, for vertical and horizontal polarization, respectively. The temperature distribution in case of horizontally po-

TABLE 6.18
Human head thermal parameters used with compound eye model.

Tissue	c_b [J/kg/°C]	λ [W/m/°C]	W_b [W/m³/°C]	Q_m [W/m³]
Brainstem	3600	0.503	35000	10000
Cerebellum	3680	0.565	35000	10000
Head skin	3500	0.420	9100	1000
Liquor	3840	0.530	0	0
Skull	1300	0.300	1000	0
Mandible	1300	0.300	1000	0
Gray Matter	3680	0.565	35000	10000

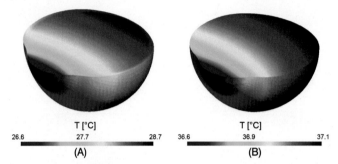

FIG. 6.52 Horizontal axial slice of the steady-state temperature distribution obtained using: (A) extracted eye model, (B) compound eye model.

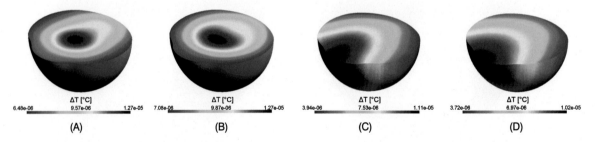

FIG. 6.53 Horizontal axial slice of the temperature increase due to 1800 MHz EM wave, obtained in the extracted eye model: (A) vertical polarization, (B) horizontal polarization; compound eye model: (C) vertical polarization, (D) horizontal polarization.

larized EM wave results in a *hot spot*, spreading in the horizontal direction, compared to the vertical spread in case of vertical polarization. Furthermore, the *hot spot* is formed in the eye vitreous region in case of the extracted model, contrary to the compound model where the *hot spot* is formed in the corneal region.

A detailed view on temperature rise along the eye visual axis can be seen on Fig. 6.54. The result from the compound model shows a steady decrease from anterior to posterior eye regions. Compared to this, the result from the extracted eye model shows first an increase in the temperature rise from cornea to roughly one-third of the vitreous region, followed by a steady decrease to sclera.

Also, as evident from Fig. 6.54, practically the same temperature distribution along the visual axis of the extracted model is obtained for both polarizations, suggesting that a symmetrical model such as extracted eye should be used cautiously if considering the incident wave polarization. The compound eye model, on the

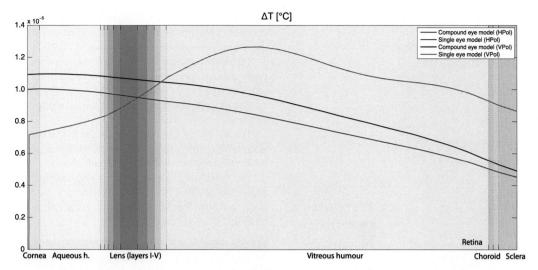

FIG. 6.54 Comparison of the temperature increase along the visual axis of the extracted and compound eye models, due to horizontally and vertically polarized 1800 MHz EM wave.

other hand, discriminates between the two polarizations, as evidenced on the same graph. A further study is currently under way to assess these effects at other frequencies.

6.2.4.4 Thermal Response of the Human Body

The thermal response of a human body exposed to HF radiation is analyzed using three human body models – cylindrical, simplified, and three-dimensional anatomically based human body models.

Cylindrical Body Model The first example deals with a simplified cylindrical body model, introduced in Chapter 4, exposed to radiation of a GSM base station antenna system mounted on a 35 m high building. The maximum value of the radiated electric field, tangential to the body, 30 m away from the antenna main beam and calculated using the ray-tracing algorithm is 15 V/m. Assuming that at 900 MHz the average conductivity of the body is approximately $\sigma = 1.4$ S/m, and the average permittivity is approximately $\varepsilon_r = 55$, the corresponding internal field in the body is 0.1 V/m [65].

The average thermal properties of the cylindrical body model (muscle properties) are given as follows [65]: $\lambda = 0.545$ W/m/°C, $W_b = 0.433$ kg/m³s, $Q_m = 703.5$ W/m³ and $C_{pb} = 3475$ J/kg°C. The arterial blood temperature and tissue density are assumed to be $T_a = 36.7$°C and $\rho = 1000$ kg/m³, respectively.

The boundary condition for the bioheat transfer equation (6.79) is to be imposed on the interface be-

tween skin and air, and is given by

$$q = H(T_s - T_a), \qquad (6.118)$$

where q denotes the heat flux density defined as

$$q = -\lambda \frac{\partial T}{\partial n}, \qquad (6.119)$$

while H, T_s and T_a denote, respectively, the convection coefficient, temperature of the skin, and temperature of the air.

The bioheat transfer equation (6.79) is solved using DR-BEM.

The obtained temperature distribution with related heat flux field is shown on Fig. 6.55.

The temperature increase obtained by the cylindrical body model, due to both the metabolic heat generation and electromagnetic power deposition, is around 37.17°C. Furthermore, the maximum calculated temperature rise due is $\Delta T = 4.6 \cdot 10^{-6}$°C and found to be rather negligible.

Simplified Body Model The second example deals with temperature increase in a human body due to radiation of GSM and UMTS base stations [84]. The height of the human model, shown on Fig. 6.56A, is 180 cm and weight is 75 kg, while the dimensions of individual body parts were chosen according to [85].

The human body is discretized using 25,658 tetrahedral elements, as shown on Fig. 6.56B. The average

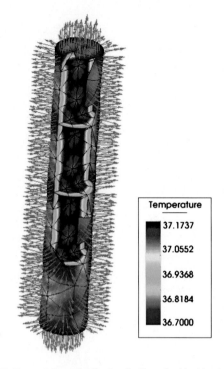

FIG. 6.55 Temperature distribution in the cylindrical body model and normal heat flux. From [65].

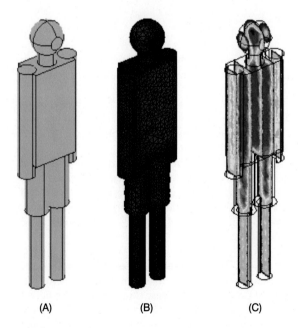

FIG. 6.56 (A) Simplified human body model, (B) Finite element mesh, (C) Temperature distribution inside the human body.

thermal parameters of the human body, assumed to be composed of muscle tissue, are taken from [86], while the arterial blood temperature is set to 36.7°C.

Dirichlet boundary conditions are imposed on the body surface, i.e., $T_s = 36.7°C$, where T_s is the surface temperature.

The bioheat equation (6.79) is solved using the finite element method (FEM).

The source of EM radiation is the GSM/UMTS base station transmitter mounted on the roof of the building, located 52 meters above the ground.

The calculated power density and average temperature rise in the human body due to radiation of GSM and UMTS base station antenna are shown in Table 6.19.

TABLE 6.19
Power density and maximum temperature rise in the simplified model of a human body.

Frequency band	Q_{EM} [W/m^3]	$\Delta T = T_{\max} - T_{bz}$ [°C]
GSM	2.45	$1.97 \cdot 10^{-3}$
UMTS	0.431	$3.46 \cdot 10^{-4}$

It is evident that the average temperature increase in the human body in both cases is negligible. Note that T_{bz} is the maximum body temperature when the human body is not exposed to electromagnetic radiation.

Temperature distribution inside the body is shown on Fig. 6.56C, where color gradation is from the lowest temperature, denoted in blue, to the hottest areas, corresponding to bright red. As the temperature rise due to induced electric field is negligible, the results from Fig. 6.56C practically represent the steady-state temperature distribution.

Anatomically Based Body Model The final example of human exposure to base station antenna radiation deals with a realistic, anatomically based, model of the human body, shown on Fig. 6.57A. The average thermal properties (of muscle tissue) are assumed as in the previous example. The finite element mesh of the homogeneous body exposed to GSM base station radiation is shown in Fig. 6.57B [87]. The corresponding value of external field is taken to be 0.16 V/m [88].

Fig. 6.57C shows the temperature distribution inside the body. The maximum temperature obtained from the finite element model of the body with muscle tissue properties, due to both metabolic heat production and electromagnetic power deposition, is around 37.178°C.

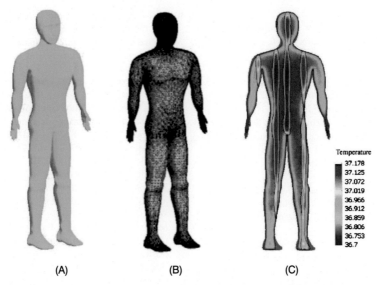

FIG. 6.57 (A) Human body model, (B) Finite element mesh, (C) Temperature distribution inside human body for $Q_{SAR} = 0$.

Consequently, the maximum temperature increase generated by power deposition due to GSM base station radiation is $\Delta T = 1.75 \cdot 10^{-9}{}^{\circ}\mathrm{C}$ which is rather negligibly small.

REFERENCES

1. Mario Cvetković, Dragan Poljak, Electromagnetic-thermal dosimetry comparison of the homogeneous adult and child brain models based on the SIE approach, Journal of Electromagnetic Waves and Applications 29 (17) (2015) 2365–2379.
2. Mario Cvetković, Dragan Poljak, Akimasa Hirata, The electromagnetic-thermal dosimetry for the homogeneous human brain model, Engineering Analysis with Boundary Elements 63 (2016) 61–73.
3. R. Harrington, Boundary integral formulations for homogeneous material bodies, Journal of Electromagnetic Waves and Applications 3 (1) (1989) 1–15.
4. W.C. Chew, M.S. Tong, B. Hu, Integral Equation Methods for Electromagnetic and Elastic Waves, Morgan & Claypol Publishers, 2009, 261 pp.
5. A.J. Poggio, E.K. Miller, Integral equation solutions of three-dimensional scattering problems, in: R. Mittra (Ed.), Computer Techniques for Electromagnetics, second edition, Hemisphere Publishing Corporation, 1987, pp. 159–264, chap. 4.
6. K. Umashankar, et al., Electromagnetic scattering by arbitrary shaped three-dimensional homogeneous lossy dielectric objects, IEEE Transactions on Antennas and Propagation 34 (6) (1986) 758–766.
7. M. Cvetković, D. Poljak, An efficient integral equation based dosimetry model of the human brain, in: Proceedings of 2014 International Symposium on Electromagnetic Compatibility, EMC EUROPE, 2014, Gothenburg, Sweden, 1–4 September 2014, 2014, pp. 375–380.
8. S. Rao, D.R. Wilton, A. Glisson, Electromagnetic scattering by surfaces of arbitrary shape, IEEE Transactions on Antennas and Propagation 30 (3) (1982) 409–418.
9. Dragan Poljak, Mario Cvetković, Andres Peratta, Cristina Peratta, Hrvoje Dodig, Akimasa Hirata, On some integral approaches in electromagnetic dosimetry, in: The Joint Annual Meeting of the Bioelectromagnetics Society and the European BioElectromagnetics Association, 2016, pp. 289–295.
10. C.T. Tai, Dyadic Green's Functions in Electromagnetic Theory, Intext Educational Publishers, 1973, 246 pp.
11. D. Livesay, K.-M. Chen, Electromagnetic fields induced inside arbitrary shaped biological bodies, IEEE Transactions on Microwave Theory and Techniques MTT-22 (12) (1974) 1273–1280.
12. Dragan Poljak, Damir Cavka, Hrvoje Dodig, Cristina Peratta, Andres Peratta, On the use of the boundary element analysis in bioelectromagnetics, Engineering Analysis with Boundary Elements 49 (2014) 2–14.
13. Hrvoje Dodig, Dragan Poljak, Andrés Peratta, Hybrid BEM/FEM edge element computation of the thermal rise in the 3D model of the human eye induced by high frequency EM waves, in: 2012 20th International Conference on Software, Telecommunications and Computer Networks, SoftCOM, IEEE, 2012, pp. 1–5.
14. Jean-Claude Nédélec, Mixed finite elements in $\mathbb{R}^3$, Numerische Mathematik 35 (3) (1980) 315–341.

15. John L. Volakis, Arindam Chatterjee, Leo C. Kempel, Finite Element Method Electromagnetics: Antennas, Microwave Circuits, and Scattering Applications, vol. 6, John Wiley & Sons, 1998.

16. Birte U. Forstmann, Max C. Keuken, Andreas Schafer, Pierre-Louis Bazin, Anneke Alkemade, Robert Turner, Multi-modal ultra-high resolution structural 7-Tesla MRI data repository, Scientific Data 1 (2014).

17. Max C. Keuken, Pierre-Louis Bazin, Andreas Schäfer, Jane Neumann, Robert Turner, Birte U. Forstmann, Ultra-high 7T MRI of structural age-related changes of the subthalamic nucleus, Journal of Neuroscience 33 (11) (2013) 4896–4900.

18. Juliane Budde, Gunamony Shajan, Klaus Scheffler, Rolf Pohmann, Ultra-high resolution imaging of the human brain using acquisition-weighted imaging at 9.4 T, Neuroimage 86 (2014) 592–598.

19. Mario Cvetković, Hrvoje Dodig, Dragan Poljak, A study on the use of compound and extracted models in the high frequency electromagnetic exposure assessment, Mathematical Problems in Engineering (2017) 2017.

20. C. Gabriel, Compilation of the Dielectric Properties of Body Tissues at RF and Microwave Frequencies, Tech. rep., Report: AL/OE-TR-1996-0037, Brooks Air Force Base, TX, 1996.

21. Av Gullstrand, The optical system of the eye, Physiological Optics 1 (1909) 350–358.

22. José Antonio Díaz, Carles Pizarro, Josep Arasa, Single dispersive gradient-index profile for the aging human lens, JOSA A 25 (1) (2008) 250–261.

23. Akimasa Hirata, Temperature increase in human eyes due to near-field and far-field exposures at 900 MHz, 1.5 GHz, and 1.9 GHz, IEEE Transactions on Electromagnetic Compatibility 47 (1) (2005) 68–76.

24. International Commission on Non-Ionizing Radiation Protection (ICNIRP), Guidelines for limiting exposure to time-varying electric, magnetic and electromagnetic fields (up to 300 GHz), Health Physics 74 (4) (1998) 494–522.

25. Krish D. Singh, Nicola S. Logan, Bernard Gilmartin, Three-dimensional modeling of the human eye based on magnetic resonance imaging, Investigative Ophthalmology & Visual Science 47 (6) (2006) 2272–2279.

26. Dragan Poljak, Human Exposure to Electromagnetic Fields, WIT Press, Southampton-Boston, 2003.

27. M.J. Ackerman, The visible human project, Proceedings of the IEEE (ISSN 0018-9219) 86 (3) (1998) 504–511, https://doi.org/10.1109/5.662875.

28. J.W. Massey, A.E. Yilmaz, AustinMan and AustinWoman: high-fidelity, anatomical voxel models developed from the VHP color images, in: 2016 38th Annual International Conference of the IEEE Engineering in Medicine and Biology Society, EMBC, ISSN 1557-170X, 2016, pp. 3346–3349.

29. E.Y.K. Ng, Jen Hong Tan, U. Rajendra Acharya, Jasjit S. Suri, Human Eye Imaging and Modeling, CRC Press, Boca Raton, 2012.

30. M. Cvetković, H. Dodig, D. Poljak, Comparison of numerical electric field and SAR results in compound and extracted eye models, in: 2nd International Multidisciplinary Conference on Computer and Energy Science, SpliTech, Split, Croatia, 12–14 July 2017, pp. 1–6.

31. Dragan Poljak, Mario Cvetković, Oriano Bottauscio, Akimasa Hirata, Ilkka Laakso, Esra Neufeld, Sylvain Reboux, Craig Warren, Antonios Giannopoulos, Fumie Costen, On the use of conformal models and methods in dosimetry for nonuniform field exposure, IEEE Transactions on Electromagnetic Compatibility 60 (2) (2018) 328–337.

32. Visual Computing Lab ISTI CNR, MeshLab, http://meshlab.sourceforge.net.

33. S.M. Blinkov, I.I. Glezer, The Human Brain in Figures and Tables. A Quantitative Handbook, Plenum Press, New York, USA, 1968, 482 pp.

34. Mario Cvetković, Dragan Poljak, Effects of electromagnetic polarization in homogeneous electromagnetic-thermal dosimetry model of human brain, in: 2015 23rd International Conference on Software, Telecommunications and Computer Networks, SoftCOM, IEEE, 2015, pp. 92–95.

35. Akimasa Hirata, Naoki Ito, Osamu Fujiwara, Influence of electromagnetic polarization on the whole-body averaged SAR in children for plane-wave exposures, Physics in Medicine & Biology 54 (4) (2009) N59.

36. A. Christ, N. Kuster, Differences in RF energy absorption in the heads of adults and children, Bioelectromagnetics 26 (S7) (2005) 31–44.

37. O.P. Gandhi, G. Lazzi, C.M. Furse, Electromagnetic absorption in the human head and neck for mobile telephones at 835 and 1900 MHz, IEEE Transactions on Microwave Theory and Techniques 44 (10) (1996) 1884–1897.

38. F. Schönborn, M. Burkhardt, N. Kuster, Differences in energy absorption between heads of adults and children in the near field of sources, IEEE Transactions on Microwave Theory and Techniques 74 (2) (1998) 160–168.

39. Independent Expert Group on Mobile Phones, Mobile Phones and Health, The Stewart Report, National Radiological Protection Board, Chilton, 2000.

40. O.P. Gandhi, G. Kang, Some present problems and a proposed experimental phantom for SAR compliance testing of cellular telephones at 835 and 1900 MHz, Physics in Medicine and Biology 47 (9) (2002) 1501–1518.

41. G. Bit-Babik, A.W. Guy, C.-K. Chou, A. Faraone, M. Kanda, A. Gessner, J. Wang, O. Fujiwara, Simulation of exposure and SAR estimation for adult and child heads exposed to radiofrequency energy from portable communication devices, Radiation Research 163 (5) (2005) 580–590.

42. J. Wang, O. Fujiwara, S. Kodera, S. Watanabe, FDTD calculation of whole-body average SAR in adult and child models for frequencies from 30 MHz to 3 GHz, Physics in Medicine and Biology 51 (1) (2006) 4119–4127.

43. E. Conil, A. Hadjem, F. Lacroux, M.F. Wong, J. Wiart, Variability analysis of SAR from 20 MHz to 2.4 GHz for different adult and child models using finite-difference time-domain, Physics in Medicine and Biology 53 (1) (2008) 1511–1525.

44. M. Fujimoto, A. Hirata, J. Wang, O. Fujiwara, T. Shiozawa, FDTD-derived correlation of maximum temperature increase and peak SAR in child and adult head models due to dipole antenna, IEEE Transactions on Electromagnetic Compatibility 48 (1) (2006) 240–247.

45. J.F. Bakker, M.M. Paulides, A. Christ, N. Kuster, G.C. van Rhoon, Assessment of induced SAR in children exposed to electromagnetic plane waves between 10 MHz and 5.6 GHz, Physics in Medicine and Biology 55 (1) (2010) 3115–3130.

46. J. Wiart, A. Hadjem, M.F. Wong, I. Bloch, Analysis of RF exposure in the head tissues of children and adults, Physics in Medicine and Biology 53 (13) (2008) 3681–3695.

47. J. Wang, O. Fujiwara, Comparison and evaluation of electromagnetic absorption characteristics in realistic human head models of adult and children for 900-MHz mobile telephones, IEEE Transactions on Microwave Theory and Techniques 51 (3) (2003) 966–971.

48. J. Wang, O. Fujiwara, S. Watanabe, Approximation of aging effect on dielectric tissue properties for SAR assessment of mobile telephones, IEEE Transactions on Electromagnetic Compatibility 48 (2) (2006) 408–413.

49. Hrvoje Dodig, Mario Cvetković, Dragan Poljak, Akimasa Hirata, Ilkka Laakso, Hybrid FEM/BEM for human heads exposed to high frequency electromagnetic radiation, WIT Transactions on The Built Environment 174 (2018) 369–378.

50. Ilkka Laakso, Satoshi Tanaka, Soichiro Koyama, Valerio De Santis, Akimasa Hirata, Inter-subject variability in electric fields of motor cortical tDCS, Brain Stimulation 8 (5) (2015) 906–913.

51. IEEE standard for safety levels with respect to human exposure to radio frequency electromagnetic fields, 3 kHz to 300 GHz, in: IEEE Std C95.1-2005 (Revision of IEEE Std C95.1-1991), 2006, pp. 1–238.

52. Matti Stenroos, Alexander Hunold, Jens Haueisen, Comparison of three-shell and simplified volume conductor models in magnetoencephalography, Neuroimage 94 (2014) 337–348.

53. Nekrasov Tanja, Measurement of Base Station Antenna Radiation, Master's thesis, Wessex Institute of Technology, 2006.

54. Eleanor R. Adair, R.C. Petersen, Biological effects of radiofrequency/microwave radiation, IEEE Transactions on Microwave Theory and Techniques 50 (3) (2002) 953–962, https://doi.org/10.1109/22.989978.

55. A. Hirata, T. Shiozawa, Correlation of maximum temperature increase and peak SAR in the human head due to handset antennas, IEEE Transactions on Microwave Theory and Techniques 51 (7) (2003) 1834–1841, https://doi.org/10.1109/TMTT.2003.814314.

56. International Commission on Non-Ionizing Radiation Protection (ICNIRP), Guidelines for limiting exposure to time-varying electric and magnetic fields (1 Hz to 100 kHz), Health Physics 99 (6) (2010) 818–836, https://doi.org/10.1097/HP.0b013e3181f06c86.

57. J. Wang, O. Fujiwara, FDTD computation of temperature rise in the human head for portable telephones, IEEE Transactions on Microwave Theory and Techniques 47 (8) (1999) 1528–1534, https://doi.org/10.1109/22.780405.

58. G.M.J. Van Leeuwen, J.J.W. Lagendijk, B.J.A.M. Van Leersum, A.P.M. Zwamborn, S.N. Hornsleth, A.N.T.J. Kotte, Calculation of change in brain temperatures due to exposure to a mobile phone, Physics in Medicine and Biology 44 (10) (1999) 2367–2379, https://doi.org/10.1088/0031-9155/44/10/301.

59. P. Bernardi, M. Cavagnaro, S. Pisa, E. Piuzzi, Specific absorption rate and temperature increases in the head of a cellular-phone user, IEEE Transactions on Microwave Theory and Techniques 48 (2000) 1118–1126, https://doi.org/10.1109/22.848494.

60. Akimasa Hirata, Masashi Morita, Toshiyuki Shiozawa, Temperature increase in the human head due to a dipole antenna at microwave frequencies, IEEE Transactions on Electromagnetic Compatibility 45 (1) (2003) 109–116.

61. W.J. Minkowycz, E.M. Sparrow, J.Y. Murthy, Handbook of Numerical Heat Transfer, second edition, John Willey & Sons, Inc., New York, 2006, 1024 pp.

62. H.H. Pennes, Analysis of tissue and arterial blood temperatures in the resting human forearm. 1948, Journal of Applied Physiology 85 (1) (1998) 5–34.

63. P.P. Silvester, R.L. Ferrari, Finite Elements for Electrical Engineers, third edition, Cambridge University Press, 1996, 516 pp.

64. M. Cvetković, D. Poljak, A. Peratta, FETD computation of the temperature distribution induced into a human eye by a pulsed laser, Progress in Electromagnetics Research, PIER 120 (2011) 403–421.

65. Dragan Poljak, Andres Peratta, Carlos A. Brebbia, The boundary element electromagnetic–thermal analysis of human exposure to base station antennas radiation, Engineering Analysis with Boundary Elements 28 (7) (2004) 763–770.

66. Paul William Partridge, Carlos Alberto Brebbia, Luiz C. Wrobel, The Dual Reciprocity Boundary Element Method, Computational Mechanics Publications, Southampton/Boston, 1992.

67. Carlos Alberto Brebbia, José Claudio Faria Telles, Luiz C. Wrobel, Boundary Element Techniques: Theory and Applications in Engineering, Springer, Berlin/Heidelberg, 1984.

68. M.A. Golberg, C.S. Chen, The theory of radial basis functions applied to the BEM for inhomogeneous partial differential equations, Boundary Elements Communications 5 (2) (1994) 57–61.

69. M.A. Golberg, C.S. Chen, A bibliography on radial basis function approximation, Boundary Elements Communications 7 (1996) 155–163.

70. M. Zhu, J.J.H. Ackerman, A.L. Sukstanskii, D.A. Yablonskiy, How the body controls brain temperature: the temperature shielding effect of cerebral blood flow, Journal of Applied Physiology 101 (5) (2006) 1481–1488, https://doi.org/10.1152/japplphysiol.00319.2006.

71. A. Sukstanskii, D. Yablonskiy, Theoretical model of temperature regulation in the brain during changes in functional activity, Proceedings of the National Academy of Sciences 103 (32) (2006) 12144–12149, https://doi.org/10.1073/pnas.0604376103.

72. D. Nelson, S. Nunneley, Brain temperature and limits on transcranial cooling in humans: quantitative modeling results, European Journal of Applied Physiology and Occupational Physiology 78 (4) (1998) 353–359, https://doi.org/10.1007/s004210050431.

73. Gilbert J. Stone, Robert R. Goodman, Kristy Z. Baker, Christopher J. Baker, Robert A. Solomon, Direct intraoperative measurement of human brain temperature, Neurosurgery 41 (1) (1997) 20–24.

74. L. Zhu, C. Diao, Theoretical simulation of temperature distribution in the brain during mild hypothermia treatment for brain injury, Medical and Biological Engineering and Computing 39 (6) (2001) 681–687, https://doi.org/10.1007/BF02345442.

75. Pekka Mellergard, Carl-Henrik Nordstrom, Epidural temperature and possible intracerebral temperature gradients in man, British Journal of Neurosurgery 4 (1) (1990) 31–38, https://doi.org/10.3109/02688699009000679.

76. Christopher M. Collins, Michael B. Smith, Robert Turner, Model of local temperature changes in brain upon functional activation, Journal of Applied Physiology 97 (6) (2004) 2051–2055, https://doi.org/10.1152/japplphysiol.00626.2004.

77. L. Mcilvoy, Comparison of brain temperature to core temperature: a review of the literature, Journal of Neuroscience Nursing 36 (1) (2004) 23–31, https://doi.org/10.1097/01376517-200402000-00004.

78. D.A. Yablonskiy, J.J.H. Ackerman, M.E. Raichle, Coupling between changes in human brain temperature and oxidative metabolism during prolonged visual stimulation, Proceedings of the National Academy of Sciences 97 (13) (2000) 7603–7608, https://doi.org/10.1073/pnas.97.13.7603.

79. R.L. McIntosh, V. Anderson, A comprehensive tissue properties database provided for the thermal assessment of a human at rest, Biophysical Reviews and Letters 5 (3) (2010) 129–151.

80. Akimasa Hirata, Improved heat transfer modeling of the eye for electromagnetic wave exposures, IEEE Transactions on Biomedical Engineering 54 (5) (2007) 959–961.

81. Mario Cvetković, Hrvoje Dodig, Dragan Poljak, Numerical comparison of compound and extracted eye models for high frequency dosimetry, International Journal for Engineering Modelling 31 (1–2) (2018) 1–13.

82. Dragan Poljak, Mario Cvetković, Hrvoje Dodig, Andres Peratta, Electromagnetic-thermal analysis for human exposure to high frequency (HF) radiation, International Journal of Design & Nature and Ecodynamics 12 (1) (2017) 55–67.

83. Mario Cvetković, Hrvoje Dodig, Dragan Poljak, Temperature increase in the extracted and compound eye models, in: 2017 25th International Conference on Software, Telecommunications and Computer Networks, SoftCOM, IEEE, 2017, pp. 1–6.

84. D. Poljak, Izloenost ljudi neionizacijskom zraenju, Kigen, Zagreb, 2006, 273 pp.

85. R. Drillis, Body segment parameters. A survey of measurement technique, Artificial Limbs 8 (1964) 329.

86. Dragan Poljak, N. Kovac, The electromagnetic-thermal analysis of human exposure to radio base station antennas, in: 17th International Conference on Applied Electromagnetics and Communications, 2003, ICECom 2003, IEEE, 2003, pp. 41–44.

87. D. Cavka, Finite element thermal model of the human exposed to electric field generated from GSM base station, in: 2006 International Conference on Software in Telecommunications and Computer Networks, 2006, pp. 22–26.

88. Damir Čavka, Mario Cvetković, Measurement of GSM base station electrical field and calculation of temperature increase inside human body due to exposure, in: SoftCOM 2005, 2005, pp. 1–6.

CHAPTER 7

Biomedical Applications of Electromagnetic Fields

7.1 TRANSCRANIAL MAGNETIC STIMULATION (TMS) TREATMENT

Transcranial magnetic stimulation (TMS) is a noninvasive and painless technique for stimulation or inhibition of particular brain regions. A stimulation coil placed in the vicinity of a patient's head is energized using a very short current pulse, generating a time varying magnetic field of high intensity that penetrates into nearby tissues such as skull and brain. According to the differential form of Faraday's law, the time varying magnetic field induces an electric field, thus causing depolarization or hyperpolarization of the neuronal cell membranes located within a few centimeters of the cortex. Reaching a certain threshold value of the membrane potential will thus result in generation of an action potential. Macroscopically, activation occurs in the brain regions with highest values of the induced electric field [1–4].

Nowadays, 1985 is considered the birth-year of TMS, when Barker et al. [5] from the University of Sheffield demonstrated the first stimulation of spinal cord having also arrived at the idea of direct and noninvasive stimulation of the brain. From the inception in mid-1980s, the basic idea of TMS remained basically the same, although a couple years after Barker, the group led by Ueno introduced the so-called figure-of-eight coil [6,7], achieving a more localized stimulation of the brain, thus laying out the path to the modern TMS system.

From the first demonstration more than three decades ago, TMS has become a very important device for various diagnostic and therapeutic purposes in the field of neurology, neurosurgery, neurophysiology, and psychiatry, as well as in studying the functional mechanism and role of specific cortical regions [5]. Furthermore, the use of navigated transcranial magnetic stimulation (nTMS) is nowadays a standard technique in preoperative mapping of eloquent brain cortices (motor, speech, language) in patients undergoing awake brain surgery [8,9], and in neurophysiologic development of nTMS methodologies for mapping these cortical areas in healthy subjects [10,11]. However, so far the most commonly studied application of TMS is related to the treatment of depression.

Although TMS is a widely used technique, considerable research is needed on the distribution of the electromagnetic fields induced in the human brain. This is corroborated by many clinical studies reporting different efficiency of TMS stimulation, primarily related to differences in relevant TMS settings such as positioning of coil, waveform of the applied pulse, frequency, number of applied stimuli and the intensity of stimulation, etc. Determination of optimal stimulation intensity is a problem many TMS studies are facing, specifically when mapping non-motor cortical areas. The single pulse TMS mapping of the primary motor cortex has a well established role in clinical neurophysiology. The generally accepted procedure is first to find the individual resting motor threshold (RMT) by stimulating the primary motor cortex for hand muscles and recording motor evoked potentials (MEP) from hand muscles. RMT is defined as the minimum TMS intensity adequate to evoke MEPs in at least 50% of trials [12]. An extensive research has been carried out on the inter-individual variability of the RMT [13,14] finding that the most important factor to this variability is the skull-to-cortex distance. Other neurophysiologic measures [15] include MEP amplitude and latency, cortical silent period duration, central motor conduction time, MEP recruitment curve, as well as cortical excitability measures (inhibition, facilitation).

7.1.1 Modeling TMS

Brain stimulation modeling is thus important not only in determining the exact location and the level of stimulation, but also in elucidating the related underlying mechanisms. The modeling can also be beneficial when interpretation of experimental results is necessary, or in the design of a new, more efficient systems like multi-coil setups for focused and deep brain stimulation [16–18].

There were many attempts to develop brain models, varying in complexity from homogeneous spherical ones [19,20] to a more recent realistic three-dimensional (3D) models using a number of various tissues [21–23].

The advantage of simple canonical models is the available analytical solution [19,24], thus making them

Human Interaction with Electromagnetic Fields. https://doi.org/10.1016/B978-0-12-816443-3.00015-7

useful for initial assessment of the distribution of the induced field. For more realistic representations, the solution can be attained using numerical methods.

Regarding the formulation of the problem, there are approaches based on the differential and integral equations. From the early days, the differential approach has become the *method of choice* in the mathematical formulation of the problems related to the determination of induced electric fields inside biological tissues, and TMS modeling is no exception.

One of the drawbacks of the differential equation approach is a need for the entire domain discretization. In case of open boundary problems, such as radiation and scattering, one needs to artificially limit the boundary of the problem or use concepts such as infinite elements, absorbing boundary conditions, etc. On the other hand, integral equation based formulations require only discretization of the boundary, allowing an additional benefit of reducing the dimensionality of the problem. Furthermore, since boundary conditions are implicit within the integral formulation, open boundary problems are treated in an exact way.

In recent decades, integral equation based approach is seeing a revival in computational electromagnetics community [25]. There is some reported work where the TMS problem is treated by the boundary element method [22,26,27]. Recently, an application of surface integral equation analysis methods to TMS has been reported [28]. The use of a rigorous TMS model based on the integral equation approach can provide a means for an alternative more accurate physical description of the problem.

The majority of TMS computational methods use the so called quasi-static approximation, where the capacitive effects and the effects of propagation are neglected. The nTMS systems currently in clinical use also rely on this approximation, as well as on the simple spherical conductor models, to determine the induced electric field [29]. Although, the quasi-static approximation results in the simplification of the governing equations, the exclusion of propagation effects at very high values [30] of tissue permittivity [31] could lead to an inaccurate assessment of the stimulated area [32–34]. A recent study using a DTI-based model has showed that neglecting the permittivity values leads to a decrease of about 72% and 24% of the maximum currents and fields, respectively [35]. Therefore, a rigorous, more accurate model, such as one reported in [28], could aid in finding out to what extent the variation of brain tissue parameters influences the induced fields and currents. Also, as the analysis on animals [36] showed the reduced TMS-efficiency in smaller brain volumes, there is

some concern on the use of adult size TMS coils in children, thus, it is interesting to investigate the influence of the brain size.

7.1.1.1 Surface Integral Equation Based Formulation for TMS

The problem of human brain exposed to a TMS coil is treated as a classical scattering problem. The electromagnetic field ($\vec{E}^{\text{inc}}, \vec{H}^{\text{inc}}$) is incident on the lossy homogeneous object representing the brain. Due to the presence of the scattering object, i.e. the brain, a scattered field denoted by ($\vec{E}^{sca}, \vec{H}^{sca}$) also exists, as illustrated on Fig. 7.1.

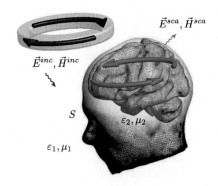

FIG. 7.1 Human brain as a lossy homogeneous dielectric (ε_2, μ_2) placed in the incident field ($\vec{E}^{\text{inc}}, \vec{H}^{\text{inc}}$) of a TMS coil.

As a first implementation of the model, we can consider a homogeneous brain compartment model only, while neglecting the skull and scalp, because the majority of the current is flowing inside the skull [37]. This will facilitate the use of a surface integral equation based approach formulated in Sect. 6.1.1.

To account for inductive and capacitive effects, the human brain is considered as a lossy material with complex permittivity and permeability (ε_2, μ_2). As biological tissues do not posses magnetic properties, μ_0 can be used as the brain permeability, i.e., the free-space permeability. The complex permittivity of the brain is given by

$$\varepsilon_2 = \varepsilon_0 \varepsilon_r - j \frac{\sigma}{\omega}, \tag{7.1}$$

where ε_0 is the permittivity of the free-space, ε_r is the relative permittivity, σ is the electrical conductivity of the brain, and $\omega = 2\pi f$ is the operating frequency.

Using the equivalence theorem, two equivalent problems are formulated in terms of equivalent electric and magnetic current densities assumed to flow on the brain surface [38,39]. Applying the boundary conditions for the electric field at surface S being the interface of the two equivalent problems

$$-\hat{n} \times \vec{E}_i^{sca}(\vec{J}, \vec{M}) = \begin{cases} \hat{n} \times \vec{E}^{inc}, & i = 1, \\ 0, & i = 2, \end{cases} \quad (7.2)$$

the frequency domain electric field integral equation (EFIE) for the lossy homogeneous human brain is obtained. In (7.2), $\vec{E}^{inc}$ is the known incident field, generated by a TMS coil, while $\vec{J}$ and $\vec{M}$ represent unknown surface currents.

The incident electric field $\vec{E}^{inc}$ in (7.2) is due to some external sources, and exists regardless of the presence of a dielectric object, i.e., the brain. In case of a plane wave incidence, it is convenient to place these sources at infinity. The TMS coil, on the other hand, placed in the vicinity of the head, can be considered as a near field source, and consequently, a coupled problem, as the induced eddy currents in the brain will in turn generate a secondary field superimposed to the primary field due to coil current.

Thus, the electric field due to TMS coil can be expressed as follows:

$$\vec{E} = -j\omega\vec{A} - \nabla\varphi, \quad (7.3)$$

where $\vec{A}$ and φ are the magnetic vector potential and electric scalar potential, respectively. The first term in (7.3) is due to current flowing in the coil, while the second one is a consequence of the accumulation of electric charge on the boundary of the medium [40].

Under the assumption that the TMS coil is decoupled from the human brain [28], i.e., its presence does not disturb the field, the electric field due to the coil can be calculated from

$$\vec{E} = -j\omega\vec{A}. \quad (7.4)$$

Assuming a uniform current density I over a coil cross-section, the magnetic vector potential at point $\vec{r}$ can be determined from the integral

$$\vec{A}(\vec{r}) = \frac{\mu_0 M I}{4\pi} \int_l \frac{d\vec{l}}{|\vec{r} - \vec{r}'|}, \quad (7.5)$$

where μ_0 is the free-space permeability, M is the number of coil windings, and $|\vec{r} - \vec{r}'|$ is the distance from the observation to the source point on the coil. The differential element of the curve $d\vec{l}$ depicts the direction of the current flow through the coil. Expression (7.5) can be solved by discretizing the coil using linear segments

[41]. Assembling the contributions from all linear segments, the magnetic vector potential can be determined at an arbitrary point in space.

On the other hand, the magnetic flux density $\vec{B}$ at point $\vec{r}$ can be determined from the Biot–Savart's law:

$$\vec{B}(\vec{r}) = \frac{\mu_0 M I}{4\pi} \int_l \frac{d\vec{l} \times \hat{r}_0}{|\vec{r} - \vec{r}'|^2}, \quad (7.6)$$

where $\hat{r}_0$ is the unit vector pointing from the source point $\vec{r}'$ to the field point $\vec{r}$.

The electric and magnetic fields at an arbitrary point inside the brain can be calculated using

$$\vec{E}_2(\vec{r}) = -j\omega\mu_2 \int_S \vec{J}(\vec{r}')G_2(\vec{r}, \vec{r}') dS'$$
$$- \frac{j}{\omega\varepsilon_2} \int_S \nabla'_S \cdot \vec{J}(\vec{r}')G_2(\vec{r}, \vec{r}') dS'$$
$$- \int_S \vec{M}(\vec{r}') \times \nabla G_2(\vec{r}, \vec{r}') dS' \quad (7.7)$$

and

$$\vec{H}_2(\vec{r}) = -j\omega\varepsilon_2 \int_S \vec{M}(\vec{r}')G_2(\vec{r}, \vec{r}') dS'$$
$$- \frac{j}{\omega\mu_2} \int_S \nabla'_S \cdot \vec{M}(\vec{r}')G_2(\vec{r}, \vec{r}') dS'$$
$$+ \int_S \vec{J}(\vec{r}') \times \nabla'G_2(\vec{r}, \vec{r}') dS', \quad (7.8)$$

where $\vec{J}$ and $\vec{M}$ are the equivalent electric and magnetic currents, respectively, determined from (6.14).

The magnetic flux density in the brain is calculated from

$$\vec{B} = \mu_0 \vec{H}. \quad (7.9)$$

From the field $\vec{E}$ inside the brain, the distribution of the current density $\vec{J}_{ind}$ for the lossy homogeneous brain is determined by

$$\vec{J}_{ind} = (\sigma + j\omega\varepsilon_0\varepsilon_r)\vec{E}, \quad (7.10)$$

where σ and ε_r are the electric conductivity and relative permittivity of the human brain, respectively. Frequency dependent parameters of the homogeneous brain tissue, taken as an average between white and gray matter from [42,43], are given in Table 7.1.

TABLE 7.1
Homogeneous human brain parameters. From CVETKOVIĆ, Mario; POLJAK, Dragan; HAUEISEN, Jens. Analysis of transcranial magnetic stimulation based on the surface integral equation formulation. IEEE Transactions on Biomedical Engineering, 2015, 62.6: 1535–1545.

Frequency	
2.44 kHz	
Relative permittivity [–]	Electrical conductivity [S/m]
ε_r	σ
46940	0.08595

It must be noted that the brain conductivity value reported in the literature varies a lot. While the value for cerebrospinal fluid (CSF) is relatively well known [44], the reported values for white and gray matter for low frequencies are ranging between 0.01 and 1 S/m [45]. The specific conductivity value chosen in Table 7.1 is well within this range.

7.1.2 Human Brain Models

The analysis of transcranial magnetic stimulation is based on the use of a human brain model introduced in Sect. 6.1.5.1. It must be pointed out that the gyrification of the brain is important for the accurate determination of maximum induced electric field [23,46], and therefore, it would be interesting to show any age-dependent changes in brain gyrification and their potential impact. The presented formulation can be used to this means, i.e., on a more detailed brain model, derived from the magnetic resonance images (MRI). But, due to difficulty in obtaining child brain MRI (due to ethical, in addition to technical reasons), as well as to facilitate the solution process, the smoothed brain surface model is featured in this work.

As already mentioned, the brain compartment model is considered, while skull and scalp are neglected. Although the homogeneous brain model does not represent the realistic scenario, as the surrounding tissues will affect the distribution of the induced fields and currents [47], the analysis in [48] showed that the inclusion of the skull or CSF would not affect the distribution of currents in the adjacent cortex. Also, the study in [26] showed that the one-compartment model, despite its simplicity, produces quite robust results, and that it is almost as accurate as the three-layered model, thus offering a good balance between accuracy and computational cost. Another study [27] presented similar trends

of the electric field distribution along investigation line in the anatomical voxel model and homogeneous head.

7.1.2.1 TMS for Pediatric Population

Although the TMS has established itself as a useful tool in adult population treatment, it has yet to prove its great potential for children [49,50]. So far, TMS is an established procedure for children in the investigation of the integrity and maturation of the motor system (corticospinal tract) [51–53]. There is also a general consensus regarding the need of posing certain safety guidelines for TMS use in children [49,50,54,55], since the child brain is considered to be significantly more plastic than the adult brain, resulting in longer lasting neuroplastic changes [49].

Compared to adults, children under 10 years of age have higher motor evoked potential thresholds (RMTs) [56], and children of 13.5 years of age have lower intracortical inhibition [57]. With the increasing age, the RMT level declines until reaching the adult levels at 13 to 16 years [58], while lower intracortical inhibition in children points to a maturation process that may have implications for greater capacity of practice-dependent neuronal plasticity in children [59]. Nowadays, it is assumed that age-related differences in TMS-evoked parameters in children reflect primarily changes during the cerebral and corticospinal myelination, intracortical synaptic and neuronal developmental process [50, 54]. Due to higher mean values and general variability between individuals, it has been suggested that determining RMTs may be less useful in children younger than 10 years [58].

When modeling TMS for children, it is reasonable to start with the adult human brain of average dimensions, and apply the linear scaling, similar to the approach used in [60], to obtain the models of the child brain. The applied scaling factors for the 10-yo and 5-yo brain can be found in Table 6.5, while Fig. 6.22 depicts a comparison between three brain models.

It is worth noting that the child brain is not simply a scaled version of the adult brain, as the surrounding tissues such as skull and skin develop at different pace [61]. Nevertheless, this scaling approach can provide some insights into the sensitivity of the results due to variable brain dimensions. The assessment of the effects these and other uncertainties have on the resulting values of interest is currently the main challenge of stochastic dosimetry [61]. Although anatomically correct children models based on MRI data should be used when available, the majority of the studies using scaled models are consistent with anatomically correct models [62].

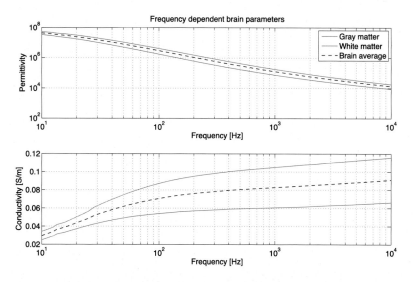

FIG. 7.2 Frequency dependent parameters of brain tissue.

TABLE 7.2
The age related parameters for the homogeneous human brain. The base (adult) parameters are taken from [42], as an average value of white and gray matter at a particular frequency of interest. The parameters for two child models are scaled based on expressions from [69]. The scaling factors are given in the third and fifth rows.

Freq. [kHz]	Permittivity ε_r			Conductivity σ [S/m]		
	2.44	3	5	2.44	3	5
adult	46940	37290	24380	0.08595	0.0867	0.0884
10-year	58535	46135	29276	0.107	0.107	0.108
10-y/adult	1.2470	1.2372	1.2193	1.2249	1.2341	1.2217
5-year	72971	57062	36234	0.133	0.132	0.131
5-y/adult	1.5546	1.5302	1.4862	1.5474	1.5225	1.4819

7.1.2.2 Brain Tissue Parameters

In addition to the obvious difference in brain size between adult and child, the biological tissue parameters such as permittivity and the electrical conductivity will significantly affect the distribution of the induced fields in the brain. These tissue parameters are often very difficult to obtain, and, in addition to being very inhomogeneous, they show variations as large as 25% from their average values [63]. Several studies showed the dielectric parameter of the brain tissue to have significant age related variations [64–66], mainly due to changes in tissue water content. Furthermore, the degree of these parameters' uncertainty at low frequencies is even more pronounced.

It is often very hard to find reliable data for the tissue parameters of interest. Moreover, parameters may vary significantly between the healthy subjects and the patients, as well as due to difference in age and sex. Significant effort has been put in [42] to acquire reliable data on biological tissues, and today it represents the most frequently used reference for the non-living human and animal tissues. The natural variation of brain tissue parameters with frequency is shown on Fig. 7.2.

The frequency dependent parameters of the homogeneous adult human brain, given in Table 7.2, are taken from [42], as an average between the white and gray matter.

The biological tissues' parameters such as the permittivity and the electrical conductivity significantly affect the distribution of the induced fields in those tissues. The work on the rat brain [67] from the early 2000s reported the higher conductivity values in young rats, thus rekindling the questions whether the same is the case in human subjects. There are some measurements of the human brain tissue parameters [68] performed 10 hours post-mortem, but no reported studies on the changing values of parameters such as permittivity and conductivity in the living subjects.

Based on the total body water concept from [69], the values for permittivity ε_r and conductivity σ of a 10- and 5-year old brain, respectively, can be obtained using expressions (6.61) and (6.62).

The age-dependent parameters used in two child models, at the three particular frequencies of interest (2.44, 3 and 5 kHz), are given in Table 7.2.

7.1.3 Numerical Results

The numerical results for three typical TMS stimulation coils (circular, figure-of-eight and butterfly) are presented in this section. All coils are discretized using 80 linear segments. The radius of the circular coil is 4.5 cm [70], while the radius of the 8-coil and the butterfly coil (10 degrees between wings) is 3.5 cm. The number of coil windings is 14 and 15, respectively, in the circular and in the other two coils.

The parameters of three typical TMS coils are given in Table 7.3.

TABLE 7.3
TMS coil parameters. A sinusoidal waveform of 2.44 kHz is assumed flowing in the coil. From CVETKOVIĆ, Mario; POLJAK, Dragan; HAUEISEN, Jens. Analysis of transcranial magnetic stimulation based on the surface integral equation formulation. IEEE Transactions on Biomedical Engineering, 2015, 62.6: 1535–1545.

	Circular	8-coil	Butterfly
Frequency	2.44 kHz	2.44 kHz	2.44 kHz
Radius of turn	4.5 cm	3.5 cm	3.5 cm
Number of turns	14	15	15
Coil current	2843 A	2843 A	2843 A

The frequency of 2.44 kHz is chosen as an operating frequency of the three coils as the maximum of the induced current normalized amplitude occurs at this frequency [70]. A comparison of the other two frequencies most often used in the TMS analysis, 3 and 5 kHz, respectively, is carried out in the following section. The

amplitude of the current is assumed 2.843 kA, similar to [70].

In all three cases considered, the coils and surface of the brain (corresponding to primary motor cortex) were separated by 1 cm. It should be noted that there are also detailed models of TMS coils [71], taking into account the geometrical spacings between the multiple coil windings. However, as the actual coil geometry, i.e., the windings separation, the thickness of the coil insulation, and the casing are neglected in this case, the chosen separation from the coil geometric center should be valid [28].

7.1.3.1 Electric Field Due to Various TMS Coils

The results obtained using SIE formulation were compared to the analytical results obtained via (7.4) and (7.5). The maximum electric field values are presented in Table 7.4 while the induced electric field in coronal cross-section is shown on Fig. 7.3.

TABLE 7.4
Comparison of maximum electric field obtained via analytical expression and proposed model. From CVETKOVIĆ, Mario; POLJAK, Dragan; HAUEISEN, Jens. Analysis of transcranial magnetic stimulation based on the surface integral equation formulation. IEEE Transactions on Biomedical Engineering, 2015, 62.6: 1535–1545 [28].

		Circular	8-coil	Butterfly
		Analytical		
E_{max}	[V/m]	161.15	321.94	328.01
		SIE model		
E_{max}	[V/m]	86.83	118.28	138.41

The results show similar distribution of the electric field in both cases. The results obtained via the SIE model show much lower values than those from the analytical ones, as expected, because (7.4) does not account for the electric field due to accumulation of charges. It was shown in [47] that the electric field decreases due to the shielding effect of the surface charges being accumulated at the surface.

Fig. 7.3 confirms the well established fact that maximum electric field of the 8-coil is directly under the coil center, compared to the circular coil where it is under the windings. Moreover, a higher value of the electric field is obtained if the windings of the 8-coil are inclined as much as 10 degrees, as in the case of the but-

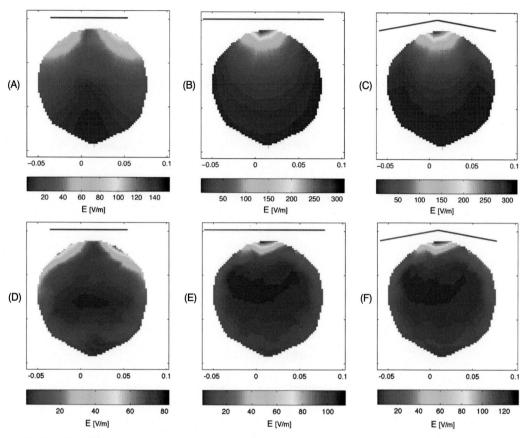

FIG. 7.3 Comparison of induced electric field in the human brain (coronal cross-section). The results on the left are obtained via analytical expressions for (A) circular, (B) 8-coil, and (C) butterfly coil, while the results on the right are obtained via proposed model for (D) circular, (E) 8-coil, and (F) butterfly coil. From [28].

terfly coil. Fig. 7.4 depicts the distribution of the electric field at the brain surface more clearly.

A comparison of the induced electric field on the coronal cross-section of the brain model, obtained using various numerical techniques such as SIE/MoM [28], FEM with cubical elements [72], complete and approximate boundary element method (BEM), complete and approximate FEM/BEM [73,74], and FEM codes using Sim4Life software, respectively, is shown on Fig. 7.5.

Table 7.5 gives a comparison of the maximum induced electric field (V/m) obtained using different numerical models for the case of the circular coil, as carried out in [75].

The comparison from Fig. 7.5 demonstrates that the results computed using the quasistatic solver and FEM method and the full wave analysis carried out via SIE/MoM do not exactly match. The electric field distribution over the cross-section is similar, but the maxi-

mum values obtained by different methods differ somewhat, as shown in Table 7.5. The interested reader can find the details on the particular formulation type and related solution method in the references found in [75].

7.1.3.2 Current Density

Knowing the distribution of the electric field inside the human brain, the induced current density J can be obtained using (7.10). The results for the maximum electric current density $J_{\max,\mathrm{ind}}$, for the homogeneous brain model, are given Table 7.6.

The calculated value of the current density given in Table 7.6 is comparable to the results reported in [70], where similar circular coil parameters were used on the four tissue brain model. The current density threshold value of 6 A/m² for the excitation of the motor cortex area was obtained in [70], while the SIE model obtained a maximum value of 7.483 A/m².

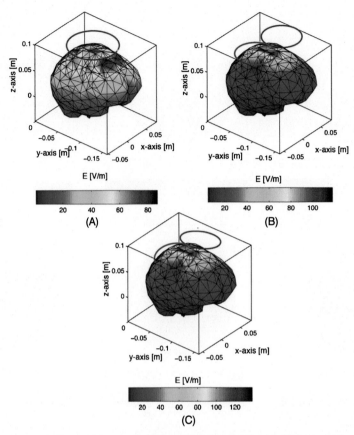

FIG. 7.4 Induced electric field on the brain surface due to: (A) circular coil, (B) figure-of-8 coil, and (C) butterfly coil. All coils are placed 1 cm over the primary motor cortex. From [28].

TABLE 7.5
Comparison of maximum induced electric field using various numerical models for the case of homogeneous human brain and circular TMS coil. From [75].

Triangles	SIE	Triangles	BEM, complete
360	115.9	696	117.1
696	122.8	1870	122.2
1224	134.5	Voxels	FEM/BEM, approx.
Triangles	BEM, approx.	5762 (6 mm)	93.6
696	108.6		FEM/BEM, complete
1870	115.6	5762 (6 mm)	93.6

7.1.3.3 Magnetic Flux Density

The results for the magnetic flux density, obtained using (7.8) and (7.9), were compared to the analytical results. The results for the maximum values are given in Table 7.7.

A comparison of the magnetic flux density in the coronal cross-section of the brain model is shown on Fig. 7.6.

The results from Table 7.7 and Fig. 7.6 indicate that the brain itself does not significantly disturb the mag-

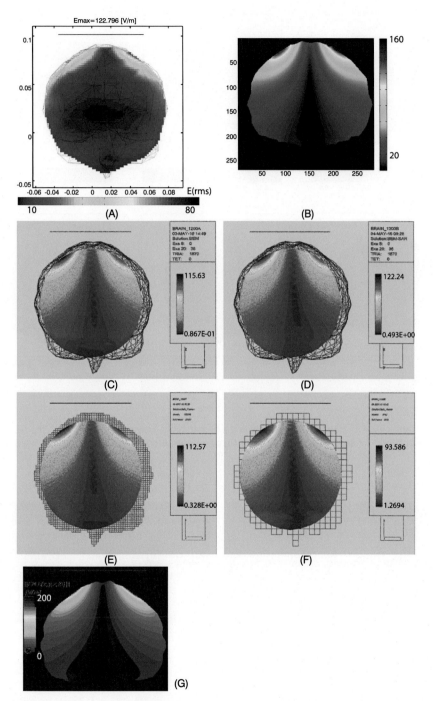

FIG. 7.5 Maps of the induced electric field due to circular TMS coil. Results obtained by: (A) SIE/MoM with 976, (B) FEM with cubical elements, (C) approximate BEM, and (D) complete BEM using 1870 triangles, respectively, (E) approximate FEM/BEM using 155546 voxels (2 mm), (F) complete FEM/BEM using 5394 voxel elements (5.5 mm), and (G) FEM using grid resolution of 0.5 mm. From [75].

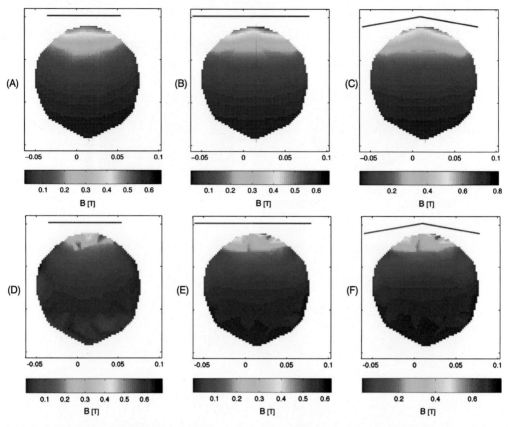

FIG. 7.6 Comparison of magnetic flux density in the human brain (coronal cross-section). The results on the left are obtained via analytical expressions for (A) circular, (B) 8-coil, and (C) butterfly coil, while the results on the right are obtained via proposed model for (D) circular, (E) 8-coil, and (F) butterfly coil. From [28].

TABLE 7.6
Comparison of the maximum value of induced electric current. From CVETKOVIĆ, Mario; POLJAK, Dragan; HAUEISEN, Jens. Analysis of transcranial magnetic stimulation based on the surface integral equation formulation. IEEE Transactions on Biomedical Engineering, 2015, 62.6: 1535–1545.

		Circular	8-coil	Butterfly
E_{max}	[V/m]	86.830	118.281	138.419
$J_{max,ind}$	[A/m^2]	7.483	10.194	11.929

TABLE 7.7
Comparison of maximum magnetic flux density. From CVETKOVIĆ, Mario; POLJAK, Dragan; HAUEISEN, Jens. Analysis of transcranial magnetic stimulation based on the surface integral equation formulation. IEEE Transactions on Biomedical Engineering, 2015, 62.6: 1535–1545 [28].

		Circular	8-coil	Butterfly
		Analytical		
B_{max}	[T]	0.679	0.672	0.826
		SIE model		
B_{max}	[T]	0.750	0.656	0.792

netic field of the coil, although a lower maximum value of the magnetic flux density was obtained for the 8-coil and butterfly coil. The distribution of the magnetic flux density in the coronal cross-section obtained using the

SIE model shows some discontinuities, which can be related to the interpolation method used. This numer-

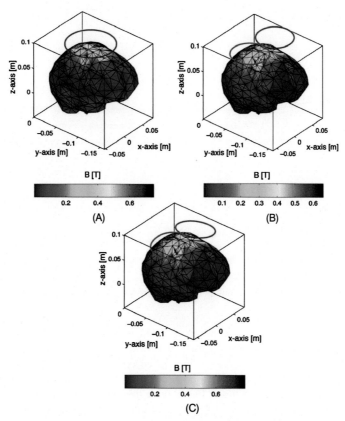

FIG. 7.7 Magnetic flux density on the brain surface due to: (A) circular coil, (B) figure-of-8 coil, and (C) butterfly coil. All coils are placed 1 cm over the primary motor cortex. From [28].

ical artifact could be overcome by calculating the field at more points before interpolating results in the neighboring area.

The magnetic flux density B on the brain surface can be clearly seen on Fig. 7.7.

It is interesting to observe the dependence of the induced electric field E and magnetic flux density B on the distance from the brain surface, as shown on Fig. 7.8.

From Fig. 7.8 the rapid decrease of both E and B fields directly under the geometric center of the stimulation coil is clearly evident in all three cases. For the circular coil, the maximum value is much lower compared to the other two coils as the maximum field will be induced under the coil windings, as shown on Fig. 7.4.

7.1.3.4 Pediatric Models Using Adult Brain Parameters

Following the analysis on the adult brain model, the induced electric field, magnetic flux density and induced current density are calculated in two models of a child brain. The calculation is first carried out using the adult tissue parameters in all three brain models, facilitating the comparison when using different brain size on the induced fields. Fig. 7.9 illustrates the effects of brain model size on the induced electric field distribution at the surface of three models, using three different coils [76].

Using the circular coil, the calculated maximum induced field is directly under the coil windings, while in case of figure-of-eight and butterfly coils, it is concentrated over a small area under the coil's geometric center.

Stimulation with same size coil in smaller brain models resulted not only in higher values of the induced electric field but also in the electric field being dispersed over a larger brain area (relative to the brain size). The higher values of the maximum induced electric field are seen in smaller brain models, except for the circular coil. This could be easily attributed to the fact

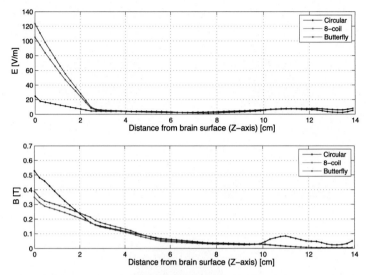

FIG. 7.8 Dependence of the induced electric field E and magnetic flux density B on the distance from the brain surface. The values given are on the points directly under the coil geometric center. From [28].

that in smaller brain models, the coil at the lateral brain parts is moved further from the surface, as seen in the first row of Fig. 7.9.

In case of the other two coils, the maximum obtained values are at the brain surface directly under the coil center, while in brain parenchyma the fields rapidly decay, as shown on Fig. 7.10. The maximum induced electric field value in two smaller models is higher compared to the adult model, at the same time showing a higher field gradient, i.e., decreasing more rapidly. Approximately 1.6 cm under the brain surface, the induced electric field values are similar for all three models.

The final comparison of the induced electric field in the adult brain model is shown in Fig. 7.11, in case of figure-of-eight coil operating at 2.44, 3 and 5 kHz, respectively.

The frequency dependent parameters of the adult brain, given in Table 7.2, are used. Stimulation by a higher frequency (sinusoidal current waveform of the stimulation coil) results in higher values of the induced electric field. The dashed line denotes the depth from the brain surface at which the electric field falls to half its maximum value. At all three frequencies, the E_{max} half value is obtained at the same depth.

7.1.3.5 Pediatric Models Using Age-Dependent Brain Parameters

The following analysis of TMS-induced fields takes into account the age dependent tissue parameters in case of two child models.

The results for the maximum values of induced electric field E_{max}, magnetic flux density B_{max} and the induced current density J_{max}, respectively, in the adult and child models, stimulated by figure of eight coil at three frequencies, are given in Table 7.8.

Table 7.8 shows that inclusion of the age dependent parameters (denoted by subscript "ε, σ") in two smaller brain models results in the increase of the induced current density J_{max}, while E_{max} and B_{max} remain practically the same.

The results suggest that the adult brain parameters are sufficient when modeling homogeneous child brains and when one is interested in the induced electric field only (e.g., in finding a region where reaching an electric field threshold value will result in the activation of that particular area). However, if one were interested in coupling the induced current and field distributions to neurophysiological equations [77–79], i.e., for calculating the transmembrane potentials of the nerve fibers in the brain [35], the age dependent parameters should not be neglected [76].

Moreover, the introduction of the age dependent parameters in two child models resulted in the change of electric field half maximum value surface distance ($0.5E_{max}$), as shown on Fig. 7.12.

In smaller brain geometries, the point with $0.5E_{max}$ is shifted closer to the surface, i.e., this value is around 7 mm for adult brain model, compared to 5.6 and 4.5 mm for 10- and 5-year old models, respectively.

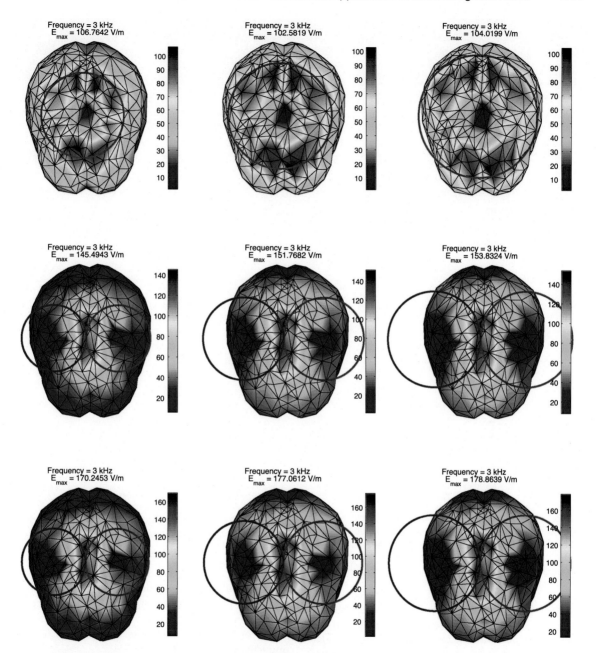

FIG. 7.9 Induced electric field at the surface of adult, 10-year old and 5-year old brain models (left to right), stimulated with three different coils at 3 kHz (top to bottom). The adult brain parameters at this particular frequency are used in the child models.

Fig. 7.13 shows the cumulative effects of the brain size and the age related parameters on the maximum induced current density J_{max} in three brain models using three different stimulation coils.

When the age dependent brain parameters (solid lines) are used, the decrease in brain size (age) is followed by the increase of the induced current values. On the other hand, when adult parameters are used in

TABLE 7.8

The maximum induced electric field E_{max}, magnetic flux density B_{max}, and induced current density J_{max} in the adult, 10-year old and 5-year old brain models, respectively, stimulated by figure-of-eight coil at three frequencies (2.44, 3, and 5 kHz).

	2.44 kHz			3 kHz			5 kHz		
	E_{max} [V/m]	B_{max} [T]	J_{max} [A/m^2]	E_{max} [V/m]	B_{max} [T]	J_{max} [A/m^2]	E_{max} [V/m]	B_{max} [T]	J_{max} [A/m^2]
adult	118.28	0.656	10.194	145.49	0.656	12.646	242.47	0.655	21.497
10-y[a]	123.37	0.773	10.633	151.76	0.776	13.192	252.84	0.774	22.417
5-y[a]	125.12	0.880	10.784	153.82	0.883	13.371	256.37	0.883	22.730
10-y$_{\varepsilon,\sigma}$[b]	123.37	0.775	13.237	151.78	0.775	16.282	252.85	0.774	27.388
5-y$_{\varepsilon,\sigma}$[b]	125.09	0.880	16.684	153.82	0.883	20.357	256.34	0.882	33.679

[a] Adult tissue parameters used in child model.
[b] Age dependent tissue parameters used in child model.

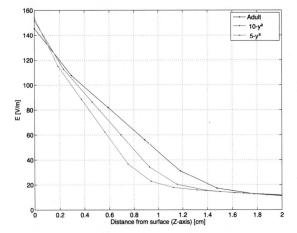

FIG. 7.10 Comparison of the induced electric field in the adult, 10-year old, and 5-year old brains stimulated by the figure-of-eight coil at 3 kHz. The adult brain parameters are used in the child models (denoted by subscript a). The location of points is directly under the geometric center of the coil.

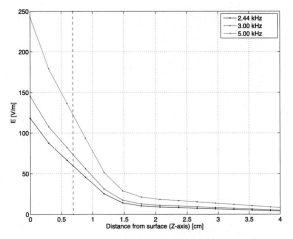

FIG. 7.11 Comparison of induced electric field in the adult brain model, due to stimulation by figure-of-eight coil at 2.44, 3 and 5 kHz.

two child brain models (dashed lines), a decrease in brain size is followed by a very small increase in the induced currents in all the cases except for the circular coils where the induced currents are decreased. This can be attributed to the smaller brains, i.e., the induced currents will *spread* over a wider area under the coil windings (as shown on Fig. 7.9), contrary to the 8-coil case where the maximum values will be obtained in a very narrow area directly under the coil.

The influence of a single brain tissue parameter such as the relative permittivity and the conductivity on the induced current density is shown in Tables 7.9 and 7.10.

Tables 7.9 and 7.10 give comparisons taking into account the following tissue properties: (a) adult tissue parameters, (b) age dependent parameters for both permittivity and the conductivity, (c) age dependent parameter only for the permittivity, and (d) only for the conductivity.

Results from Tables 7.9 and 7.10 show a very small effect on the induced current density if age dependence of permittivity is considered. Contrary, the age dependence of the conductivity will have a very significant effect on the calculated current values. Moreover, the induced current density is increased by the same factor used to scale the age dependent parameters, as evident from Tables 7.2, 7.9 and 7.10.

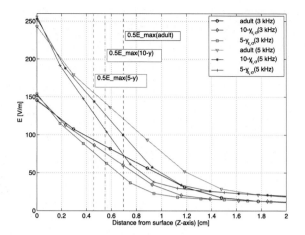

FIG. 7.12 Comparison of the induced electric field in the adult, 10-year old, and 5-year old brains stimulated by the figure-of-eight coil at 3 and 5 kHz. The age related parameters are used in the child models (denoted by subscript "ε, σ"). Dashed vertical lines represent the distance from the brain surface where the maximum electric field value E_{max} (from the surface) falls to half its value. This distance is the same for both frequencies.

7.2 NERVE FIBER EXCITATION

The widespread neurological disorders such as dementia, Alzheimer's, Parkinson's, epilepsy, multiple sclerosis, etc., are affecting hundreds of millions of people worldwide. These conditions pose an actual public health problems in developing countries, as estimated by high occurrences and the high numbers of un-

treated people. Therapeutic procedures involving electrical stimulation of nerves such as percutaneous electrical nerve stimulation (PENS), electro-acupuncture or transcutaneous electrical nerve stimulation (TENS) [79–81], are therefore widely used in the treatment of these neurological disorders.

The studies on the electrical excitation of nerves, among other aspects, involve: nerve excitation using stimulating electrodes, nerve conduction velocity tests or non-invasive stimulation of nerves.

The need for better understanding the complex functioning of the nervous system, and the nerve cells as its basis, has been continuously motivating researchers to carry out an efficient and accurate nerve fiber modeling. Thus, computational models of nerve fibers could facilitate a way to study the nerve fiber response to different stimulus waveforms, as often used within various electrotherapy techniques. Moreover, such a mathematical model can be beneficial and versatile tool for interpreting experimental results and analyzing variables for which there could be difficulties in the laboratory implementation.

7.2.1 Nerve Fiber Models

Basically, two types of electric potential occur in the stimulated nerve cells: the electrotonic potential and the action potential. The electrotonic potential, which is due to the local changes in the ion conductivity, decays along the length of the fiber, in which case the passive membrane shows linear nature satisfying Ohm's law. The action potential, on the other hand, is initiated

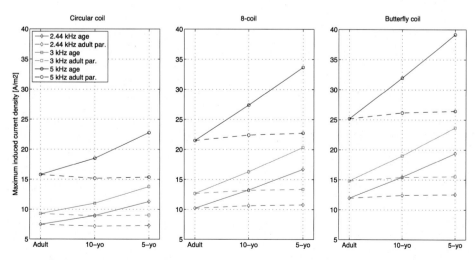

FIG. 7.13 Comparison of a maximum induced current density J in three brain models using three different stimulation coils at 2.44, 3 and 5 kHz.

TABLE 7.9
The maximum induced current density J [A/m^2] in the 10-year old brain model for three different coils at three frequencies.

Freq. [kHz]	Circular			8-coil			Butterfly		
	2.44	3	5	2.44	3	5	2.44	3	5
10-y[a]	7.1901	8.9167	15.157	10.633	13.192	22.417	12.404	15.390	26.153
10-y$_{\varepsilon,\sigma}$[b]	1.2450	1.2336	1.2218	1.2449	1.2343	1.2217	1.2448	1.2342	1.2219
10-y$_\varepsilon$[c]	1.0015	1.0013	1.0015	1.0013	1.0011	1.0014	1.0012	1.0011	1.0014
10-y$_\sigma$[d]	1.2432	1.2331	1.2206	1.2433	1.2327	1.2206	1.2436	1.2327	1.2205

[a] Adult tissue parameters used in child model.
[b] Age dependent tissue parameters used in child model. Induced currents normalized with respect to the values obtained using adult tissue parameters.
[c] Age dependent relative permittivity, adult parameters for conductivity.
[d] Adult parameters for relative permittivity, age dependent conductivity.

TABLE 7.10
The maximum induced current density J [A/m^2] in the 5-year old brain model for three different coils at three frequencies.

Freq. [kHz]	Circular			8-coil			Butterfly		
	2.44	3	5	2.44	3	5	2.44	3	5
5-y[a]	7.2911	9.0417	15.370	10.784	13.371	22.730	12.538	15.547	26.431
5-y$_{\varepsilon,\sigma}$[b]	1.5472	1.5222	1.4823	1.5471	1.5225	1.4817	1.5476	1.5225	1.4819
5-y$_\varepsilon$[c]	1.0045	1.0035	1.0038	1.0039	1.0035	1.0036	1.0039	1.0035	1.0034
5-y$_\sigma$[d]	1.5467	1.5203	1.4798	1.5448	1.5203	1.4769	1.5452	1.5204	1.4794

[a] Adult tissue parameters used in child model.
[b] Age dependent tissue parameters used in child model. Induced currents normalized with respect to the values obtained using adult tissue parameters.
[c] Age dependent relative permittivity, adult parameters for conductivity.
[d] Adult parameters for relative permittivity, age dependent conductivity.

when the threshold potential to which the membrane potential must be depolarized is achieved.

Most nerve fiber models are based on the idea of the neural membrane as a leaky capacitor [77,79,82–84]. In this case, nerve fibers are described by the partial differential equation named cable equation. This equation takes into account the leakiness and capacitance of the nerve fiber membrane, in addition to the finite resistance of the intracellular fluid. This model, in its most general form, assumes a very long electrically conducting core, surrounded by a membrane. Usually, the additional simplification is used by assuming a sufficiently small cross-section, thus enabling the problem to be treated as one-dimensional. Although the cable theory is rather simplified and usually assumes that neurons do not interact significantly, it lies at the core of our understanding how an individual neuron works.

One of the most revolutionary nerve fiber models, able to describe the action potential initiation and prop-

agation, as well as the voltage-gated ion channels, in the unmyelinated squid giant axon, was first published in 1952 by Hodgkin and Huxley [77], and latter extended to modeling the myelinated nerve fibers [85]. This model is expressed by a set of five ordinary nonlinear equations representing the reaction of the frog nerve fiber when stimulated by a current pulse. Latter work by McNeal in 1976 included the stimulation by placing the electrodes in the extracellular space [82], while in 1985 Reilly modified this model by including nonlinear Frankenhaeuser–Huxley equations at all Ranvier nodes [86]. This model, based on the amphibian data, is often known as Spatially Extended Nonlinear Node Model (SENN). Several years later, the first model based on the experimentally obtained data from myelinated fibers of rabbits was published, nowadays known after the authors as Chiu, Ritchie, Rogert, Stagg and Sweeney, or CRRSS, model [83,84].

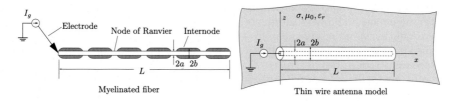

FIG. 7.14 Thin wire antenna model of the myelinated nerve fiber.

More recent models mainly tend to improve the cable theory. For example, Schnabel and Struijk analytically modeled the electric field distribution in a volume conductor which is presented as a set of concentric, homogeneous and infinitely long cylinders immersed in an unbounded domain [87], while Einziger et al. analytically derived the cable equation for the myelinated axon by extending the well-known cable equation for the unmyelinated type of axon [88]. In 2005 Heimburg and Jackson developed the soliton model which proposes the impulses to travel along the nerve fiber in the form of sound pulses known as solitons [89]. It overcomes several problems that were not explained by the previous models and as such represents a direct challenge to the widely accepted Hodgkin–Huxley model.

7.2.1.1 Nerve Fiber Antenna Model

Recently, a completely new approach of the nerve fiber modeling, based on the wire antenna theory, was introduced [81,90,91]. This approach, based on the antenna theory, relies on the corresponding Pocklington integro-differential equation in the frequency domain, whose solution results in the intracellular current distribution along the nerve fiber. While previous models and measurements were mainly oriented on determining the action potential, the nerve fiber antenna model, on the other hand, provides insight into the intracellular current distribution along the nerve fiber, which is another important parameter when determining the nerve fiber activation.

The myelinated nerve fiber with an arbitrary number of Ranvier's nodes and the corresponding straight thin wire antenna model [81,91,92] are shown in Fig. 7.14.

According to Fig. 7.14, L denotes the length of the nerve fiber, while a and b are the inner axon radius and the radius including the myelin coating, respectively. Electrode nerve fiber stimulation is taken into account by means of the equivalent current source I_g located at $L = 0$ (fiber beginning). The current generator I_g represents the nerve fiber stimulation used in electro-acupuncture or percutaneous electrical nerve stimulation (PENS), usually realized by inserting thin needles

through the skin. The fiber, oriented along the x axis, is assumed to be immersed in an unbounded homogeneous lossy medium, defined by electrical conductivity σ, permeability μ_0, and relative permittivity ε_r.

The mathematical formulation is based on the homogeneous Pocklington integro-differential equation for the unknown intracellular current I_a along the nerve fiber [91,92] given by

$$-\frac{1}{j4\pi\omega\varepsilon_{\text{eff}}} \int_0^L \left(\frac{\partial^2}{\partial x^2} - \gamma^2\right) g(x, x') I_a(x') \, dx' = 0,$$

(7.11)

while $g(x, x')$ is the Green's function for the lossy medium defined as

$$g(x, x') = \frac{e^{-\gamma R}}{R},$$

(7.12)

where R denotes the distance from the source to the observation point, respectively, γ is the complex propagation constant given by

$$\gamma = \sqrt{j\omega\mu\sigma - \omega^2\mu\varepsilon_0\varepsilon_r},$$

(7.13)

and ε_{eff} is the complex permittivity of a lossy medium given by

$$\varepsilon_{\text{eff}} = \varepsilon_0\varepsilon_r - j\frac{\sigma}{\omega},$$

(7.14)

with ω being the angular frequency.

The current generator I_g is taken into account via following boundary conditions for the nerve fiber:

$$I_a(0) = I_g, \quad I_a(L) = 0.$$

(7.15)

The boundary condition (7.15) at the nerve fiber end L is the so-called *sealed end boundary condition*, corresponding to the physical situation such as when the nerve fiber is transected during some injury or when performing some experiment on the isolated nerve fiber.

It is worth mentioning that the properties of the lossy medium, nerve fiber membrane and myelin sheath are considered via the conductivity and relative permittivity which are obtained from the cable and transmission line equations, respectively, as described in [91,92].

The Helmholtz equation for the corresponding transmembrane voltage is given by [91]

$$\frac{\partial^2 V}{\partial x^2} - \gamma_m^2 V = 0, \qquad (7.16)$$

where

$$\gamma_m = \sqrt{\frac{1 + j\omega\tau}{\lambda^2}} \qquad (7.17)$$

and V is a transmembrane voltage, while λ is the length constant defined by [92]

$$\lambda = \sqrt{\frac{r_m a}{2\rho_a}}, \qquad (7.18)$$

and τ is the time constant given by [8]

$$\tau = r_m c_m. \qquad (7.19)$$

The resistivity of axoplasm, as featured in (7.18), is denoted by ρ_a, while r_m and c_m, given in (7.19), are the myelin layer resistance for unit area and the capacitance of the Ranvier's node membrane per unit area, respectively.

After some rearranging, the following relations for the conductivity and relative permittivity can be obtained:

$$\sigma = \frac{2\rho_a c_m}{\mu a}, \qquad (7.20)$$

$$\varepsilon_r = \frac{2\rho_a}{a\omega^2 \mu \varepsilon_0 r_m}. \qquad (7.21)$$

7.2.1.2 Numerical Solution

Homogeneous Pocklington integro-differential equation (7.11) is solved via the Galerkin–Bubnov Indirect Boundary Element Method (GB-IBEM). The details could be found elsewhere, e.g., in [90]. A more general treatment of a thin-wire embedded in a lossy medium can be found in [93,94]. For the sake of brevity, the procedure is only briefly outlined in the following.

Therefore, using the boundary element formalism, (7.11) can be transformed into a set of linear equations, which can be written by the following matrix notation [91,95]:

$$\sum_{i=1}^{n} [Z]_{ji}^e \{\alpha\}_i^e = 0, \quad j = 1, 2, \ldots, n, \qquad (7.22)$$

where n is the number of wire segments, $[Z]_{ji}^e$ is the impedance matrix representing the interaction between the observation segment j and the source segment i, and $\{\alpha\}_i^e$ is the solution vector, expressed using the global nodes. The impedance matrix element is expressed by

$$[Z]_{ji}^e = -\frac{1}{j4\pi\omega\varepsilon_{\text{eff}}} \left[\int\limits_{\Delta l_j} \int\limits_{\Delta l_i} \{D\}_j \{D'\}_i^T g(x, x') \, dx \, dx' \right. $$
$$\left. - \gamma^2 \int\limits_{\Delta l_j} \int\limits_{\Delta l_i} \{f\}_j \{f'\}_i^T g(x, x') \, dx \, dx' \right], $$
$$(7.23)$$

where matrices $\{f\}$ and $\{f'\}$ contain the linear shape functions and matrices $\{D\}$ and $\{D'\}$ contain their derivatives. The length of the observation and source segment is given by Δl_j and Δl_i, respectively. The nerve fiber stimulation, taken into account via equivalent current generator, is incorporated into matrix system (7.22) using the boundary conditions (7.15).

7.2.2 Passive Nerve Fiber

When modeling the passive nerve fiber, the ionic current at each Ranvier's node has to be considered. This can be achieved by introducing the additional current sources at the non-activated Ranvier's nodes. The example of the passive nerve fiber model, having three Ranvier's nodes and four internodes, is depicted in Fig. 7.15.

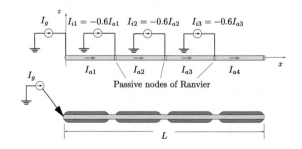

FIG. 7.15 Passive nerve fiber model.

According to Fig. 7.15 and the CRRSS (Chiu, Ritchie, Rogert, Stagg and Sweeney) model [79,84], the ionic current at each non-activated Ranvier's node can be considered to be approximately equal to $-0.6I_a$, where I_a denotes the intracellular current flowing into the observed Ranvier's node. This results in intracellular current I_{a2} leaving the first Ranvier's node having the value

TABLE 7.11
Values obtained for the ionic current analytical expression.

Constant	A	B	D	E	G	H	K
Unit	μA/ms	$(ms)^{-1}$	$(ms)^{-1}$	μA/ms	$(ms)^{-2}$	μA	$(ms)^{-2}$
Value	20.75	199.9	46.91	0.664	23	0.676	5000

of $0.4I_{a1}$. Similarly, the intracellular current I_{a3} flowing out of the second Ranvier's node is equal to $0.4I_{a2}$ while the intracellular current I_{a4} leaving the third Ranvier's node is equal to $0.4I_{a3}$.

7.2.3 Active Nerve Fiber

The model of an active nerve fiber, representing a myelinated nerve fiber using three active Ranvier's nodes and four internodes, is shown in Fig. 7.16.

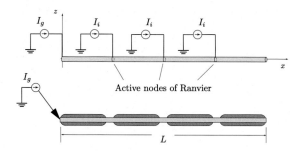

FIG. 7.16 Active nerve fiber model.

According to Fig. 7.16, each node is represented by a wire junction of two thin wires, representing an active Ranvier's node. The nerve fiber is stimulated by a current generator I_g imposed as a boundary condition at the fiber beginning. The three additional current sources introduced at the active Ranvier nodes represent an ionic current I_i of the particular activated node.

This ionic current could be determined by analyzing the CRRSS model and the resulting analytical expression for the ionic current could be obtained by curve fitting procedure [91]. This expression is given by

$$I_i(t) = -Au(t-t_1)\left[e^{-B(t-t_1)} - e^{-D(t-t_1)}\right]$$
$$+ Eu(t-t_1)e^{-G(t-t_2)^2} + He^{-K(t-t_3)^2}, \quad (7.24)$$

where $t_1 = 31$ μs, $t_2 = 252$ μs, $t_3 = 26.25$ μs, while constants $A, B, D, E, G, H,$ and $K,$ and their respectively units, are given in Table 7.11.

An illustration of the ionic current, determined for the nerve fiber having diameter of 20 μm, is depicted in Fig. 7.17.

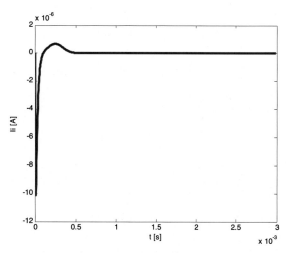

FIG. 7.17 Ionic current of activated node of Ranvier.

Using a thin wire junction model, an active node, on the other hand, can be represented by two thin wires, separated by some infinitesimally small distance Δx, as shown in Fig. 7.18.

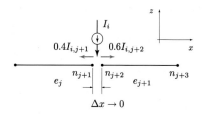

FIG. 7.18 Two wire junction representation of the active node.

Although the thin wires do not have direct electrical contact, it can be said that nodes n_{j+1} and n_{j+2} are at the same position. In order for this geometry to behave like the real wire junction, Kirchhoff's law has to be sat-

TABLE 7.12			
Nerve fiber parameters according to [81].			
Parameter	**Description**	**Value**	**Unit**
b	Radius including the myelin sheath	10	µm
a	Inner axon radius	0.64·b	µm
l_{int}	Internode length	100·2b	µm
L	Nerve fiber length	l_{int}·10	µm
ρ_a	Resistivity of the axoplasm	1.1	Ωm
r_m	Myelin resistance for unit area	10	Ωm²
c_m	Membrane capacitance per unit area	7.3	µF/m

isfied at the junction [96]:

$$I_i = 0.4I_{i,j+1} + 0.6I_{i,j+2}, \qquad (7.25)$$

where $I_{i,j+1}$ represents the ionic current value flowing out of the junction in one direction and $I_{i,j+2}$ is current flowing out of the junction in the opposite direction.

Determining the intracellular current in the activated nerve fiber is carried out in several steps. First, the nerve fiber is stimulated by a current generator I_g at the fiber beginning, similar to the case of the passive nerve fiber. If the stimulating current exceeds the threshold value at the corresponding Ranvier's node, the second current source, representing the ionic current, activates. It should be noted here that the threshold for the nerve fiber activation depends on the strength and duration of the stimulus [97–99]. The intracellular current for the activated fiber is obtained as a sum of the node activation current and the ionic current of the activated node. The resulting current, flowing out of the activated node, then activates the next node. The same procedure is repeated for each node along the nerve fiber.

7.2.4 Computational Examples

Some illustrative computational examples for the intracellular current along the nerve fiber due to excitation to equivalent current source are presented in the following.

7.2.4.1 Numerical Results for Passive Nerve Fiber

The intracellular current is first calculated for the passive nerve fiber. The length of the nerve fiber is 2 cm, represented by 9 Ranvier's nodes and 10 corresponding internodes. The parameters used in the antenna model of the fiber nerve are given in Table 7.12 [81].

Using a subthreshold rectangular current pulse, as shown in Fig. 7.19, the fiber is stimulated at its be-

ginning ($L = 0$). The location of Ranvier's nodes is at $x = 2, 4, 6, 8, 10, 12, 14, 16$, and 18 mm along the fiber length.

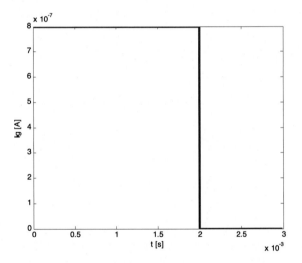

FIG. 7.19 Rectangular subthreshold current pulse.

Using the Inverse Fast Fourier Transform (IFFT), the frequency domain results are transformed to the time domain. The intracellular current distribution, at time $t = 1$ ms, along the passive nerve fiber is shown in Fig. 7.20. From the same figure, a comparison to the CRRSS model [79,83,84] can be seen.

Figs. 7.21 and 7.22 show the subthreshold intracellular current versus time in the Ranvier's nodes 2, and 6, respectively.

The results for the intracellular current obtained by the antenna model seem to be in a rather satisfactory agreement to the results obtained with the CRRSS model. From Figs. 7.21 and 7.22 it can be seen that the amplitude difference between the two models in-

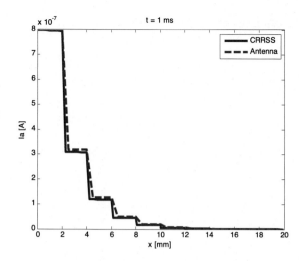

FIG. 7.20 Intracellular current along the passive nerve fiber, $t = 1$ ms, $L = 2$ cm.

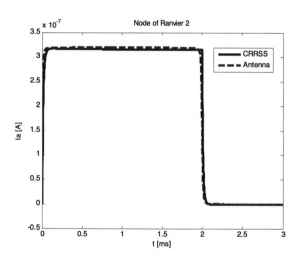

FIG. 7.21 Intracellular current in the passive Ranvier's node 2, $L = 2$ cm.

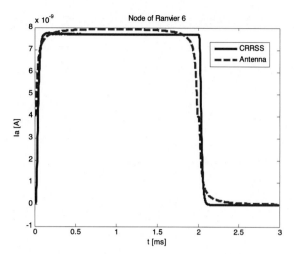

FIG. 7.22 Intracellular current in the passive Ranvier's node 6, $L = 2$ cm.

internodes. A rectangular current pulse used to depolarize the nerve fiber membrane is plotted in Fig. 7.23.

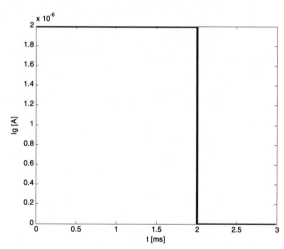

FIG. 7.23 Rectangular superthreshold current pulse.

creases at nodes positioned towards the nerve fiber end. The reason for this is that in the antenna model, the signal attenuates slower along the fiber compared to CRRSS model. Also, note that the intracellular current waveform in Ranvier's nodes follows the form of the excitation pulse, owing to the subthreshold current stimulation, the voltage gated channels are not activated.

7.2.4.2 Numerical Results for Active Nerve Fiber

The intracellular current for the active nerve fiber is calculated again for a model with 9 Ranvier's nodes and 10

The current pulse shown in Fig. 7.23 excites the nerve fiber at the fiber beginning. Once a current threshold value of 1.5 µA reaches the first Ranvier's node, the other current source activates. Summing the intracellular current activating the node and the ionic current of the following current source yields the resulting intracellular current in the given activated Ranvier's node. The resulting current then activates the following Ranvier's node. A plot of the obtained intracellular current is shown in Fig. 7.24.

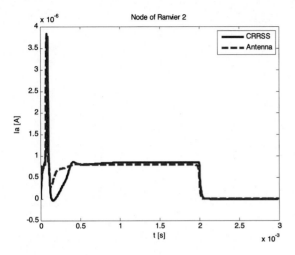

FIG. 7.24 Intracellular current in the active Ranvier's node 2, $L = 2$ cm.

The results for the intracellular current obtained via antenna fiber model and CRRSS model show a rather satisfactory agreement to a certain extent. Some minor discrepancies appear due to the ionic current approximation. Intracellular current component related to the rectangular current pulse is in a very good agreement.

The intracellular current for the sixth Ranvier's node is shown in Fig. 7.25.

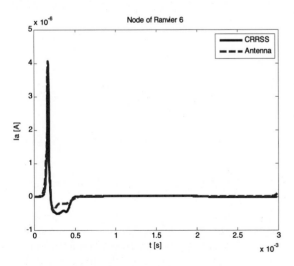

FIG. 7.25 Intracellular current in the active Ranvier's node 6, $L = 2$ cm.

Obviously, the rectangular current pulse decays along the fiber, while the intracellular current part keeps almost the same amplitude at the activated node. The

signal of the activated nerve propagates along the fiber almost without attenuation.

Finally, Fig. 7.26 shows the intracellular current distribution along the nerve fiber, plotted for the time instant $t = 0.2$ ms and obtained using the nerve fiber antenna and CRRSS models.

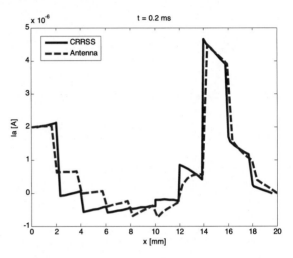

FIG. 7.26 Intracellular current along the active nerve fiber, $t = 0.2$ ms, $L = 2$ cm.

There is a satisfactory agreement between the results, with some visible discrepancies in amplitude of the intracellular current traveling along the fiber, mainly due to the instability of numerical procedure in the antenna model. Moreover, the discrepancy is due to the approximate way that the fiber active state is taken into account. Namely, the additional current source should take into account all the biological effects occurring during the activation of the fiber. Nonetheless, the wire antenna model offers a completely new approach to nerve fiber modeling, in addition to a promising perspective of facilitating a simple way of coupling the nerve fiber model to an external electromagnetic field excitation.

7.3 LASER RADIATION

Since the first working demonstration of a ruby laser by Teodore Maiman in 1960, lasers have become a very important tool in many industries such as entertainment, communications and science. It is therefore of no surprise that from the early days medicine has also embraced this electromagnetic (EM) source as a form of treatment [100], particularly in ophthalmology and dermatology. Nowadays, laser systems, being promoted as a sort of 21st century medical knife, are implemented in eye surgery procedures covering the whole EM spectrum from ultraviolet (UV) to infrared (IR) wavelengths [100–102].

Modern ophthalmological procedures include the ArF (argon fluoride) excimer systems used in laser in situ keratomileusis (LASIK) and photorefractive keratectomy (PRK) procedures, ruby lasers used in welding of detached retina, Nd:YAG (neodymium-doped yttrium aluminum garnet) laser used in after-cataract and glaucoma treatment, Ho:YAG (holmium:YAG) laser for contactless thermal keratoplasty, to name only a few.

The widespread use of lasers and the application in various eye surgeries have resulted in the need to understand and quantify the basic interaction between the laser light and the human eye tissues. Electromagnetic wave incident on the human eye will result either in a transmission, reflection, scattering or absorption of the EM energy. Among these processes, absorption is the most important as photon energy absorbed by a tissue results in reemitted radiant energy, or in EM energy transformed into heat [102], in turn increasing the temperature of the tissue.

The blood flow is a key element in the temperature regulation of a living organism. However, the human eye is a special case, since the blood flow inside an eye is not sufficient to regulate the heating. Consequently, one of the most important tasks in surgical application of lasers is the assessment of temperature variation in tissues subjected to high intensity laser radiation before the operation takes place. Due to this fact, several models were developed to predict the various processes inside the human eye, with the ultimate goal of helping the medical doctors in minimizing the possible damage to intraocular tissue while operating with laser.

Some early work on quantifying the effects of EM radiation on the temperature rise in eye tissues has been carried out by Taflove and Brodwin [103], and Emery et al. [104], who determined the intraocular temperature distribution due to microwave radiation using the finite difference and the finite element method (FEM), respectively. During the 1980s, Scott [105,106] studied the cataract formation due to IR radiation, using a two-dimensional (2D) FEM model of heat transfer inside the human eye, while Lagendijk [107] used finite differences to determine the temperature rise inside the human and animal eyes undergoing hyperthermia treatment.

However, the earliest effort on the analysis of laser–tissue interaction and the eye thermal response to IR radiation has been carried out by Mainster [108]. Following this pioneer effort, various models have been developed in the following decades, including the ther-

mal model of the eye exposed to visible and infrared lasers by Amara [109], and a three-dimensional (3D) model of the eye irradiated by argon laser by Sbirlea and L'Huillier [110], to mention just a few.

In the last decade or so, many human eye models have been developed, ranging from a geometrically simple ones [111–113], to a detailed 2D [114,115] and 3D models [116,117], taking into account numerous eye tissues. Although some methods of solving heat transfer problems in a human eye include semianalytical approach [112], generally, numerical techniques based on the use of finite volume [111,114], finite element [118] or boundary element methods [116], respectively, are employed.

7.3.1 Laser–Eye Interaction

Before going into details of modeling laser–eye interaction, it is useful to briefly outline the necessary steps and to specify the required parameters that need to be taken into account. As shown on Fig. 7.27, the first step is the calculation of heat generated inside eye tissues caused by the endogenous processes but also due to external sources such as laser radiation.

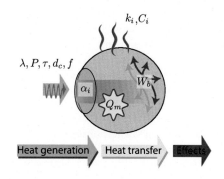

FIG. 7.27 Required parameters for modeling laser–eye interaction.

Heat generated due to laser radiation depends on parameters such as the wavelength λ, power density P, exposure time τ, spot size d_c, and the frequency of the applied pulses f. Moreover, generated heat is dependent on the tissue optical properties, such as the absorption coefficient α, which is not readily available.

Finally, to account for the endogenous heat sources, parameters such as the volumetric perfusion rate of blood W_b and the internal volumetric heat generation Q_m, i.e., metabolic rate, are necessary. These are always present in the organism, and are responsible for maintaining the constant body temperature.

TABLE 7.13
Thermal properties of human eye tissues.

	W_b [kg/m^{3}s]	Q_m [W/m^3]	k [J/m s K]	C [J/kg K]	ρ [kg/m^3]
Vitreous humour	0	0	0.594	3997	1009
Lens	0	0	0.400	3000	1100
Aqueous humour	0	0	0.578	3997	1003
Cornea	0	0	0.580	4178	1076
Sclera	0	0	0.580	4178	1170
Cilliary body	2700	690	0.498	3340	1040
Choroid	0	0	0.530	3840	1060
Retina	35000	10000	0.565	3680	1039

On the other hand, the heat transfer to surrounding tissues is characterized by the tissue thermal properties such as the thermal conductivity k and the specific heat capacity C. The values of these parameters are readily available in the literature, e.g., [119], and are given in Table 7.13.

7.3.2 Tissue Optical Parameters

The absorption of laser light by a specific target tissue is the fundamental goal of clinical lasers [102]. In biological tissues, absorption is mainly caused by either water molecules or macromolecules such as proteins and pigments. Absorption in the IR region of the spectrum can be primarily attributed to water molecules, while proteins and pigments are main absorbers in UV and visible range of the spectrum [102,120,121].

The tissue components responsible for photon absorption, known commonly as chromophores, are primarily dependent on the wavelength. Human eye chromophores that absorb light in the visible spectrum are melanin, located in the retinal pigment epithelium, iris pigment epithelium, uvea, and trabecular meshwork, hemoglobin, located in red blood cells within blood vessels, and xanthophyll, located in the macular region of the retina [100].

The strong wavelength dependence is the reason for absorption coefficient being the most important tissue optical parameter, and the fact that different eye tissues are strongly absorbing at specific wavelengths while weakly absorbing at others is the fundamental principle medical lasers are based on [102].

When a beam of light from the visible spectrum range is incident on the eye, on its way to photoreceptors in the retina, it passes relatively weakly absorbing tissues such as cornea, aqueous humour, lens, and vitre-

ous humour. On the other hand, as cornea is strongly absorbing in the ultraviolet, it will protect the light-sensitive cells on the retina from this high energy radiation.

Unfortunately, the data for the absorption coefficient are not readily available for various eye tissues. A number of methods are known for measuring the optical properties of tissues, as outlined in [122]. Generally, one can classify them as direct techniques, using the Lambert–Beer's law, and indirect techniques, where one needs to implement a complex theoretical model of light scattering in order to obtain the parameter.

The method based on Lambert–Beer's law uses the values of transmittances to obtain the absorption coefficient α as

$$\alpha = \frac{1}{d} \ln |T_{\text{tran}}|, \qquad (7.26)$$

where d is the thickness of a specific tissue and T_{tran} is the tissue total transmittance. The values for transmittances can be found in [123,124].

For the choroid, where the data on the reflectivity of the posterior tissue (sclera) and the anterior tissue (fundus) is available, the absorption coefficient can be calculated according to:

$$\alpha = \frac{1}{d} \ln(\frac{I_0'}{I_0}), \qquad (7.27)$$

$$I_0' = I_0[T_{\text{tran}}(1 + R_{pi}(1 + R_{ai}))], \qquad (7.28)$$

where I_0 and I_0' are the intensities of the incident and the transmitted beams, respectively, R_{pi} is reflectance (%) of the posterior tissue and R_{ai} is reflectance (%) of the anterior tissue.

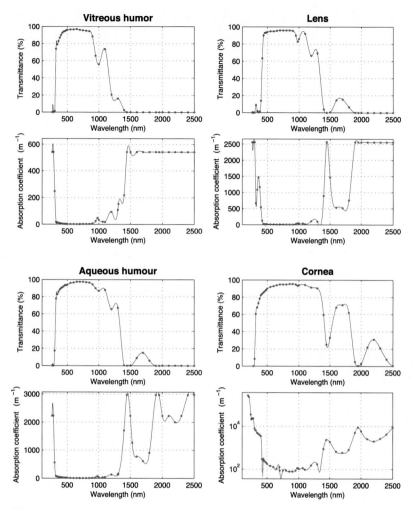

FIG. 7.28 Wavelength dependent transmittances and absorption coefficients for vitreous humour, lens, aqueous humour, and cornea.

Absorption coefficient values reported in a number of respected papers [109,110,120,121,125–132] were added to the calculated ones. The wavelength dependent transmittances and the calculated absorption coefficients of several human eye tissues are shown on Fig. 7.28, while absorption coefficients for specific laser wavelengths are given in Table 7.14, as reported in our previous work [115,133,134].

7.3.3 Model of the Human Eye

Although relatively small in comparison to other organs, measuring circa 24 mm along pupillary axis and 23 mm in diameter, the human eye is a complex optical system comprising many different tissues. In order to model the eye as accurately as possible, various tissues need to be taken into account, as well as their thermal, optical and electrical parameters.

We have assumed our model of the eye to be a solid structure of given dimensions, consisting of total of eight homogeneous tissues, namely, cornea, aqueous humour, cilliary body, lens, vitreous humour, retina, choroid, and sclera. To facilitate the model development at this stage, some simplifications regarding the anatomy have been made. For example, omitting the optic nerve provides a possibility of extending the simple two-dimensional (2D) model to three-dimensions (3D) by a simple rotation around the pupillary axis.

TABLE 7.14
Absorption coefficient α [1/m] of eye tissues.

Tissue/wavelength	193 nm	694.3 nm	1053 nm	1064 nm	2090 nm
Vitreous humour	542.7	2	22.4	20	542.7
Lens	2558.4	9.5	43.2	43.5	2558.4
Aqueous humour	2228.3	8.4	36.7	35	2228.3
Cornea	270000	124	111.7	113	2923.8
Sclera	28800	358.5	560	634.8	2923.8
Cilliary body	2228.3	8.2	42.9	42.5	2228.3
Choroid	48475	16848	6671.7	6615	6398.2
Retina	6526.2	44000	594.1	10000	4370.3

Fig. 7.29 shows the geometry of our 3D model of the eye, with denoted eye tissues, obtained by rotating one half of a 2D model around the central axis by 360 degrees.

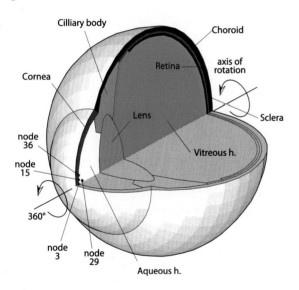

FIG. 7.29 Three-dimensional model of the human eye and the associated eye tissues.

7.3.4 Laser Source Modeling

When modeling laser source irradiating a human eye, one can use a simplifying fact that many laser beams take the form of a Gaussian profile. This corresponds to the fundamental transverse mode, or a TEM$_{00}$ mode, of a laser. This consideration is valid when the beam divergence is very small. In that case, the solutions for the electric field and intensity can be represented in the

form of a Gaussian function [101], e.g., in cylindrical coordinates (r, z), the intensity distribution at a time t can be written as follows:

$$I(r, z, t) = I_0 e^{-\alpha z} e^{-\frac{2r^2}{w^2}} e^{-\frac{8t^2}{\tau^2}}, \qquad (7.29)$$

where I_0 is the incident value of intensity, w and τ are the Gaussian beam waist and pulse duration, respectively.

Fig. 7.30 illustrates the laser beam propagating in the z-direction with a typical Gaussian profile depicted above. Along each transversal plane of the propagation axis, the beam is of the same spatial profile.

The same Fig. 7.30 shows the temporal profile of the pulse power or intensity, denoted by the second exponential term in Eq. (7.29).

The absorbed laser energy $H(r, z, t)$, by some specific eye tissue located at cylindrical coordinates (r, z), can be calculated from

$$H(r, z, t) = \alpha I(r, z, t), \qquad (7.30)$$

where α is the wavelength dependent absorption coefficient of the specific tissue and $I(r, z, t)$ is the intensity at that particular coordinates, at time t.

Thus, the first step is to calculate the laser beam intensity at corneal surface using

$$I_c = \frac{4P}{d_c^2 \pi}, \qquad (7.31)$$

where P denotes the laser power and d_c is the beam diameter on the cornea.

This is followed by determining the beam diameter being formed on the retina. This diameter will be much smaller than the area of the pupillary opening due to

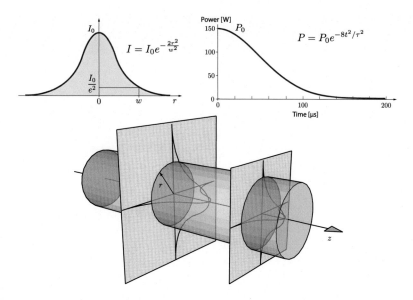

FIG. 7.30 Spatial and temporal profile of a Gaussian beam propagating in the z-direction.

the diffraction on this aperture [135]. If we take into account the focusing action of the lens, the retinal beam diameter can be related to the wavelength and the size of a pupillary aperture by

$$d_r = 2.44 \frac{\lambda f}{d_p}, \tag{7.32}$$

where λ is wavelength of the laser beam, f is the focal distance of the lens taken to be 17 mm, and d_p is the diameter of a pupillary opening.

The beam intensity on the retina can then be determined using

$$I_r = I_c \frac{d_p^2}{d_r^2}. \tag{7.33}$$

From the calculated values of d_r and I_r, we can now interpolate the values for the beam width and intensity at each intermediate position along the laser beam path. Thus, we have assumed the beam propagating through all tissues along the pupillary axis. As a result, the tissue along this same axis will have absorbed the energy of a laser pulse depending on the tissue absorption coefficient, given in Table 7.14.

We must note here that Eq. (7.33) represents the worst case scenario where all optical power of the stable laser beam is focused onto a very small spot on the retina. In a realistic situation, the human eye is not a such a good optical device. Additionally, due to tiny eye movements, the size of the focused image on the retina

will be much larger than that expressed by Eq. (7.32) [136].

7.3.5 Heat Transfer in the Human Eye

The starting point for all mathematical analysis of bioheat transfer in the living tissues is the Pennes' equation [137]. Although the equation has been criticized over the years [138], it is nonetheless one of the most frequently used approaches for the analysis of bioheat transfer phenomena. The Pennes' equation, given by

$$\rho C \frac{\partial T}{\partial t} = \nabla (k \nabla T) + W_b C_{pb} (T_a - T) + Q_m + H, \tag{7.34}$$

represents the energy balance between the conductive heat transfer, the endogenous heat generated by the metabolic processes and the effects due to the blood vessels acting as a thermal sinks or sources.

The left-hand side of Eq. (7.34) features ρ and C, denoting the density and specific heat capacity of the tissue, respectively, W_b represents the volumetric perfusion rate, C_{pb} is the specific heat capacity of blood, and Q_m is the volumetric heat generated by specific tissue.

The bioheat equation (7.34) features an additional term, H, representing the heat deposited in the tissue due to some external source. In this case, H is the deposited energy due to a laser radiation, but it could be any other source, e.g., of electromagnetic radiation [139].

TABLE 7.15
Various parameters used in the eye thermal model.

Parameter	Label	Value	Unit
Heat transfer coefficient of cornea	h_c	14	W/m^{2}K
Heat transfer coefficient of sclera	h_s	65	W/m^{2}K
Stefan–Boltzmann constant	σ	$5.67 \cdot 10^{-8}$	W/m^{2}K^4
Emissivity of cornea	ϵ	0.975	–
Evaporation rate of cornea	E_{vap}	40	W/m^2
Ambient air temperature	T_{amb}	25	°C
Arterial blood temperature	T_a	36.7	°C

The bioheat transfer equation is supplemented with natural boundary conditions for the surface of cornea and sclera, denoted by Γ_1 and Γ_2, respectively:

$$-k\frac{\partial T}{\partial n} = h_c\left(T - T_{amb}\right) + \sigma\epsilon\left(T^4 - T^4_{amb}\right) + E_{vap} \in \Gamma_1, \tag{7.35}$$

$$-k\frac{\partial T}{\partial n} = h_s\left(T - T_a\right) \in \Gamma_2, \tag{7.36}$$

where k represents the specific thermal conductivity of tissue, given in Table 7.13, h_c and h_s are the heat transfer coefficients of cornea and sclera, respectively, σ is the Stefan–Boltzmann constant, ϵ is the corneal surface emissivity, E_{vap} is the corneal surface evaporation rate, T_{amb} is the ambient air temperature, and T_a is the temperature of the arterial blood. The typical values of these parameters used in our model are given in Table 7.15.

A representation of the complete eye domain including the associated boundaries is depicted in Fig. 7.31.

It should be noted that Eq. (7.35) represents the heat exchange between the corneal surface and surrounding air taking into account the convection, radiation and evaporation processes. Eq. (7.36), on the other hand, represents the heat exchange between the scleral tissue and surrounding ocular globe due to convection only.

Some simplifications are considered related to Eq. (7.35). Namely, to avoid the iterative procedure required when dealing with nonlinear terms, such as the second term representing the heat released to the surroundings q_{rad}, we use the following binomial expansion:

$$\left(T^4 - T^4_{amb}\right) = \left(T - T_{amb}\right)\left(T + T_{amb}\right)\left(T^2 + T^2_{amb}\right). \tag{7.37}$$

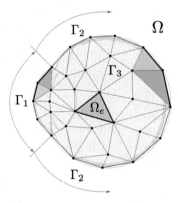

FIG. 7.31 Illustration of the human eye domain Ω meshed with triangular finite elements Ω_e; Γ_1 and Γ_2 are boundaries on cornea and sclera, respectively.

This results in q_{rad} being expressed as

$$q_{rad} = \sigma\epsilon T^3_{app}\left(T - T_{amb}\right), \tag{7.38}$$

where T^3_{app} is given by

$$T^3_{app} = \left(T + T_{amb}\right)\left(T^2 + T^2_{amb}\right). \tag{7.39}$$

Usually, the following step, as in [109], is to approximate Eq. (7.39) by T^3_F, where a constant value of $T_F = 100°C$ is used. However, this way the values for q_{rad} are underestimated, particularly at higher temperatures resulting when the eye is radiated by a laser [117].

A much better approach would thus be to approximate the value of T_{app} by 10°C segments, while taking into account different values of ambient temperature (20, 25, 30°C), as done in [117].

7.3.6 Numerical Solution of the Heat Transfer

The numerical solution of bioheat equation (7.34) is carried out using the finite element method (FEM). First, the steady state variant of Eq. (7.34) is solved, where the left-hand side is equated to zero and the last term from the right-hand side, representing the source term, is omitted. Thus, the bioheat equation reduces to

$$\nabla(k\nabla T) + W_b C_{pb}(T_a - T) + Q_m = 0. \quad (7.40)$$

More details on this can be found in Sect. 6.2, where the solution is also carried out using the dual reciprocity boundary element method (BEM).

In the following transient analysis, i.e., when the laser is irradiating the eye, the results obtained from the steady state case are utilized as the initial conditions.

Multiplying Eq. (7.34) using the weighted functions W_j and integrating over the complete domain Ω, the following integral formulation is obtained:

$$\int_\Omega \rho C \frac{\partial T}{\partial t} W_j d\Omega = \int_\Omega \nabla(k\nabla T) W_j d\Omega$$
$$+ \int_\Omega W_b c_b (T_a - T) W_j d\Omega$$
$$+ \int_\Omega (Q_m + H) W_j d\Omega. \quad (7.41)$$

Integration by parts followed by an application of generalized Gauss' theorem yields:

$$\int_\Omega \nabla(k\nabla T) W_j d\Omega = \oint_\Gamma k \frac{\partial T}{\partial n} W_j d\Gamma - \int_\Omega k\nabla T \nabla W_j d\Omega, \quad (7.42)$$

$$\int_\Omega \nabla \vec{A} d\Omega = \oint_\Gamma \vec{A}\vec{n} d\Gamma. \quad (7.43)$$

Substituting (7.42) and (7.43) into Eq. (7.41) leads to the weak formulation of the Pennes' bioheat equation:

$$\int_\Omega \rho C \frac{\partial T}{\partial t} W_j d\Omega = \oint_\Gamma k \frac{\partial T}{\partial n} W_j d\Gamma - \int_\Omega k\nabla T \nabla W_j d\Omega$$
$$+ \int_\Omega W_b c_b (T_a - T) W_j d\Omega$$
$$+ \int_\Omega (Q_m + H) W_j d\Omega. \quad (7.44)$$

The expressions for the boundary conditions of the eye model are given by:

$$-\int_{\Gamma_1} k \frac{\partial T}{\partial n} W_j d\Gamma = \int_{\Gamma_1} (h_c + \sigma\epsilon T_{app}^3) T W_j d\Gamma$$
$$- \int_{\Gamma_1} (h_c T_{amb} + \sigma\epsilon T_{app}^3 T_{amb}$$
$$+ E_{vap}) W_j d\Gamma, \quad (7.45)$$

$$-\int_{\Gamma_2} k \frac{\partial T}{\partial n} W_j d\Gamma = \int_{\Gamma_2} h_s T W_j d\Gamma - \int_{\Gamma_2} h_s T_a W_j d\Gamma. \quad (7.46)$$

Expanding the unknown temperature over a finite element using a linear combination of shape functions N_j gives

$$T^e = \sum_{j=1}^M \zeta_j N_j, \quad (7.47)$$

where M represents the number of local nodes of the finite element, i.e., in case of the tetrahedral elements used in 3D model, $M = 4$.

Eq. (7.47) can be written in the matrix form

$$T^e = \{N\}^T \{\zeta\} = \begin{bmatrix} N_1 & N_2 & N_3 & N_4 \end{bmatrix} \begin{bmatrix} \zeta_1 \\ \zeta_2 \\ \zeta_3 \\ \zeta_4 \end{bmatrix}, \quad (7.48)$$

where vector $\{\zeta\}$ represents the unknown coefficients to be solved for.

The expansion of the temperature gradient in terms of the shape functions is given by the following:

$$\nabla T = \begin{bmatrix} \frac{\partial T}{\partial x} \\ \frac{\partial T}{\partial y} \\ \frac{\partial T}{\partial z} \end{bmatrix} = \begin{bmatrix} \frac{\partial N_1}{\partial x} & \frac{\partial N_2}{\partial x} & \frac{\partial N_3}{\partial x} & \frac{\partial N_4}{\partial x} \\ \frac{\partial N_1}{\partial y} & \frac{\partial N_2}{\partial y} & \frac{\partial N_3}{\partial y} & \frac{\partial N_4}{\partial y} \\ \frac{\partial N_1}{\partial z} & \frac{\partial N_2}{\partial z} & \frac{\partial N_3}{\partial z} & \frac{\partial N_4}{\partial z} \end{bmatrix} \begin{bmatrix} \zeta_1 \\ \zeta_2 \\ \zeta_3 \\ \zeta_4 \end{bmatrix}. \quad (7.49)$$

Inserting (7.48) and (7.49) into Eq. (7.44) and using the same weighting functions as the shape functions, i.e., $W_j = N_j$, leads to

$$\frac{\rho C}{\Delta t} \int_\Omega T^i W_j d\Omega + \int_\Omega (k\nabla T^i \nabla W_j + W_b c_b T^i W_j) d\Omega$$
$$+ \int_{\Gamma_1} (h_c + \sigma\epsilon T_{app}^3) T^i W_j d\Gamma_1 + \int_{\Gamma_2} h_s T^i W_j d\Gamma_2$$
$$= \int_{\Gamma_1} (h_c T_{amb} + \sigma\epsilon T_{app}^3 T_{amb} + E_{vap}) W_j d\Gamma_1$$
$$+ \int_{\Gamma_2} h_s T_a W_j d\Gamma_2$$

$$+ \int_{\Omega} (W_b c_b T_a + Q_m + H) W_j d\Omega$$

$$+ \frac{\rho C}{\Delta t} \int_{\Omega} T^{i-1} W_j d\Omega. \tag{7.50}$$

The time derivative from Eq. (7.44) is expressed as the finite difference $(T^i - T^{i-1})/\Delta t$ in Eq. (7.50), where T^i and T^{i-1} represent the nth node temperature at the current and previous time, respectively, while Δt is the time step.

Rewriting above equation in matrix form, and solving the matrix system, the temperature distribution inside the eye is obtained.

7.3.7 Numerical Results

The first set of the numerical results are obtained using the 2D and 3D models of the human eye when there is no laser radiation. The obtained results from the steady state analysis are then utilized as initial conditions for the transient analysis. As there are many combinations of laser parameters that can be utilized in this sort of analysis, due to many different laser systems currently in use, the results are presented only for the more commonly used ones, e.g., Nd:YAG, Ho:YAG, Ruby, ArF excimer and Nd:YLF laser.

The laser–tissue interaction of the last two laser systems is based on different mechanism than for the first three lasers. If not applied at high enough frequency, these lasers' pulses, shorter than 1 µs, do not cause thermal effects [101]. Instead, the interaction mechanisms, on which ArF excimer and Nd:YLF lasers are based, are the photoablation and the plasma-induced ablation, respectively.

The photoablation features the decomposition of a very thin surface layer of tissue, on the order of several micrometers, without thermal damage to the adjacent tissue [101,140]. When determining the dependence of the ablated tissue depth on the laser intensity, the authors base their assumptions on the Lambert–Beer's law [141], as utilized in the previously presented model. In order to do a more realistic analysis in this case, however, the corneal layers would have to be discretized more finely, taking into account several different corneal layers. Moreover, in addition to the heat transfer, one would have to take into account the mass transfer as well.

Hence, the results for the ArF excimer laser presented here should be considered for the illustrative purposes only, as the presented model takes into account only the thermal effects.

In case of Nd:YLF laser, where the interaction mechanism is based on the plasma-induced ablation, the most important parameter is the electric field induced in the eye tissue [101]. Therefore, the results for this parameter, similar to an excimer case, are presented without any inference.

The relation between the electric field value and the laser intensity is given by

$$I = \frac{1}{2} \epsilon_0 c E^2, \tag{7.51}$$

where ϵ_0 is the dielectric constant of free-space, and c is the speed of light. The results for the electric field are obtained using Eq. (7.51).

7.3.7.1 Steady-State Temperature Distribution

The 2- and 3-dimensional eye models have been discretized using 21,595 triangular elements (with 11,079 nodes), and 148,664 tetrahedral elements (with 27,625 nodes), respectively. The steady state temperature distributions for both models are shown in Fig. 7.32. One-half of sagittal and transverse cross-sections are shown for the 3D model, facilitating the temperature distribution view inside the eye.

As seen from Fig. 7.32, the two-dimensional model obtained lower temperature values in the cornea, aqueous humour and lens tissues compared to the three-dimensional model. Some previous studies [142,143] attributed this difference between 2D and 3D steady state results to the lack of heat transfer in the direction perpendicular to the plane of the 2D model.

Fig. 7.33 depicts the temperature distribution along the eye pupillary axis for both models, obtained using varying temperatures of the ambient air. As is obvious from Fig. 7.33, higher values of this parameter lead to a smaller discrepancy between the results from the two models.

Several papers [105,109,144,145] reported similar distribution of temperature along the pupillary axis. The largest discrepancy those papers reported was in the corneal surface temperature, due to the different values of the ambient temperature [105], corneal heat transfer coefficient [105,144,145], and exclusion of the evaporation term in the boundary condition [109].

The interested reader might find helpful the overview of the corneal surface temperature values, obtained using various techniques, in the work by Ng and Ooi [118]. These values range from 32°C to 36.6°C, where the mean reported value is 34.66°C.

Using 25°C for the ambient air temperature, our 2D and 3D models obtained 32.34°C and 35.54°C, respectively, thus showing that steady-state results are in a good agreement with other results published in the relevant literature.

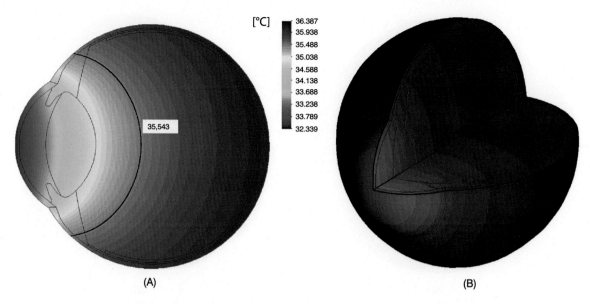

FIG. 7.32 Steady state temperature in the human eye: (A) 2D model, (B) 3D model.

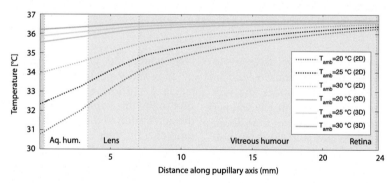

FIG. 7.33 Steady state temperature along the eye pupillary axis (2D model – full line, 3D model – dotted line) for ambient air temperature: 20°C, 25°C, and 30°C.

7.3.7.2 Nd:YAG 1064 nm Laser

Neodymium-YAG laser is a versatile laser used in procedures such as the posterior capsulotomy (after cataract), the peripheral iridotomy (treatment of glaucoma), for retinal photocoagulation, and many more [100].

Nd:YAG parameters used in our model are: single pulse of 1 ms duration, laser power 0.16 W, corneal beam diameter 2 mm, diameter of papillary opening 7.3 mm, similar to [109].

Fig. 7.34 shows the temperature distribution in 2D and 3D eye models following application of a laser pulse. The numerical results for a 3D model are shown as sagittal and transverse cross-sections along the pupillary axis. Additionally, detailed temperature fields around the vitreous humour, retina and choroid are given in the inset of Fig. 7.34. The maximum temperature is achieved in the vitreous humour.

Compared to the results from a 3D model, the 2D model obtained significantly lower temperatures, as clearly seen for a node on the retina, where the temperatures are 103.66°C and 120.38°C for 2D and 3D models, respectively.

Some additional differences between the two results are related to the location of maximum temperature achieved. It was expected to obtain the maximum temperature on the retina, as this has been the goal of a laser beam focusing. However, as seen from Fig. 7.34, the three-dimensional model obtained this maximum at the point in the vitreous humour.

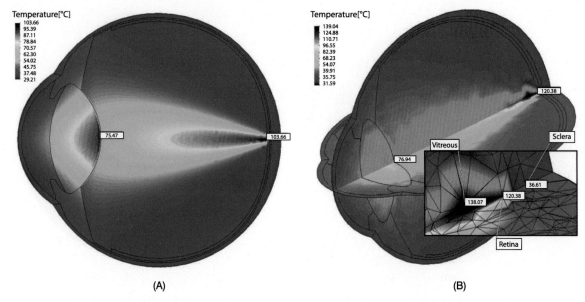

FIG. 7.34 Temperature distribution due to Nd:YAG 1064 nm laser. Comparison between: (A) 2D model, (B) 3D model (sagittal and transversal slice). Nd:YAG laser parameters: $\tau = 1$ ms, $P = 0.16$ W, $d_c = 2$ mm, $d_p = 7.3$ mm.

To overcome this issue, it would be necessary to refine the mesh in and around the retina region. In the current implementation of this model, the very thin retinal layer comprises only one finite element per thickness. Since the retinal thickness varies from about 0.15 mm anteriorly [146], to around 0.4 mm posteriorly, this would necessitate the use of a very fine mesh grid, hence it is left for the future work.

The final remark is related to the temperature distribution in the crystalline lens tissue. The 2D model obtained a lower value on a node at the lens–vitreous boundary compared to the 3D model, 75.47°C and 76.94°C, respectively, however, it is interesting that the 2D model obtained a significantly higher temperature gradient throughout the lens tissue. Again, this can be easily attributed to the absence of heat transfer in the perpendicular direction in the 2D model.

7.3.7.3 Ho:YAG 2090 nm Laser

Holmium:YAG laser is used in the laser thermo-keratoplasty (L-TKP) for the vision correction [147,148]. A typical treatment using this laser consists of delivering several pulses in a circular pattern to induce the local shrinkage of the corneal collagen.

Parameters used for this laser are: 7 pulses of 200 μs duration, 2 seconds switch-off time between pulses,

laser power 150 W, beam diameter on the cornea 0.6 mm, and pupil diameter 1.5 mm.

Fig. 7.35 shows the obtained temperature distribution in the three-dimensional model of the eye during the application of the final laser pulse, as well as 2 seconds after the laser has been switched off.

Fig. 7.36 shows the temperature evolution of several selected nodes (nodes location depicted in Fig. 7.29) against calculation step. Also, the nodes' respective distances from the cornea center are given.

A comparison of 2D and 3D model temperature evolution on the selected cornea node n3 (0.35 mm, 0 mm) can be seen in the same Fig. 7.36.

The three-dimensional model again obtained a higher maximum temperature at the cornea node n3 compared to the 2D model, i.e., 230.14°C and 187.44°C, respectively.

The transient analysis of our models obtained significantly higher temperatures following the application of seven laser pulses, compared to other authors [112,116]. The study from [116] reported a temperature around 110°C, and in that respect our present model obtained much higher values.

In order to initiate the shrinkage of corneal tissue, the temperatures on the order of 100°C are required, according to [149]. As both our numerical models obtained higher than expected temperatures, the analysis

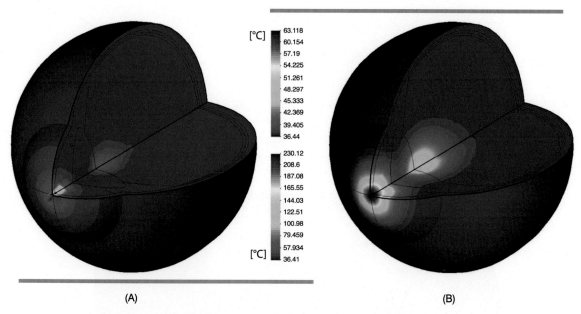

FIG. 7.35 Temperature distribution due to Ho:YAG 2090 nm laser: (A) following the application of the last pulse, (B) 2 seconds after the last pulse. Ho:YAG laser parameters: number of pulses = 7, $\tau = 200$ μs, $\tau_{off} = 2$ s, $P = 150$ W, $d_c = 0.6$ mm, $d_p = 1.5$ mm.

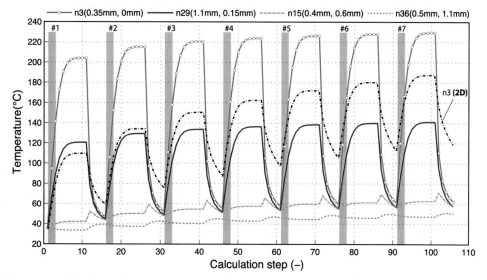

FIG. 7.36 Ho:YAG laser: Temperature evolution at selected nodes. Numbers in parenthesis represent the distance from the corneal geometric center. Pulse application showed only illustratively.

has been carried out to assess the importance of some of the modeling parameters. Fig. 7.37 shows the effects of the pupillary opening, pulse power and cornea absorption coefficient on the maximum obtained temperature.

The parameter analysis showed pupillary opening diameter to have the highest impact on the obtained temperature, as reported in [109,115,134]. In the normal human eye, this value varies from 1.5 to about 8 mm,

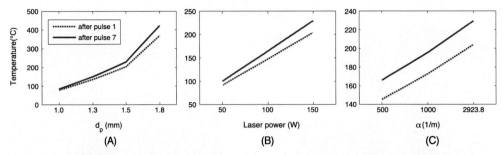

FIG. 7.37 The effects of various model parameters on the maximum temperature: (A) pupillary opening, (B) applied laser power, and (C) cornea absorption coefficient.

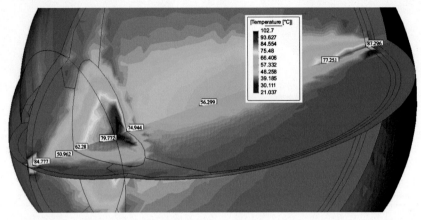

FIG. 7.38 Temperature distribution along pupillary axis due to 694.3 nm ruby laser. Ruby laser parameters: $\tau = 10$ ms, $P = 0.15$ W, $d_c = 8$ mm, $d_p = 7.3$ mm.

and is related to the ambient light conditions. Using the pupillary diameter value of 1.0 mm yielded results closer to the expected values.

It is worth mentioning that the position of the laser beam itself is very important. In the current implementation of the eye model, all seven pulses are applied to the same spot, i.e., at cornea center. A more realistic modeling would be to take into account the annular distribution of the pulses, thus resulting in much lower temperatures.

7.3.7.4 694.3 nm Ruby Laser

Ruby laser was the first laser introduced to the ophthalmology in the early 1960s [100]. Since then, it has been extensively used as a selective heat source to coagulate the retinal tissue in welding detached retinal segments to the underlying choroid. In order to prevent unnecessary vaporization and carbonization of the tissue, the temperatures achieved using this laser should not exceed 80°C [101].

Parameters used in our calculation are: pulse duration 10 ms, laser power 0.15 W, beam diameter on the cornea 8 mm, and pupil diameter 7.3 mm.

Fig. 7.38 shows the temperature distribution along eye pupillary axis, obtained from the three-dimensional model. The temperature of 87.3°C is obtained on the boundary between the vitreous humour and retina, while the maximum temperature of 102.7°C is obtained in the crystalline lens. Using similar laser parameters, Amara [109] obtained a maximum temperature increase of 97°C.

Parameter analysis similar to that carried out for Ho:YAG laser showed that laser beam diameter on the cornea has the most significant effect on the obtained temperature, as previously reported by [109]. This is due to a fact that the same amount of laser energy concentrated on a smaller spot surface will result in more absorbed power, and hence, higher temperatures. Further analysis showed again that wider pupillary opening results in higher temperatures. Finally, the analysis of the ambient temperature showed some small effect on the

TABLE 7.16
Effect of the ambient temperature on the temperature at selected nodes in the eye. Nodes situated on the boundary between: n.3 cornea–aqueous, n.290 aqueous–lens, n.1925 lens–vitreous, n.26589 vitreous–retina, n.26925 retina–choroid, and n.27245 choroid–sclera.

T_{amb} (°C)	n.3	n.290	n.1925	n.26589	n.26925	n.27245	T_{max}
20	84.54	62.1	74.83	87.25	33.86	38.54	106.44
25	84.78	62.28	74.94	87.3	33.9	38.58	106.56
30	85.02	62.46	75.05	87.34	33.94	38.62	106.68

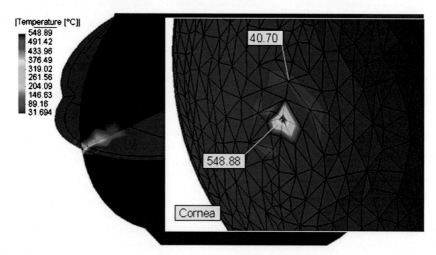

FIG. 7.39 Temperature around the anterior eye region and detail of the corneal surface due to excimer laser. ArF excimer parameters: $\tau = 15$ ns, laser energy 450 µJ, $d_c = 0.2$ mm, $d_p = 1$ mm.

anterior parts of the eye, while almost no direct effect on the posterior parts, as seen in Table 7.16.

7.3.7.5 ArF 193 nm Excimer Laser

The two most widely used laser eye surgery methods nowadays are the photorefractive keratectomy (PRK) and the laser in situ keratomileusis (LASIK), both performed using the ultraviolet ArF excimer laser, due to its ability to provide clean, thin cuts in the cornea by means of ablation [150,151].

As the interaction between ArF laser and eye tissues is based on a different mechanism compared to the previous three lasers, the results given here are only for illustrative purposes.

Parameters for the excimer laser used in the model are: pulse duration 15 ns, laser energy 450 µJ, beam diameter on the cornea 0.2 mm, and pupil diameter 1 mm.

Fig. 7.39 shows the temperature distribution in the anterior regions of the 3D eye model, with inset showing a more detailed view of the temperature field around the central part of the cornea.

7.3.7.6 Nd:YLF 1053 nm Laser

Neodimium:YLF is a versatile laser as it can, in addition to thermal effects, induce the effects of plasma ablation and photodisruption.

The interested reader can find more details on the physical parameters of plasma formation in [101], the reported parameters are: 10 ps pulse duration, power density $8 \cdot 10^{11}$ W/cm^2, and electric field $2.5 \cdot 10^7$ V/cm.

The following parameters are used for the Nd:YLF laser simulation: pulse duration 30 ps, pulse energy 100 µJ, beam diameter on the cornea 1 mm, and pupil diameter 7.3 mm. Fig. 7.40 shows the distribution of the electric field, deposited laser energy and temperature around the posterior part of the eye.

The values obtained using the three-dimensional model are: $1.3 \cdot 10^{11}$ W/cm^2 for the power density and

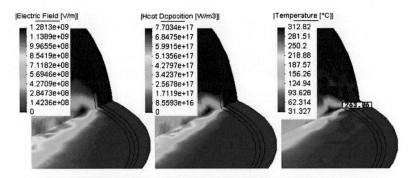

FIG. 7.40 Distribution of the electric field, deposited laser energy and the temperature due to Nd:YLF laser. Laser parameters: $\tau = 30$ ps, pulse energy 100 μJ, $d_c = 1$ mm, $d_p = 7.3$ mm.

$1.28 \cdot 10^7$ V/cm for the electric field. Maximum temperature achieved on the retina is around 263°C.

REFERENCES

1. E. Wassermann, C. Epstein, U. Ziemann (Eds.), Oxford Handbook of Transcranial Stimulation, first edition, Oxford University Press, New York, 2008, 580 pp.
2. A. Pascual-Leone, N.J. Davey, J. Rothwell, E.M. Wassermann, B.K. Puri (Eds.), Handbook of Transcranial Magnetic Stimulation, first edition, A Hodder Arnold Publication, 2002, 416 pp.
3. M. Hallett, Transcranial magnetic stimulation and the human brain, Nature 406 (6792) (2000) 147–150.
4. R. Plonsey, R.C. Barr, Bioelectricity: A Quantitative Approach, third edition, Springer Science+Business Media, LLC, 2007, 542 pp.
5. A.T. Barker, et al., Non-invasive magnetic stimulation of the human motor cortex, The Lancet (1985) 1106–1107.
6. S. Ueno, et al., Localized stimulation of neural tissues in the brain by means of a paired configuration of time varying magnetic fields, Journal of Applied Physics 64 (10) (1988) 5862–5864.
7. S. Ueno, et al., Functional mapping of the human motor cortex obtained by focal and vectorial magnetic stimulation of the brain, IEEE Transactions on Magnetics 26 (5) (1990) 1539–1544.
8. Thomas Picht, Sandro M. Krieg, Nico Sollmann, Judith Rösler, Birat Niraula, Tuomas Neuvonen, Petri Savolainen, Pantelis Lioumis, Jyrki P. Mäkelä, Vedran Deletis, et al., A comparison of language mapping by preoperative navigated transcranial magnetic stimulation and direct cortical stimulation during awake surgery, Neurosurgery 72 (5) (2013) 808–819.
9. J. Rösler, B. Niraula, V. Strack, A. Zdunczyk, S. Schilt, P. Savolainen, P. Lioumis, J. Mäkelä, P. Vajkoczy, D. Frey, et al., Language mapping in healthy volunteers and brain tumor patients with a novel navigated TMS system: evidence of tumor-induced plasticity, Clinical Neurophysiology 125 (3) (2014) 526–536.
10. Vedran Deletis, Maja Rogić, Isabel Fernández-Conejero, Andreu Gabarrós, Ana Jerončić, Neurophysiologic markers in laryngeal muscles indicate functional anatomy of laryngeal primary motor cortex and premotor cortex in the caudal opercular part of inferior frontal gyrus, Clinical Neurophysiology 125 (9) (2014) 1912–1922.
11. Maja Rogić, Vedran Deletis, Isabel Fernández-Conejero, Inducing transient language disruptions by mapping of Broca's area with modified patterned repetitive transcranial magnetic stimulation protocol, Journal of Neurosurgery 120 (5) (2014) 1033–1041.
12. Paolo M. Rossini, A.T. Barker, A. Berardelli, M.D. Caramia, G. Caruso, R.Q. Cracco, M.R. Dimitrijević, M. Hallett, Y. Katayama, C.H. Lücking, Non-invasive electrical and magnetic stimulation of the brain, spinal cord and roots: basic principles and procedures for routine clinical application. Report of an IFCN committee, Electroencephalography and Clinical Neurophysiology 91 (2) (1994) 79–92.
13. Mark G. Stokes, Christopher D. Chambers, Ian C. Gould, Tracy R. Henderson, Natasha E. Janko, Nicholas B. Allen, Jason B. Mattingley, Simple metric for scaling motor threshold based on scalp-cortex distance: application to studies using transcranial magnetic stimulation, Journal of Neurophysiology 94 (6) (2005) 4520–4527.
14. S. Knecht, J. Sommer, M. Deppe, O. Steinsträter, Scalp position and efficacy of transcranial magnetic stimulation, Clinical Neurophysiology 116 (8) (2005) 1988–1993.
15. P.M. Rossini, D. Burke, R. Chen, L.G. Cohen, Z. Daskalakis, R. Di Iorio, V. Di Lazzaro, F. Ferreri, P.B. Fitzgerald, M.S. George, et al., Non-invasive electrical and magnetic stimulation of the brain, spinal cord, roots and peripheral nerves: basic principles and procedures for routine clinical and research application. An updated report from an IFCN committee, Clinical Neurophysiology 126 (6) (2015) 1071–1107.
16. J. Ruohonen, R.J. Ilmoniemi, Focusing and targeting of magnetic brain stimulation using multiple coils, Medical and Biological Engineering and Computing 36 (3) (1998) 297–301.

17. J. Ruohonen, et al., Theory of multichannel magnetic stimulation: toward functional neuromuscular rehabilitation, IEEE Transactions on Biomedical Engineering 46 (6) (1999) 646–651.

18. L. Gomez, et al., Numerical analysis and design of single-source multicoil TMS for deep and focused brain stimulation, IEEE Transactions on Biomedical Engineering 60 (10) (2013) 2771–2782, https://doi.org/10.1109/TBME.2013.2264632.

19. F. Grandori, P. Ravazzani, Magnetic stimulation of the motor cortex-theoretical considerations, IEEE Transactions on Biomedical Engineering 38 (2) (1991) 180–191.

20. N.M. Branston, P.S. Tofts, Analysis of the distribution of currents induced by a changing magnetic field in a volume conductor, Physics in Medicine and Biology 36 (2) (1991) 161–168.

21. M. Nadeem, et al., Computation of electric and magnetic stimulation in human head using the 3-D impedance method, IEEE Transactions on Biomedical Engineering 50 (7) (2003) 900–907.

22. F.S. Salinas, et al., 3D modeling of the total electric field induced by transcranial magnetic stimulation using the boundary element method, Physics in Medicine and Biology 54 (2009) 3631–3647.

23. M. Chen, D.J. Mogul, Using increased structural detail of the cortex to improve the accuracy of modeling the effects of transcranial magnetic stimulation on neocortical activation, IEEE Transactions on Biomedical Engineering 57 (5) (2010) 1216–1226.

24. H. Eaton, Electric field induced in a spherical volume conductor from arbitrary coils: application to magnetic stimulation and MEG, Medical and Biological Engineering and Computing 30 (4) (1992) 433–440.

25. W.C. Chew, M.S. Tong, B. Hu, Integral Equation Methods for Electromagnetic and Elastic Waves, Morgan & Claypol Publishers, 2009, 261 pp.

26. A. Nummenmaa, et al., Comparison of spherical and realistically shaped boundary element head models for transcranial magnetic stimulation navigation, Clinical Neurophysiology 124 (10) (2013) 1995–2007, https://doi.org/10.1016/j.clinph.2013.04.019.

27. Oriano Bottauscio, Mario Chiampi, Luca Zilberti, Mauro Zucca, Evaluation of electromagnetic phenomena induced by transcranial magnetic stimulation, IEEE Transactions on Magnetics 50 (2) (2014) 1033–1036.

28. M. Cvetković, D. Poljak, J. Haueisen, Analysis of transcranial magnetic stimulation based on the surface integral equation formulation, IEEE Transactions on Biomedical Engineering (ISSN 0018-9294) 62 (6) (2015) 1535–1545, https://doi.org/10.1109/TBME.2015.2393557.

29. J. Ruohonen, J. Karhu, Navigated transcranial magnetic stimulation, Neurophysiologie Clinique/Clinical Neurophysiologie 40 (1) (2010) 7–17.

30. T. Wagner, et al., Three-dimensional head model simulation of transcranial magnetic stimulation, IEEE Transactions on Biomedical Engineering 51 (9) (2004) 1586–1598.

31. N.S. Stoykov, et al., Frequency- and time-domain FEM models of EMG: capacitive effects and aspects of dispersion, IEEE Transactions on Biomedical Engineering 49 (8) (2002) 763–772.

32. C.R. Butson, C.C. McIntyre, Tissue and electrode capacitance reduce neural activation volumes during deep brain stimulation, Clinical Neurophysiology 116 (10) (2005) 2490–2500.

33. B. Tracey, M. Williams, Computationally efficient bioelectric field modeling and effects of frequency-dependent tissue capacitance, Journal of Neural Engineering 8 (2011) 1–7.

34. C.A. Bossetti, et al., Analysis of the quasi-static approximation for calculating potentials generated by neural stimulation, Journal of Neural Engineering 5 (2008) 44–53.

35. Nele De Geeter, Guillaume Crevecoeur, Luc Dupré, Wim Van Hecke, Alexander Leemans, A DTI-based model for TMS using the independent impedance method with frequency-dependent tissue parameters, Physics in Medicine and Biology 57 (8) (2012) 2169.

36. J.D. Weissman, C.M. Epstein, K.R. Davey, Magnetic brain stimulation and brain size: relevance to animal studies, Electroencephalography and Clinical Neurophysiology 85 (3) (1992) 215–219, https://doi.org/10.1016/0168-5597(92)90135-X.

37. M. Stenroos, et al., Comparison of three-shell and simplified volume conductor models in magnetoencephalography, NeuroImage (2014) 1–12, https://doi.org/10.1016/j.neuroimage.2014.01.006.

38. R. Harrington, Boundary integral formulations for homogeneous material bodies, Journal of Electromagnetic Waves and Applications 3 (1) (1989) 1–15.

39. A.J. Poggio, E.K. Miller, Integral equation solutions of three-dimensional scattering problems, in: R. Mittra (Ed.), Computer Techniques for Electromagnetics, second edition, Hemisphere Publishing Corporation, 1987, pp. 159–264, chap. 4.

40. B.J. Roth, J.M. Saypol, M. Hallett, L.G. Cohen, A theoretical calculation of the electric field induced in the cortex during magnetic stimulation, Electroencephalography and Clinical Neurophysiology 81 (1) (1991) 47–56.

41. L.G. Cohen, et al., Effects of coil design on delivery of focal magnetic stimulation. Technical considerations, Electroencephalography and Clinical Neurophysiology 75 (4) (1990) 350–357.

42. C. Gabriel, S. Gabriel, Compilation of the Dielectric Properties of Body Tissues at RF and Microwave Frequencies, Tech. rep., AL/OE-TR-1996-0037, King's College London, 1996.

43. S. Gabriel, et al., The dielectric properties of biological tissues: II. Measurements in the frequency range 10 Hz to 20 GHz, Physics in Medicine and Biology 41 (11) (1996) 2251–2269.

44. S.B. Baumann, et al., The electrical conductivity of human cerebrospinal fluid at body temperature, Bioelectromagnetics 44 (3) (1997) 220–223.

45. C. Gabriel, et al., Electrical conductivity of tissue at frequencies below 1 MHz, Physics in Medicine and Biology 54 (16) (2009) 4863.

46. Alexander Opitz, Mirko Windhoff, Robin M. Heidemann, Robert Turner, Axel Thielscher, How the brain tissue shapes the electric field induced by transcranial magnetic stimulation, Neuroimage 58 (3) (2011) 849–859.

47. P.C. Miranda, et al., The electric field induced in the brain by magnetic stimulation: a 3-D finite-element analysis of the effect of tissue heterogeneity and anisotropy, IEEE Transactions on Biomedical Engineering 50 (9) (2003) 1074–1085.

48. Laleh Golestanirad, Michael Mattes, Juan R. Mosig, Claudio Pollo, Effect of model accuracy on the result of computed current densities in the simulation of transcranial magnetic stimulation, IEEE Transactions on Magnetics 46 (12) (2010) 4046–4051.

49. R.E. Frye, A. Rotenberg, M. Ousley, A. Pascual-Leon, Transcranial magnetic stimulation in child neurology: current and future directions, Journal of Child Neurology 23 (1) (2008) 79–96, https://doi.org/10.1177/0883073807307972.

50. T. Rajapakse, A. Kirton, Non-invasive brain stimulation in children: applications and future directions, Translational Neuroscience 4 (2) (2013) 1–29, https://doi.org/10.2478/s13380-013-0116-3.

51. K. Müller, V. Hömberg, Development of speed of repetitive movements in children is determined by structural changes in corticospinal efferents, Neuroscience Letters 144 (1) (1992) 57–60.

52. K. Müller, V. Hömberg, A. Aulich, H.G. Lenard, Magnetoelectrical stimulation of motor cortex in children with motor disturbances, Electroencephalography and Clinical Neurophysiology/Evoked Potentials Section 85 (2) (1992) 86–94.

53. J.A. Eyre, S. Miller, G.J. Clowry, E.A. Conway, C. Watts, Functional corticospinal projections are established prenatally in the human foetus permitting involvement in the development of spinal motor centres, Brain 123 (1) (2000) 51–64.

54. M.A. Garvey, V. Mall, Transcranial magnetic stimulation in children, Clinical Neurophysiology 119 (5) (2008) 973–984, https://doi.org/10.1016/j.clinph.2007.11.048.

55. Chandramouli Krishnan, Luciana Santos, Mark D. Peterson, Margaret Ehinger, Safety of noninvasive brain stimulation in children and adolescents, Brain Stimulation 8 (1) (2015) 76–87.

56. Marjorie A. Garvey, Donald L. Gilbert, Transcranial magnetic stimulation in children, European Journal of Paediatric Neurology 8 (1) (2004) 7–19.

57. V. Mall, S. Berweck, U.M. Fietzek, F.X. Glocker, U. Oberhuber, M. Walther, J. Schessl, J. Schulte-Monting, R. Korinthenberg, F. Heinen, Low level of intracortical inhibition in children shown by transcranial magnetic stimulation, Neuropediatrics 35 (2) (2004) 120–125.

58. Kuang-Lin Lin, Alvaro Pascual-Leone, Transcranial magnetic stimulation and its applications in children, Chang Gung Medical Journal 25 (7) (2002) 424–436.

59. Ulf Ziemann, Wolf Muellbacher, Mark Hallett, Leonardo G. Cohen, Modulation of practice-dependent plasticity in human motor cortex, Brain 124 (6) (2001) 1171–1181.

60. G. Bit-Babik, A.W. Guy, C.-K. Chou, A. Faraone, M. Kanda, A. Gessner, J. Wang, O. Fujiwara, Simulation of exposure and SAR estimation for adult and child heads exposed to radiofrequency energy from portable communication devices, Radiation Research 163 (5) (2005) 580–590.

61. J. Wiart, A. Hadjem, M.F. Wong, I. Bloch, Analysis of RF exposure in the head tissues of children and adults, Physics in Medicine and Biology 53 (13) (2008) 3681–3695.

62. A. Christ, N. Kuster, Differences in RF energy absorption in the heads of adults and children, Bioelectromagnetics 26 (S7) (2005) 31–44.

63. S. Gabriel, R.W. Lau, Camelia Gabriel, The dielectric properties of biological tissues: II. Measurements in the frequency range 10 Hz to 20 GHz, Physics in Medicine and Biology 41 (11) (1996) 2251.

64. Gernot Schmid, et al., Age dependence of dielectric properties of bovine brain and ocular tissues in the frequency range of 400 MHz to 18 GHz, Physics in Medicine and Biology 50 (19) (2005) 4711.

65. A. Peyman, S.J. Holden, S. Watts, R. Perrott, C. Gabriel, Dielectric properties of porcine cerebrospinal tissues at microwave frequencies: in vivo, in vitro and systematic variation with age, Physics in Medicine and Biology 52 (8) (2007) 2229.

66. A. Peyman, C. Gabriel, E.H. Grant, Günter Vermeeren, Luc Martens, Variation of the dielectric properties of tissues with age: the effect on the values of SAR in children when exposed to walkie-talkie devices, Physics in Medicine and Biology 54 (2) (2008) 227.

67. A. Peyman, A.A. Rezazadeh, C. Gabriel, Changes in the dielectric properties of rat tissue as a function of age at microwave frequencies, Physics in Medicine and Biology 46 (6) (2001) 1617–1629.

68. G. Schmid, G. Neubauer, P.R. Mazal, Dielectric properties of human brain tissue measured less than 10 h postmortem at frequencies from 800 to 2450 MHz, Bioelectromagnetics 24 (6) (2003) 423–430.

69. J. Wang, O. Fujiwara, S. Watanabe, Approximation of aging effect on dielectric tissue properties for SAR assessment of mobile telephones, IEEE Transactions on Electromagnetic Compatibility 48 (2) (2006) 408–413.

70. T. Kowalski, et al., Current density threshold for the stimulation of neurons in the motor cortex area, Bioelectromagnetics 23 (6) (2002) 421–428.

71. F.S. Salinas, et al., Detailed 3D models of the induced electric field of transcranial magnetic stimulation coils, Physics in Medicine and Biology 52 (2007) 2879–2892.

72. Ilkka Laakso, Akimasa Hirata, Fast multigrid-based computation of the induced electric field for transcranial magnetic stimulation, Physics in Medicine and Biology 57 (23) (2012) 7753.

73. Oriano Bottauscio, Mario Chiampi, Luca Zilberti, Boundary element solution of electromagnetic and bioheat

equations for the simulation of SAR and temperature increase in biological tissues, IEEE Transactions on Magnetics 48 (2) (2012) 691–694.

74. Oriano Bottauscio, Mario Chiampi, Jeff Hand, Luca Zilberti, A GPU computational code for eddy-current problems in voxel-based anatomy, IEEE Transactions on Magnetics 51 (3) (2015) 1–4.

75. Dragan Poljak, Mario Cvetković, Oriano Bottauscio, Akimasa Hirata, Ilkka Laakso, Esra Neufeld, Sylvain Reboux, Craig Warren, Antonios Giannopoulos, Fumie Costen, On the use of conformal models and methods in dosimetry for nonuniform field exposure, IEEE Transactions on Electromagnetic Compatibility 60 (2) (2018) 328–337.

76. Mario Cvetković, Dragan Poljak, Electromagnetic-thermal dosimetry comparison of the homogeneous adult and child brain models based on the SIE approach, Journal of Electromagnetic Waves and Applications 29 (17) (2015) 2365–2379.

77. Alan L. Hodgkin, Andrew F. Huxley, A quantitative description of membrane current and its application to conduction and excitation in nerve, The Journal of Physiology 117 (4) (1952) 500–544.

78. Bradley J. Roth, Peter J. Basser, A model of the stimulation of a nerve fiber by electromagnetic induction, IEEE Transactions on Biomedical Engineering 37 (6) (1990) 588–597.

79. F. Rattay, The basic mechanism for the electrical stimulation of the nervous system, Neuroscience 89 (2) (1999) 335–346.

80. J.H.M. Frijns, J.H. Ten Kate, A model of myelinated nerve fibres for electrical prosthesis design, Medical and Biological Engineering and Computing 32 (4) (1994) 391–398.

81. Ivana Zulim, Vicko Dorić, Dragan Poljak, Khalil El Khamlichi Drissi, Antenna model for passive myelinated nerve fiber, in: 2015 23rd International Conference on Software, Telecommunications and Computer Networks, SoftCOM, IEEE, 2015, pp. 87–91.

82. Donald R. McNeal, Analysis of a model for excitation of myelinated nerve, IEEE Transactions on Biomedical Engineering 23 (4) (1976) 329–337.

83. S.Y. Chiu, J.M. Ritchie, R.B. Rogart, D. Stagg, A quantitative description of membrane currents in rabbit myelinated nerve, The Journal of Physiology 292 (1) (1979) 149–166.

84. J.D. Sweeney, J.T. Mortimer, Dominique M. Durand, Modeling of mammalian myelinated nerve for functional neuromuscular stimulation, in: IEEE/Engineering in Medicine and Biology Society Annual Conference, IEEE, 1987, pp. 1577–1578.

85. B. Frankenhaeuser, A.F. Huxley, The action potential in the myelinated nerve fibre of Xenopus laevis as computed on the basis of voltage clamp data, The Journal of Physiology 171 (2) (1964) 302–315.

86. J. Patrick Reilly, Vanda T. Freeman, Willard D. Larkin, Sensory effects of transient electrical stimulation-evaluation with a neuroelectric model, IEEE Transactions on Biomedical Engineering 32 (12) (1985) 1001–1011.

87. Veit Schnabel, Johannes J. Struijk, Evaluation of the cable model for electrical stimulation of unmyelinated nerve fibers, IEEE Transactions on Biomedical Engineering 48 (9) (2001) 1027–1033.

88. Pinchas D. Einziger, Leonid M. Livshitz, Joseph Mizrahi, Generalized cable equation model for myelinated nerve fiber, IEEE Transactions on Biomedical Engineering 52 (10) (2005) 1632–1642.

89. Thomas Heimburg, Andrew D. Jackson, On soliton propagation in biomembranes and nerves, Proceedings of the National Academy of Sciences 102 (28) (2005) 9790–9795.

90. D. Poljak, S. esni, A simple antenna model of the human nerve, in: 2013 21st International Conference on Software, Telecommunications and Computer Networks, SoftCOM 2013, 2013, pp. 1–4.

91. Dragan Poljak, Mario Cvetković, Vicko Dorić, Ivana Zulim, Zoran ogaš, Maja Rogić Vidaković, Jens Haueisen, Khalil El Khamlichi Drissi, Integral equation formulations and related numerical solution methods in some biomedical applications of electromagnetic fields: Transcranial Magnetic Stimulation (TMS), nerve fiber stimulation, International Journal of E-Health and Medical Communications (IJEHMC) 9 (1) (2018) 65–84.

92. Ivana Zulim, Vicko Dorić, Dragan Poljak, Myelinated nerve fiber antenna model activation, in: 2016 24th International Conference on Software, Telecommunications and Computer Networks, SoftCOM, IEEE, 2016, pp. 1–5.

93. Dragan Poljak, Vicko Doric, Wire antenna model for transient analysis of simple grounding systems, part I: the vertical grounding electrode, Progress in Electromagnetics Research 64 (2006) 149–166.

94. Dragan Poljak, Vicko Doric, Wire antenna model for transient analysis of simple grounding systems, part II: the horizontal grounding electrode, Progress in Electromagnetics Research 64 (2006) 167–189.

95. Dragan Poljak, Advanced Modeling in Computational Electromagnetic Compatibility, John Wiley & Sons, Inc., 2007.

96. D. Cerdic, Advanced Transient Analysis of Complex Grounding Systems, Master's thesis, University of Wales, Wessex Institute of Technology, UK, 2008.

97. L.A. Geddes, J.D. Bourland, The strength-duration curve, IEEE Transactions on Biomedical Engineering 32 (6) (1985) 458–459.

98. David Boinagrov, Jim Loudin, Daniel Palanker, Strength–duration relationship for extracellular neural stimulation: numerical and analytical models, Journal of Neurophysiology 104 (4) (2010) 2236–2248.

99. Zoe Ashley, Hazel Sutherland, Hermann Lanmuller, Ewald Unger, Feng Li, Winfried Mayr, Helmut Kern, Jonathan C. Jarvis, Stanley Salmons, Determination of the chronaxie and rheobase of denervated limb muscles in conscious rabbits, Artificial Organs 29 (3) (2005) 212–215.

100. K. Thompson, Q. Ren, J. Parel, Therapeutic and diagnostic application of lasers in ophthalmology, Proceedings of the IEEE 80 (6) (1992) 838–860.

101. M.H. Niemz, Laser-Tissue Interactions: Fundamentals and Applications, third, enlarged edition, Springer-Verlag, Berlin, 2003, 66 pp.

102. Lisa Carroll, Tatyana R. Humphreys, LASER-tissue interactions, Clinics in Dermatology 24 (2006) 2–7.

103. A. Taflove, M. Brodwin, Computation of the electromagnetic fields and induced temperatures within a model of the microwave-irradiated human eye, IEEE Transactions on Microwave Theory and Techniques 23 (11) (1975) 888–896.

104. A. Emery, P. Kramar, A. Guy, J. Lin, Microwave induced temperature rises in rabbit eyes in cataract research, Journal of Heat Transfer 97 (1975) 123–128.

105. J. Scott, A finite element model of heat transport in the human eye, Physics in Medicine and Biology 33 (2) (1988) 227–241.

106. J. Scott, The computation of temperature rises in the human eye induced by infrared radiation, Physics in Medicine and Biology 33 (2) (1988) 243–257.

107. J. Lagendijk, A mathematical model to calculate temperature distributions in human and rabbit eyes during hyperthermic treatment, Physics in Medicine and Biology 27 (11) (1982) 1301–1311.

108. M.A. Mainster, Ophthalmic applications of infrared lasers – thermal considerations, Investigative Ophthalmology & Visual Science 18 (4) (1979) 414–420.

109. E. Amara, Numerical investigations on thermal effects of laser–ocular media interaction, International Journal of Heat and Mass Transfer 38 (13) (1995) 2479–2488.

110. G. Sbirlea, J.P. L'Huillier, A powerful finite element for analysis of argon laser iridectomy – influence of natural convection on the human eye, Transactions on Biomedicine and Health 4 (1997) 67–79.

111. K.J. Chua, J.C. Ho, S.K. Chou, M.R. Islam, On the study of the temperature distribution within a human eye subjected to a laser source, International Communications in Heat and Mass Transfer 32 (2005) 1057–1065.

112. F. Manns, D. Borja, J.M. Parel, W. Smiddy, W. Culbertson, Semianalytical thermal model for subablative laser heating of homogeneous nonperfused biological tissue: application to laser thermokeratoplasty, Journal of Biomedical Optics 8 (2) (2003) 288–297.

113. A. Podol'tsev, G. Zheltov, Photodestructive effect of IR laser radiation on the cornea, Optics and Spectroscopy 102 (1) (2007) 142–146.

114. A. Narasimhan, K. Jha, L. Gopal, Transient simulations of heat transfer in human eye undergoing laser surgery, International Journal of Heat and Mass Transfer 53 (2010) 482–490.

115. M. Cvetkovic, D. Poljak, A. Peratta, Thermal modelling of the human eye exposed to laser radiation, in: 2008 International Conference on Software, Telecommunications and Computer Networks, vol. 10, 2008, pp. 16–20.

116. E.H. Ooi, W.T. Ang, E.Y.K. Ng, A boundary element model of the human eye undergoing laser thermokeratoplasty, Computers in Biology and Medicine 38 (2008) 727–737.

117. M. Cvetković, D. Poljak, A. Peratta, FETD computation of the temperature distribution induced into a human eye by a pulsed laser, Progress in Electromagnetics Research 120 (2011) 403–421.

118. E.Y.K. Ng, E.H. Ooi, Ocular surface temperature: a 3D FEM prediction using bioheat equation, Computers in Biology and Medicine 37 (2007) 829–835.

119. S.C. DeMarco, G. Lazzi, W. Liu, J.D. Weiland, M.S. Humayun, Computed SAR and thermal elevation in a 0.25-mm 2-D model of the human eye and head in response to an implanted retinal stimulator – part I: models and methods, IEEE Transactions on Antennas and Propagation 51 (9) (2003) 2274–2285.

120. W. Makous, J. Gould, Effects of lasers on the human eye, IBM Journal of Research and Development 12 (3) (1968) 257–271.

121. J. Krauss, C. Puliafito, W. Lin, J. Fujimoto, Interferometric technique for investigation of laser thermal retinal damage, Investigative Ophthalmology & Visual Science 28 (8) (1987) 1290–1297.

122. W. Cheong, S. Prahl, A. Welch, A review of the optical properties of biological tissues, IEEE Journal of Quantum Electronics 26 (12) (1990) 2166–2185.

123. E.A. Boettner, Spectral Transmission of the Eye, Tech. rep., Report of the University of Michigan Ann Arbor Contract AF41(609)-2966, 1967.

124. A. Takata, L. Zaneveld, W. Richter, Laser-Induced Thermal Damage of Skin, Tech. rep., Final rept. Sep 76-Apr 77, 1977.

125. Dhiraj K. Sardar, Guang-Yin Swanland, Raylon M. Yow, Robert J. Thomas, Andrew T.C. Tsin, Optical properties of ocular tissues in the near infrared region, Lasers in Medical Science 22 (1) (2007) 46–52.

126. A. Lembares, X.H. Hu, G.W. Kalmus, Absorption spectra of corneas in the far ultraviolet region, Investigative Ophthalmology & Visual Science 38 (6) (1997) 1283–1287.

127. Lajos Kolozsvári, Antal Nógrádi, Béla Hopp, Zsolt Bor, UV absorbance of the human cornea in the 240- to 400-nm range, Investigative Ophthalmology & Visual Science 43 (7) (2002) 2165–2168.

128. E. Chan, B. Sorg, D. Protsenko, M. O'Neil, et al., Effects of compression on soft tissue optical properties, IEEE Journal of Selected Topics in Quantum Electronics 2 (4) (1996) 943–949.

129. A. Vogel, C. Dlugos, R. Nuffer, R. Birngruber, Optical properties of human sclera, and their consequences for transscleral laser applications, Lasers in Surgery and Medicine 11 (4) (1991) 331–340.

130. W. Weinberg, R. Birngruber, B. Lorenz, The change in light reflection of the retina during therapeutic laser photocoagulation, IEEE Journal of Quantum Electronics 20 (12) (1984) 1481–1489.

131. C. Cain, A. Welch, Measured and predicted laser-induced temperature rises in the rabbit fundus, Investigative Ophthalmology & Visual Science 13 (1) (1974) 60–70.

132. Walter J. Geeraets, R.C. Williams, Guy Chan, William T. Ham, DuPont Guerry, Frederick H. Schmidt, The relative absorption of thermal energy in retina and choroid, Investigative Ophthalmology 1 (3) (1962) 340–347.

133. M. Cvetkovic, A. Peratta, D. Poljak, Thermal modelling of the human eye exposed to infrared radiation of 1064 nm Nd:YAG and 2090 nm Ho:YAG lasers, WIT Transactions in Biomedicine and Health 14 (2009) 221–231.

134. M. Cvetkovic, D. Cavka, D. Poljak, A. Peratta, 3D FEM temperature distribution analysis of the human eye exposed to laser radiation, in: The Proceedings of Eight International Conference on Modelling in Medicine and Biology, BIOMED 2009, Crete, 2009.

135. H. Cember, Introduction to Health Physics, third edition, McGraw Hill, Inc., New York, 1996, 733 pp.

136. Frank Träger (Ed.), Springer Handbook of Lasers and Optics, Springer Science+Business Media, LLC, New York, 2007.

137. H.H. Pennes, Analysis of tissue and arterial blood temperatures in the resting human forearm. 1948, Journal of Applied Physiology 85 (1) (1998) 5–34.

138. E. Wissler, Pennes' 1948 paper revisited, Journal of Applied Physiology 85 (1) (1998) 35–41.

139. D. Poljak, A. Peratta, C.A. Brebbia, The boundary element electromagnetic–thermal analysis of human exposure to base station antennas radiation, Engineering Analysis with Boundary Elements 28 (2004) 763–770.

140. M. Kitai, V. Popkov, V. Semchischen, A. Kharizov, The physics of UV laser cornea ablation, IEEE Journal of Quantum Electronics 27 (2) (1991) 302–307.

141. R. Srinivasan, V. Mayne-Banton, Self developing photoetching of poly (ethylene terephthalate) films by far ultraviolet excimer laser radiation, Applied Physics Letters 41 (1982) 576–578.

142. R.U. Acharya, E.Y.K. Ng, J.S. Suri (Eds.), Image Modeling of the Human Eye, chap. 11, Artech House, Inc., 2008, pp. 229–252.

143. E.Y.K. Ng, E.H. Ooi, U.R. Archarya, A comparative study between the two-dimensional and three-dimensional human eye models, Mathematical and Computer Modelling 48 (5–6) (2008) 712–720.

144. E.Y.K. Ng, E.H. Ooi, FEM simulation of the eye structure with bioheat analysis, Computer Methods and Programs in Biomedicine 82 (2006) 268–276.

145. E.H. Ooi, W.T. Ang, E.Y.K. Ng, Bioheat transfer in the human eye: a boundary element approach, Engineering Analysis with Boundary Elements 31 (6) (2007) 494–500.

146. Deborah Pavan-Langston, Manual of Ocular Diagnosis and Therapy, 6th edition, Lippincott Williams & Wilkins, 2008, 560 pp.

147. S. Esquenazi, V. Bui, O. Bibas, Surgical correction of hyperopia, Survey of Ophthalmology 51 (4) (2006) 381–418.

148. H. Stringer, J. Parr, Shrinkage temperature of eye collagen, Nature 204 (4965) (1964) 1307.

149. R. Brinkmann, J. Kampmeier, U. Grotehusmann, A. Vogel, N. Koop, M. Asiyo-Vogel, R. Birngruber, Corneal collagen denaturation in laser thermokeratoplasty, in: Laser–Tissue Interaction VII, in: Proceedings of the SPIE, vol. 2681 (56), 1996, pp. 56–63.

150. J. Parrish, T. Deutsch, Laser photomedicine, IEEE Journal of Quantum Electronics 20 (12) (1984) 1386–1396.

151. M.W. Berns, L. Chao, A.W. Giebel, L.H. Liaw, J. Andrews, B. VerSteeg, Human corneal ablation threshold using the 193-nm ArF excimer laser, Investigative Ophthalmology & Visual Science 40 (5) (1999) 826–830.

The Generalized Symmetric Form of Maxwell's Equations

For a linear, homogeneous and isotropic medium, the time-harmonic, symmetric form of Maxwell's equations, i.e., the form in which both electric and fictitious magnetic charges and currents are taken into account, can be written as follows:

$$\nabla \times \vec{E} = -j\omega\mu\vec{H} - \vec{M}, \tag{A.1}$$

$$\nabla \times \vec{H} = j\omega\varepsilon\vec{E} + \vec{J}, \tag{A.2}$$

$$\nabla \cdot \vec{E} = \frac{1}{\varepsilon}\rho_e, \tag{A.3}$$

$$\nabla \cdot \vec{H} = \frac{1}{\mu}\rho_m. \tag{A.4}$$

Note that $\vec{M}$ and ρ_m are the fictitious magnetic surface current and charge density, respectively. The time-harmonic factor $e^{j\omega t}$, which is implied, has been omitted from the equations.

As it is well-known and widely accepted, the electric current $\vec{J}$ and charge ρ_e give rise to electric field $\vec{E}$ and magnetic field $\vec{H}$ which may be expressed in terms of two auxiliary quantities: the magnetic vector potential $\vec{A}$ and electric scalar potential φ,

$$\vec{E} = -j\omega\vec{A} - \nabla\varphi, \tag{A.5}$$

$$\vec{H} = \frac{1}{\mu}\nabla \times \vec{A}. \tag{A.6}$$

Similarly, the effects of the fictitious magnetic current $\vec{M}$ and charge ρ_m may be taken into account by means of the electric vector potential $\vec{F}$ and the magnetic scalar potential ψ:

$$\vec{H} = -j\omega\vec{F} - \nabla\psi, \tag{A.7}$$

$$\vec{E} = -\frac{1}{\varepsilon}\nabla \times \vec{F}. \tag{A.8}$$

Two generalized equations for the electric and magnetic fields may be obtained by combining the effects of $\vec{J}$, $\vec{M}$, ρ_e and ρ_m in Eqs. (A.5)–(A.8) as follows:

$$\vec{E} = -j\omega\vec{A} - \nabla\varphi - \frac{1}{\varepsilon}\nabla \times \vec{F}, \tag{A.9}$$

$$\vec{H} = -j\omega\vec{F} - \nabla\psi - \frac{1}{\mu}\nabla \times \vec{A}. \tag{A.10}$$

To express the fields in Eqs. (A.9) and (A.10) in terms of the sources $\vec{J}$ and $\vec{M}$, first the non-homogeneous Helmholtz's equations relating to $\vec{J}$ and ρ_e, derived from the Maxwell equations, are considered:

$$\nabla^2\vec{A} + k^2\vec{A} = -\mu\vec{J}, \tag{A.11}$$

$$\nabla^2\varphi + k^2\varphi = -\frac{\rho_e}{\varepsilon}, \tag{A.12}$$

where

$$k = \omega\sqrt{\mu\varepsilon} = \frac{\omega}{c} \tag{A.13}$$

is the wave number and c the velocity of light.

The magnetic sources counterparts of (A.11) and (A.12) then can be written as:

$$\nabla^2\vec{F} + k^2\vec{F} = -\varepsilon\vec{M}, \tag{A.14}$$

$$\nabla^2\psi + k^2\psi = -\frac{\rho_m}{\mu}. \tag{A.15}$$

Now, the particular solutions of Helmholtz type equations (A.11)–(A.12) and (A.14)–(A.15) are given in terms of the four potentials:

$$\vec{A}(\vec{r}) = \frac{\mu}{4\pi} \iint_S \vec{J}(\vec{r}')\, G(\vec{r}, \vec{r}')\, d\vec{S}(\vec{r}'), \tag{A.16}$$

$$\vec{F}(\vec{r}) = \frac{\varepsilon}{4\pi} \iint_S \vec{M}(\vec{r}')\, G(\vec{r}, \vec{r}')\, d\vec{S}(\vec{r}'), \tag{A.17}$$

$$\varphi(\vec{r}) = -\frac{1}{j4\pi\omega\varepsilon} \iint_S \nabla'_S \cdot \vec{J}(\vec{r}')\, G(\vec{r}, \vec{r}')\, d\vec{S}(\vec{r}'), \tag{A.18}$$

$$\psi(\vec{r}) = -\frac{1}{j4\pi\omega\mu} \iint_S \nabla'_S \cdot \vec{M}(\vec{r}')\, G(\vec{r}, \vec{r}')\, d\vec{S}(\vec{r}'), \tag{A.19}$$

where the corresponding Green's function of the lossless homogeneous medium is

$$G(\vec{r}, \vec{r}') = \frac{e^{-jk|\vec{r}-\vec{r}'|}}{|\vec{r} - \vec{r}'|}, \tag{A.20}$$

where $\vec{r}$ is the position vector at the observation point and $\vec{r}\,'$ is the source point.

The corresponding equations of continuity expressing the conservation of electric and magnetic charges, respectively, in time-harmonic form are given by:

$$\nabla' \cdot \vec{J}(\vec{r}\,') = -j\,\omega\rho_e, \qquad (A.21)$$

$$\nabla' \cdot \vec{M}(\vec{r}\,') = -j\,\omega\rho_m. \qquad (A.22)$$

Eqs. (A.21) and (A.22) may be substituted by expressions (A.17) and (A.19) to express the potentials in terms of fictitious magnetic field sources.

A Note on Integral Equations

An integral equation contains an unknown function u under the integral sign. Integral equations represent one of the most powerful mathematical tools in mathematical physics and engineering. Integral equations appear in a variety of applications, usually being obtained from a corresponding differential equation as integral formulation often makes solution of the problem easier.

Linear integral equations are often analyzed and can be classified as Fredholm or Volterra equations. The Fredholm equations of the first and second kind can be written in the form:

$$f(x) = \int_a^b K(x,t)\,u(t)\,dt, \tag{B.1}$$

$$f(x) = u(x) - \lambda \int_a^b K(x,t)\,u(t)\,dt, \tag{B.2}$$

where λ is a parameter, $u(x)$ is a function to be determined, $f(x)$ is the right-hand side and pertains to the excitation, and $K(x,t)$ is the integral equation kernel.

Furthermore, the Volterra equations of the first and second kind, having a variable upper limit of integration, are formulated as follows:

$$f(x) = \int_a^x K(x,t)\,u(t)\,dt, \tag{B.3}$$

$$f(x) = u(x) - \lambda \int_a^x K(x,t)\,u(t)\,dt. \tag{B.4}$$

Note that integral equations (B.1) to (B.4) become homogeneous for $f(x) = 0$. An integral equation is singular when either integration limits or kernel $K(x,t)$ have infinite value. Furthermore, the kernel $K(x,t)$ is symmetric when a condition $K(x,t) = K(t,x)$ is satisfied.

There is a close connection between the integral and differential formulation of a given problem. In general, differential equations can be expressed in terms of integral equations, but not vice versa. Whereas boundary conditions are imposed externally in differential equations, they are incorporated within an integral equation.

Thus, if one considers a first order differential equation and related boundary condition (Cauchy problem):

$$\frac{du}{dx} = f(x,u), \quad a \le x \le b, \tag{B.5}$$

$$u(x_0) = U_0, \tag{B.6}$$

having performed the integration, Eq. (B.5) can be written as the Volterra integral of the second kind,

$$u(x) = \int_a^x f[t,u(\xi)]\,d\xi + U_0. \tag{B.7}$$

Since Eqs. (B.5) and (B.7) are equivalent, any solution of Eq. (B.7) satisfies both Eq. (B.5) and the prescribed boundary condition (B.6). Therefore, an integral equation formulation includes both the differential equation and the boundary condition(s).

Thus, the corresponding integral formulation contains a complete description of the given problem without necessity of defining additional boundary conditions.

Similarly, when integrating a second order ordinary differential equation (accompanied with corresponding boundary conditions),

$$\frac{d^2u}{dx^2} = f(x,u), \quad a \le x \le b, \tag{B.8}$$

the following formulation is obtained:

$$\frac{du}{dx} = \int_a^x f[x,u(t)]\,dt + \frac{du}{dx}\Big|_{x=a}. \tag{B.9}$$

Integrating Eq. (B.9) by parts provides

$$u(x) = u(a) - \frac{du}{dx}\Big|_{x=a} \cdot a + \frac{du}{dx}\Big|_{x=a} \cdot x$$

$$+ \int_a^x (x-t)\,f[x,u(t)]\,dt. \tag{B.10}$$

Thus, (B.10) finally becomes

$$u(x) = u(a) + (x - a) \left.\frac{du}{dx}\right|_{x=a} + \int_a^x (x - t) \, f\,[x, u(t)]\, dt.$$

$$(B.11)$$

Obviously, the integral equation (B.11) represents both the differential equation (B.8) and the boundary conditions. Only one-dimensional integral equations have been discussed, but the approach could be extended to integral equations containing unknowns in two or more dimensions.

It is also worth noting that a majority of integral equations can be solved only by numerical methods.

APPENDIX C

Scalar Green's Function and the Solution to Helmholtz Equation

Solutions to the vector Helmholtz equations (A.11) and (A.14), and the scalar Helmholtz equations (A.12) and (A.15), respectively, could be found by using the solution to the scalar Helmholtz equation [1,2]. If the Laplace operator is written using the Cartesian coordinate system,

$$\nabla^2 \vec{A} = \nabla^2 A_x \hat{e}_x + \nabla^2 A_y \hat{e}_y + \nabla^2 A_z \hat{e}_z, \qquad \text{(C.1)}$$

and using the fact that within homogeneous space the magnetic and electric vector potentials $\vec{A}$ and $\vec{F}$ are collinear with electric and magnetic current densities $\vec{J}$ and $\vec{M}$, respectively, it follows that the vector Helmholtz equation can be separated into three scalar equations of the same form for each of the components.

Thus, Eqs. (A.11)–(A.15) can be expressed using the following general form:

$$\left(\nabla^2 + k^2\right) \Psi\left(\vec{r}\right) = -s\left(\vec{r}\right), \qquad \text{(C.2)}$$

where Ψ represents any of the scalar potentials, φ and ψ, or the nth component of the vector potentials, $\vec{A}$ and $\vec{F}$, while s denotes the right-hand side terms, μJ_n, εM_n, ρ/ϵ, ρ_m/μ, from the corresponding equations.

Solution to the scalar Helmholtz equation (C.2) can be found using the well known method of Green's functions [2], where the Green's function represents the solution or the system response due to point source excitation $\delta(\vec{r} - \vec{r}')$. Using the property of linearity, the solution due to some arbitrary distribution of sources can be thus obtained by superposition of point sources.

C.1 SCALAR GREEN'S FUNCTION

The scalar Green's function for the scalar Helmholtz equation is the solution to the following equation:

$$\left(\nabla^2 + k^2\right) G(\vec{r}, \vec{r}') = -\delta\left(\vec{r} - \vec{r}'\right), \qquad \text{(C.3)}$$

where $\vec{r}$ and $\vec{r}'$ denote the observation and the source point, respectively.

To be uniquely defined, it is necessary for the Green's function to satisfy the boundary conditions as well. In

case of unbounded space, the boundary conditions are given in terms of a so-called radiation or Sommerfeld's condition [1].

To facilitate the procedure, the source point $\vec{r}'$ is positioned at the origin of the coordinate system, resulting in spherically symmetric $G(\vec{r})$, dependent only on the radial distance r.

Rewriting the Laplace operator in the spherical coordinate system, we get

$$\left[\frac{1}{r^2}\frac{d}{dr}\left(r^2\frac{d}{dr}\right) + k^2\right] G(r) = -\delta(r). \qquad \text{(C.4)}$$

When $\vec{r} \neq 0$, we have $\delta(r) = 0$, and the differential equation (C.4) becomes homogeneous. Choosing $u(r)/r$ for $G(r)$ and multiplying (C.4) with r yields

$$\frac{1}{r}\frac{d}{dr}\left[r^2\frac{d}{dr}\left(\frac{u(r)}{r}\right)\right] + k^2 u(r) = 0. \qquad \text{(C.5)}$$

Expanding (C.5), followed by some cancelation, a homogeneous differential equation of the second order is obtained:

$$\left(\frac{d^2}{dr^2} + k^2\right) u(r) = 0 \qquad \text{(C.6)}$$

whose characteristic equation is given by

$$\lambda^2 + k^2 = 0. \qquad \text{(C.7)}$$

Solution to (C.7) is $\lambda_{1,2} = \pm jk$, and the corresponding solution to differential equation (C.6) is

$$u(r) = C_1 e^{-jkr} + C_2 e^{jkr}. \qquad \text{(C.8)}$$

The exponent of the first term represents the forward traveling wave from the source positioned at the origin, while the second term is the backward traveling wave from the source positioned at infinity. Because of physical validity, as there are no sources at infinity, $C_2 = 0$. Moreover, only the first term in (C.8) satisfies the Sommerfeld boundary condition [3]:

$$\lim_{r \to \infty} r\left(\frac{\partial G}{\partial r} + jkG\right) = 0, \qquad \text{(C.9)}$$

245

yielding

$$G(r) = C_1 \frac{e^{-jkr}}{r}. \tag{C.10}$$

Constant C_1 can be determined if the region around the origin is considered. Eq. (C.3) is thus

$$\nabla \cdot (\nabla G) + k^2 G = -\delta(r), \tag{C.11}$$

where integration over the small spherical volume V_0 of radius r_0 is performed in the limiting case, i.e., when $r_0 \to 0$, resulting in

$$\lim_{r_0 \to 0} \int_{V_0} \left[\nabla \cdot (\nabla G) + k^2 G \right] dV = - \lim_{r_0 \to 0} \int_{V_0} \delta(r) \, dV, \tag{C.12}$$

where the right-hand side of (C.12) is equal to -1 due to the property of Dirac delta function.

The second term from the left hand side of (C.12) is equal to zero:

$$\lim_{r_0 \to 0} \int_{V_0} k^2 G \, dV = \lim_{r_0 \to 0} \int_{V_0} k^2 C_1 \frac{e^{-jkr}}{r} r^2 \sin\theta \, dr \, d\theta \, d\varphi$$

$$= 0, \tag{C.13}$$

while for the first integrand of (C.12), application of (E.1) and (E.36) results in

$$\lim_{r_0 \to 0} \oint_{S_0} \hat{r} \cdot \hat{r} \, (1 + jkr) \frac{G}{r} r^2 \sin\theta \, d\theta \, d\varphi$$

$$= \lim_{r_0 \to 0} \oint_{S_0} (1 + jkr) C_1 e^{-jkr} \sin\theta \, d\theta \, d\varphi$$

$$= \lim_{r_0 \to 0} \int_0^{2\pi} \int_0^{\pi} C_1 \sin\theta \, d\theta \, d\varphi = 4\pi C_1 \tag{C.14}$$

and $C_1 = 1/4\pi$.

Moving the source point from the origin, the Green's function $G(\vec{r}, \vec{r}')$ satisfying Eq. (C.3), as well as Sommerfeld boundary condition, is obtained as

$$G(\vec{r}, \vec{r}') = \frac{e^{-jk|\vec{r} - \vec{r}'|}}{4\pi |\vec{r} - \vec{r}'|}, \tag{C.15}$$

representing a particular solution of the Helmholtz equation (C.3).

C.2 SCALAR HELMHOLTZ EQUATION SOLUTION

The solution to the inhomogeneous Helmholtz equation can be demonstrated on the example for the magnetic vector potential $\vec{A}$. Using the Cartesian coordinate system to write the Laplace operator, the scalar equation for each vector potential component can be written using similar form, e.g., for the x-component we get

$$\nabla^2 A_x (\vec{r}) + k^2 A_x (\vec{r}) = -\mu J_x (\vec{r}). \tag{C.16}$$

Using the operator notation, (C.16) can be written as [1]

$$\mathcal{L} A_x = -\mu J_x (\vec{r}), \tag{C.17}$$

where $\mathcal{L}$ denotes the linear vector differential operator, $\mathcal{L} = \nabla^2 + k^2$.

Introducing the point excitation source placed at $\vec{r}'$ on the right-hand side of (C.17), the solution representing the response to this point source is a scalar function $G(\vec{r}, \vec{r}')$ satisfying the following equation:

$$\mathcal{L} G(\vec{r}, \vec{r}') = -\delta(\vec{r} - \vec{r}'), \tag{C.18}$$

where $\mathcal{L}$ operates only on the unprimed coordinates.

Multiplying (C.18) with $J_x(\vec{r}')$, followed by integration over source, results in

$$\int_V \mathcal{L} G(\vec{r}, \vec{r}') J_x(\vec{r}') \, dV' = - \int_V \delta(\vec{r} - \vec{r}') J_x(\vec{r}') \, dV'. \tag{C.19}$$

Using the sampling property of the Dirac delta function,

$$\int_V \delta(\vec{r} - \vec{r}') f(\vec{r}) \, dV = f(\vec{r}'), \tag{C.20}$$

the right-hand side of (C.19) results in $J_x(\vec{r})$.

Operator $\mathcal{L}$ can be taken out of the integral, leading to

$$\mathcal{L} \int_V G(\vec{r}, \vec{r}') J_x(\vec{r}') \, dV' = -J_x(\vec{r}). \tag{C.21}$$

Comparing (C.21) and (C.17), it follows that

$$A_x(\vec{r}) = \mu \int_V G(\vec{r}, \vec{r}') J_x(\vec{r}') \, dV', \tag{C.22}$$

and, since all Cartesian components satisfy the equation of the same form, the general vector solution to the inhomogeneous Helmholtz equation (A.11) is

$$\vec{A}(\vec{r}) = \mu \int_V G(\vec{r}, \vec{r}\,') \vec{J}(\vec{r}\,')\, dV'. \tag{C.23}$$

A similar result can be obtained for the other potentials as well.

REFERENCES

1. G.W. Hanson, A.B. Yakovlev, Operator Theory for Electromagnetics: An Introduction, Springer-Verlag New York, Inc., 2002, 634 pp.
2. Mathew N.O. Sadiku, Numerical Techniques in Electromagnetics, second edition, CRC Press LLC, 2001, 760 pp.
3. A. Ishimaru, Electromagnetic Wave Propagation, Radiation and Scattering, Prentice-Hall, Inc., Englewood Cliffs, New Jersey, USA, 1991, 637 pp.

APPENDIX D

Derivation of EFIE From the Vector Analog of Green's Theorem

In case of unbounded free-space, the electric and magnetic fields due to some radiating sources are relatively straightforward to determine using (A.9) and (A.10), respectively. However, due to the presence of a material body, the fields are perturbed, resulting in a significantly more complex problem, as it is required to satisfy the boundary conditions at the interface between the two regions.

One possible way of solving such a problem is to use the integral approach, where the field at a particular point in space is linked to all sources enclosed within the problem domain but also to some external sources taken into account via their resulting fields on the boundary surface. To facilitate this connection, the vector analog of Green's theorem, or Green's second identity, is used [1], named after the self-taught English genius George Green, who formulated them during the 1820s.

D.1 THE VECTOR ANALOG OF GREEN'S THEOREM

The vector analog of Green's theorem, or Green's second identity, is derived using the divergence theorem (E.1), applied on vector $\vec{P} \times \nabla \times \vec{Q}$:

$$\int_V \nabla \cdot \left(\vec{P} \times \nabla \times \vec{Q} \right) dV = \int_S \left(\vec{P} \times \nabla \times \vec{Q} \right) \cdot \hat{n} \, dS.$$

(D.1)

Applying (E.18) on the left-hand side of (D.1) results in Green's first identity:

$$\int_V \left(\nabla \times \vec{Q} \cdot \nabla \times \vec{P} - \vec{P} \cdot \nabla \times \nabla \times \vec{Q} \right) dV$$

$$= \int_S \left(\vec{P} \times \nabla \times \vec{Q} \right) \cdot \hat{n} \, dS.$$

(D.2)

Following the same procedure on vector $\vec{Q} \times \nabla \times \vec{P}$ yields

$$\int_V \nabla \cdot \left(\vec{Q} \times \nabla \times \vec{P} \right) dV = \int_S \left(\vec{Q} \times \nabla \times \vec{P} \right) \cdot \hat{n} \, dS,$$

(D.3)

which, using (E.18), results in

$$\int_V \left(\nabla \times \vec{P} \cdot \nabla \times \vec{Q} - \vec{Q} \cdot \nabla \times \nabla \times \vec{P} \right) dV$$

$$= \int_S \left(\vec{Q} \times \nabla \times \vec{P} \right) \cdot \hat{n} \, dS.$$

(D.4)

Subtracting (D.2) from (D.4), results in Green's second identity, i.e., the vector analog of Green's theorem:

$$\int_V \left(\vec{Q} \cdot \nabla \times \nabla \times \vec{P} - \vec{P} \cdot \nabla \times \nabla \times \vec{Q} \right) dV$$

$$= \int_S \left(\vec{P} \times \nabla \times \vec{Q} - \vec{Q} \times \nabla \times \vec{P} \right) \cdot \hat{n} \, dS. \quad (D.5)$$

D.2 ON THE USE OF GREEN'S SECOND IDENTITY

Let us consider the problem depicted in Fig. D.1.

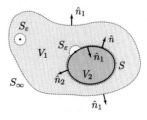

FIG. D.1 Problem description with denoted regions and boundaries.

The lossy material body V_2 of complex parameters (ε_2, μ_2), bounded by surface S, is placed in an infinite space V_1, with complex parameters (ε_1, μ_1), bounded by S_1, where $S_1 = S \cup S_\varepsilon \cup S_\infty$, with electric and magnetic source currents $(\vec{J}_1, \vec{M}_1)$. Here S_∞ and S_ε represent the boundary at infinity, and of an infinitesimal space V_ε considered when $\vec{r} \to \vec{r}\,'$, respectively. The unit vector $\hat{n}$ is directed from V_2 to V_1, while unit vectors $\hat{n}_1$ and $\hat{n}_2$, respectively, are outward directed for each region.

Let $\vec{P}$ and $\vec{Q}$ be vector functions with continuous first and second derivatives within V_1, satisfying Helmholtz equation (C.3), chosen as

$$\vec{P} = \hat{a} G(\vec{r}, \vec{r}\,'), \quad \vec{Q} = \vec{E}(\vec{r}), \tag{D.6}$$

where $\vec{E}$ is the electric field, $\hat{a}$ is the constant unit vector, and $G(\vec{r}, \vec{r}\,')$ denotes the free-space scalar Green's function

$$G(\vec{r}, \vec{r}\,') = \frac{e^{-jkR}}{4\pi R}, \quad R = |\vec{r} - \vec{r}\,'|, \tag{D.7}$$

where R is the distance from the observation point $\vec{r}$ to the source point $\vec{r}\,'$.

Inserting (D.6) into (D.5), after some manipulation, the first term on the left-hand side of (D.5) can be written as

$$\nabla \times \nabla \times \vec{P} = \nabla \times \nabla \times \hat{a} G = \nabla \left(\hat{a} \cdot \nabla G \right) + k^2 G, \tag{D.8}$$

where $k = \omega \sqrt{\mu \varepsilon}$ is the free-space wave number, μ and ε are the permittivity and permeability, respectively, and $\omega = 2\pi f$ is the operating frequency.

Multiplying (D.8) with $\vec{E}$ and integrating over volume V_1 leads to the following expression:

$$I = \int_{V_1} \vec{E} \cdot \left[\nabla \left(\hat{a} \cdot \nabla G \right) + \hat{a} k^2 G \right] dV, \tag{D.9}$$

which, after some mathematical manipulations, results in

$$I = \int_{S_1} \vec{E} \left(\hat{a} \cdot \nabla G \right) \hat{n} \, dS - \int_{V_1} \left(\hat{a} \cdot \nabla G \right) \frac{\rho}{\varepsilon} \, dV + \int_{V_1} \vec{E} \cdot \hat{a} k^2 G \, dV, \tag{D.10}$$

where ρ is the electric charge density.

After some rearranging, the constant vector can be moved outside the integrals:

$$I = \hat{a} \cdot \int_{S_1} \nabla G \left(\hat{n} \cdot \vec{E} \right) dS + \hat{a} \cdot \int_{V_1} \left(k^2 G \vec{E} - \frac{\rho}{\varepsilon} \nabla G \right) dV. \tag{D.11}$$

Following the same procedure, the second term on the left-hand side of (D.5) can be written as

$$\vec{P} \cdot \nabla \times \nabla \times \vec{Q} = \hat{a} \cdot G \left[k^2 \vec{E} - j\omega \mu \vec{J} - \nabla \times \vec{M} \right], \tag{D.12}$$

where the following substitution has been used:

$$\nabla \times \nabla \times \vec{E} = k^2 \vec{E} - j\omega \mu \vec{J} - \nabla \times \vec{M}. \tag{D.13}$$

Integrating (D.12) over volume V_1, after some rearranging, the following expression is obtained:

$$II = \hat{a} \cdot \int_{V_1} G \left(k^2 \vec{E} - j\omega \mu \vec{J} \right) dV - \hat{a} \cdot \int_{V_1} G \left(\nabla \times \vec{M} \right) dV. \tag{D.14}$$

Partial integration of the last term in (D.14), followed by an application of vector form of Stokes theorem, leads to

$$II = \hat{a} \cdot \int_{V_1} G \left(k^2 \vec{E} - j\omega \mu \vec{J} \right) dV - \hat{a} \cdot \int_{S_1} \hat{n} \times \left(G \vec{M} \right) dS + \hat{a} \cdot \int_{V_1} \nabla G \times \vec{M} \, dV. \tag{D.15}$$

Finally, subtracting (D.15) from (D.11), the left-hand side of Green's second identity (D.5) can be written as follows:

$$\text{LHS} \equiv \hat{a} \cdot \int_{S_1} \nabla G \left(\hat{n} \cdot \vec{E} \right) dS$$
$$+ \hat{a} \cdot \int_{V_1} \left(k^2 G \vec{E} - \frac{\rho}{\varepsilon} \nabla G \right) dV$$
$$- \hat{a} \cdot \int_{V_1} G \left(k^2 \vec{E} - j\omega \mu \vec{J} \right) dV$$
$$- \hat{a} \cdot \int_{S_1} \hat{n} \times \left(G \vec{M} \right) dS + \hat{a} \cdot \int_{V_1} \nabla G \times \vec{M} \, dV, \tag{D.16}$$

which, after some rearrangement, results in

$$\text{LHS} \equiv \int_{V_1} \left(j\omega \mu \vec{J} G - \frac{\rho}{\varepsilon} \nabla G - \nabla G \times \vec{M} \right) dV$$
$$+ \int_{S_1} \left[\nabla G \left(\hat{n} \cdot \vec{E} \right) + \hat{n} \times \left(G \vec{M} \right) \right] dS. \tag{D.17}$$

Treatment of the right-hand side of (D.5) is practically identical. The first term can be written as

$$\vec{P} \times \nabla \times \vec{Q} = \hat{a}G \times \nabla \times \vec{E}, \qquad (D.18)$$

where the use of $\nabla \times \vec{E} = -j\omega\mu\vec{H} - \vec{M}$ will lead to

$$\left(\vec{P} \times \nabla \times \vec{Q}\right) \cdot \hat{n} = \hat{a} \cdot \left[\hat{n} \times \left(j\omega\mu G\vec{H} + G\vec{M}\right)\right].$$

$$(D.19)$$

Similar, the second term from the right-hand side of (D.5) can be written as

$$\left(\vec{Q} \times \nabla \times \vec{P}\right) \cdot \hat{n} = \left[\vec{E} \times \nabla \times \left(\hat{a}G\right)\right] \cdot \hat{n}$$

$$= \hat{a} \cdot \left(\hat{n} \cdot \vec{E}\right) \times \nabla G. \qquad (D.20)$$

Integrating (D.19) and (D.20) over the surface S_1, followed by a subtraction, the following is obtained:

$$\text{RHS} \equiv \int\limits_{S_1} \hat{a} \cdot \left[\hat{n} \times \left(j\omega\mu G\vec{H} + G\vec{M}\right)\right] dS$$

$$- \int\limits_{S_1} \hat{a} \cdot \left(\hat{n} \cdot \vec{E}\right) \times \nabla G \, dS$$

$$= \hat{a} \cdot \int\limits_{S_1} \left[-\left(\hat{n} \times \vec{E}\right) \times \nabla G + j\omega\mu G\left(\hat{n} \times H\right)\right.$$

$$\left. + \hat{n} \times \left(G\vec{M}\right)\right] dS. \qquad (D.21)$$

Equating (D.17) and (D.21), after some additional mathematical manipulations, results in the following formula:

$$\int\limits_{V_1} \left[j\omega\mu\vec{J}G + \vec{M} \times \nabla'G - \frac{\rho}{\varepsilon}\nabla'G\right] dV'$$

$$= \int\limits_{S_1} \left[j\omega\mu G\left(\hat{n} \times H\right) - \left(\hat{n} \times \vec{E}\right) \times \nabla'G\right.$$

$$\left. - \left(\hat{n} \cdot \vec{E}\right)\nabla'G\right] dS', \qquad (D.22)$$

where a substitution between the source and the observation points, respectively, has been done.

Expression (D.22) is of a suitable form, as all sources ($\vec{J}$, $\vec{M}$, ρ) within volume V_1 are on the left-hand side, while the right-hand side features electric and magnetic fields on the boundary $S_1 = S \cup S_\varepsilon \cup S_\infty$ of volume V_1. In the following, the behavior of surface integral in (D.22) is considered separately for each boundary segment.

D.3 REGION AROUND SINGULARITY

When $\vec{r} \to \vec{r}'$, it is necessary to determine surface integral of (D.22) for two special cases: when the observation point is placed inside V_1, and on the boundary surface S_1.

In the former case, a small spherical region S_ε of radius ε is excluded around the observation point, as shown on Fig. D.2A.

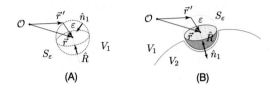

FIG. D.2 Singularity extraction when the observation point is: (A) in V_1, (B) on S_1.

Inserting (D.7) into (D.22), the first integral on the right-hand side of (D.22) vanishes, when $R \to 0$:

$$\lim_{R \to 0} \int\limits_{S_\varepsilon} j\omega\mu \frac{e^{-jkR}}{4\pi R}\left(\hat{n} \times \vec{H}\right)\varepsilon^2 \sin\theta \, d\theta \, d\varphi = 0, \quad (D.23)$$

resulting in

$$\lim_{R \to 0} \int\limits_{S_\varepsilon} \left[-\left(\hat{n} \times \vec{E}\right) \times \nabla'G - \left(\hat{n} \cdot \vec{E}\right)\nabla'G\right] dS. \quad (D.24)$$

The gradient of Green's function,

$$\nabla G(\vec{r}, \vec{r}') = -(1 + jkR)\frac{G(\vec{r}, \vec{r}')}{4\pi R^2}\vec{R}, \qquad (D.25)$$

inserted into (D.24) yields

$$\lim_{R \to 0} \int\limits_{S_\varepsilon} \left[\vec{E}\left(\hat{n} \cdot \hat{n}\right) - \left(\hat{n} \cdot \vec{E}\right)\hat{n} + \left(\hat{n} \cdot \vec{E}\right)\hat{n}\right]$$

$$\times (1 + jkR)\frac{e^{-jkR}}{4\pi R^2}$$

$$= \lim_{R \to 0} \int\limits_{S_\varepsilon} \left[\vec{E}\right](1 + jkR)\frac{e^{-jkR}}{4\pi R^2} dS$$

$$= \frac{\vec{E}}{4\pi} \int\limits_{\varphi=0}^{2\pi} \int\limits_{\theta=0}^{\pi} \sin\theta \, d\theta \, d\varphi = \vec{E}(\vec{r}). \qquad (D.26)$$

In case when the observation point is on the boundary surface S, the hemispherical region around the observation point needs to be excluded as shown in Fig. D.2B, and the corresponding value of the surface integral (D.26) is

$$\frac{\vec{E}}{4\pi} \int\limits_{\varphi=0}^{2\pi} \int\limits_{\theta=0}^{\pi/2} \sin\theta \, d\theta \, d\varphi = \frac{1}{2}\vec{E}(\vec{r}). \qquad (D.27)$$

D.4 SOMMERFELD BOUNDARY CONDITIONS

The boundary conditions on the surface S_∞, as shown on Fig. D.3, where $R \to \infty$, represent the so-called Sommerfeld radiating conditions.

FIG. D.3 Sommerfeld boundary conditions.

Selecting the spherical surface of infinite radius R, whose center is positioned at the source point $\vec{r}'$ and radius vector $\hat{R}$ is pointing outwards from V_1, leads to

$$\lim_{R\to\infty} \int\limits_{S_\infty} \left\{ j\omega\mu G\left(\hat{R}\times\vec{H}\right) \right.$$

$$\left. + \left(\frac{1}{R}+jk\right)\underbrace{\left[\left(\hat{R}\times\vec{E}\right)\times\hat{R}+\left(\hat{R}\cdot\vec{E}\right)\hat{R}\right]}_{\vec{E}}\frac{e^{-jkR}}{4\pi R} \right\}$$

$$\times R^2 \sin\theta \, d\theta \, d\varphi$$

$$= \lim_{R\to\infty} \int\limits_{S_\infty} \left[jkR\left(\eta\hat{R}\times\vec{H}+\vec{E}\right)\frac{e^{-jkR}}{4\pi} \right] \sin\theta \, d\theta \, d\varphi$$

$$= 0, \qquad (D.28)$$

easily demonstrating that the surface integral of (D.22) vanishes over surface S_∞.

D.5 STRATTON–CHU EXPRESSION

Inserting (D.26) into (D.22), while replacing the surface integral designation (S_1 changing to S), yields the

following integral expression for the electric field:

$$\vec{E}(\vec{r}) = \int\limits_{V_1} \left[j\omega\mu\vec{J}G + \vec{M}\times\nabla'G - \frac{\rho}{\varepsilon}\nabla'G \right] dV'$$

$$+ \int\limits_{S} \left[-j\omega\mu G\left(\hat{n}\times H\right) + \left(\hat{n}\times\vec{E}\right)\times\nabla'G \right.$$

$$\left. + \left(\hat{n}\cdot\vec{E}\right)\nabla'G \right] dS'. \qquad (D.29)$$

The integral expression (D.29) is the so-called Stratton–Chu [2,3] solution for the electric field. The first term from (D.29) represents the electric field due to the sources placed inside the region V_1, while the second one is the integral representation of the exterior sources expressed in terms of the field over the boundary surface S.

Eq. (D.29) enables one to consider the contribution of the sources placed outside the region of interest via their field components, or the equivalent sources on the boundary surface, and represents the mathematical form of the equivalence principle [4].

D.6 APPLICATION OF THE EQUIVALENCE PRINCIPLE

Introducing the equivalent electric and magnetic current densities $\vec{J}_S$ and $\vec{M}_S$, respectively, on the surface S,

$$\vec{J}_S = \hat{n}\times\vec{H}_1 = -\hat{n}_1\times\vec{H}_1, \quad \vec{M}_S = -\hat{n}\times\vec{E}_1 = \hat{n}_1\times\vec{E}_1, \qquad (D.30)$$

the electric field in V_1 can be expressed in the following way:

$$\vec{E}_1(\vec{r}) = \int\limits_{V_1} \left[j\omega\mu_1\vec{J}_1 G_1 + \vec{M}_1\times\nabla'G_1 - \frac{\rho_1}{\varepsilon_1}\nabla'G_1 \right] dV'$$

$$+ \int\limits_{S} \left[-j\omega\mu_1\vec{J}_S G_1 - \vec{M}_S\times\nabla'G_1 \right.$$

$$\left. + \frac{\rho_S}{\varepsilon_1}\nabla'G_1 \right] dS'. \qquad (D.31)$$

The last term of (D.31) featured the use of a continuity equation and the following substitution:

$$\hat{n}_1 \cdot \vec{E}_1 = \frac{j}{\omega\varepsilon_1}\nabla\cdot\left(\hat{n}_1\times\vec{H}_1\right). \qquad (D.32)$$

According to (D.31), the electric field in V_1 is due to the sources inside this volume ($\vec{J}_1$, $\vec{M}_1$), and also due to the equivalent surface currents ($\vec{J}_S$, $\vec{M}_S$), introduced to the surface S. Since there are no external sources in (D.31), the entire space is homogenized, facilitating the use of a free-space Green's function G_1.

The left-hand side of (D.31) results from integral (D.22) evaluated over the infinitesimal surface S_ε, while it is equal to zero when (D.22) is evaluated at S_∞. When the observation point is on the boundary S, the expression needs to be modified, resulting in

$$\frac{1}{2}\vec{E}_1(\vec{r}) = \int_{V_1}\left[j\omega\mu_1\vec{J}_1G_1 + \vec{M}_1 \times \nabla'G_1 \right.$$
$$\left. - \frac{\rho_1}{\varepsilon_1}\nabla'G_1\right]dV'$$
$$+ \int_S\left[-j\omega\mu_1\vec{J}_SG_1 - \vec{M}_S \times \nabla'G_1 \right.$$
$$\left. + \frac{\rho_S}{\varepsilon_1}\nabla'G_1\right]dS'. \tag{D.33}$$

Finally, when the observation point is outside V_1, i.e., inside V_2, it can be readily shown that the left-hand side of (D.31) vanishes [5]. Namely, when the source point and the observation point are in V_1 and V_2, respectively, it is not required to exclude the infinitesimal region around singularity, leading to

$$\int_{V_1}\left[j\omega\mu_1\vec{J}_1G_1 + \vec{M}_1 \times \nabla'G_1 - \frac{\rho_1}{\varepsilon_1}\nabla'G_1\right]dV'$$
$$= -\int_S\left[-j\omega\mu_1\vec{J}_SG_1 - \vec{M}_S \times \nabla'G_1 + \frac{\rho_S}{\varepsilon_1}\nabla'G_1\right]dS'. \tag{D.34}$$

Subsequently inserting (D.34) into (D.31), it follows that $\vec{E}_1(\vec{r}) = 0$.

On the other hand, the electric field in V_2 can be expressed as

$$\vec{E}_2(\vec{r}) = \int_{V_2}\left[j\omega\mu_2\vec{J}_2G_2 + \vec{M}_2 \times \nabla'G_2 - \frac{\rho_2}{\varepsilon_2}\nabla'G_2\right]dV'$$
$$+ \int_S\left[-j\omega\mu_2\vec{J}_SG_2 - \vec{M}_S \times \nabla'G_2 \right.$$
$$\left. + \frac{\rho_S}{\varepsilon_2}\nabla'G_2\right]dS', \tag{D.35}$$

where equivalent electric and magnetic currents are introduced on the surface S:

$$\vec{J}_S = \hat{n} \times \vec{H}_2 = \hat{n}_2 \times \vec{H}_2, \quad \vec{M}_S = -\hat{n} \times \vec{E}_2 = -\hat{n}_2 \times \vec{E}_2, \tag{D.36}$$

and we are using $\hat{n}_1 = -\hat{n}_2$.

Again, (D.35) states that the electric field in V_2 is due to sources inside this region ($\vec{J}_2$, $\vec{M}_2$), as well as some external sources taken into account via equivalent surface currents. Furthermore, the whole space is homogenized again.

Similar to region 1, it is easily shown that the left-hand side of (D.35) vanishes when the observation point is outside region 2 [5], while when the observation point is on surface S, it follows that

$$\frac{1}{2}\vec{E}_2(\vec{r}) = \int_{V_2}\left[j\omega\mu_2\vec{J}_2G_2 + \vec{M}_2 \times \nabla'G_2 \right.$$
$$\left. - \frac{\rho_2}{\varepsilon_2}\nabla'G_2\right]dV'$$
$$+ \int_S\left[-j\omega\mu_2\vec{J}_SG_2 - \vec{M}_S \times \nabla'G_2 \right.$$
$$\left. + \frac{\rho_S}{\varepsilon_2}\nabla'G_2\right]dS'. \tag{D.37}$$

The boundary conditions for the electric field on the surface S require the tangential components of vector $\vec{E}$ to be continuous. One can assume there are no field sources inside region 2, while the sources inside region 1 (volume integral over V_1) can be taken into account via the incident field vector $\vec{E}^{\text{inc}}$. The electric field at point $\vec{r}$ on the boundary surface S from the viewpoint of region 1 and expression (D.33) becomes

$$\frac{1}{2}\vec{E}_1(\vec{r}) = \vec{E}^{\text{inc}} + \int_S\left[-j\omega\mu_1\vec{J}_SG_1 - \vec{M}_S \times \nabla'G_1 \right.$$
$$\left. + \frac{\rho_S}{\varepsilon_1}\nabla'G_1\right]dS'. \tag{D.38}$$

The electric field at the same point $\vec{r}$ on the boundary surface S from the viewpoint of region 2 and expression (D.35) is

$$\frac{1}{2}\vec{E}_2(\vec{r}) = \int_S\left[-j\omega\mu_2\vec{J}_SG_2 - \vec{M}_S \times \nabla'G_2 \right.$$
$$\left. + \frac{\rho_S}{\varepsilon_2}\nabla'G_2\right]dS'. \tag{D.39}$$

Equating the left-hand sides of (D.38) and (D.39), i.e., satisfying the boundary conditions at surface S, $\hat{n} \times$

$\vec{E}_1 = \hat{n} \times \vec{E}_2$, the right-hand sides are

$$-\hat{n} \times \int_S \left[-j\omega\mu_1 \vec{J}_S G_1 - \vec{M}_S \times \nabla' G_1 + \frac{\rho_S}{\varepsilon_1} \nabla' G_1 \right] dS'$$

$$= \hat{n} \times \vec{E}^{\mathrm{inc}} - \hat{n} \times \int_S \left[j\omega\mu_2 \vec{J}_S G_2 + \vec{M}_S \times \nabla' G_2 \right.$$

$$\left. - \frac{\rho_S}{\varepsilon_2} \nabla' G_2 \right] dS' = 0. \qquad (D.40)$$

Furthermore, substituting the continuity equation into (D.40) yields the coupled set of surface integral equations:

$$\vec{E}^{\mathrm{inc}} = \int_S j\omega\mu_1 \vec{J} G_1 \, dS - \frac{j}{\omega\varepsilon_1} \int_S \nabla'_S \cdot \vec{J} \nabla' G_1 \, dS$$

$$+ \int_S \vec{M} \times \nabla' G_1 \, dS,$$

$$0 = \int_S j\omega\mu_2 \vec{J} G_2 \, dS - \frac{j}{\omega\varepsilon_2} \int_S \nabla'_S \cdot \vec{J} \nabla' G_2 \, dS$$

$$+ \int_S \vec{M} \times \nabla' G_2 \, dS, \qquad (D.41)$$

representing the electric field integral equation (EFIE) formulation.

REFERENCES

1. J.D. Jackson, Classical Electrodynamics, third edition, John Wiley & Sons, Inc., Hoboken, New Jersey, 1998, 808 pp.
2. J.A. Stratton, L.J. Chu, Diffraction theory of electromagnetic waves, Physical Review 56 (1) (1939) 99–107.
3. J.A. Stratton, Electromagnetic Theory, McGraw-Hill, New Jersey, 1941, 640 pp.
4. W.C. Chew, M.S. Tong, B. Hu, Integral Equation Methods for Electromagnetic and Elastic Waves, Morgan & Claypol Publishers, 2009, 261 pp.
5. K.-M. Chen, A mathematical formulation of the equivalence principle, IEEE Transactions on Microwave Theory and Techniques 37 (10) (1989) 1576–1581.

APPENDIX E

Useful Identities

E.1 INTEGRAL THEOREMS

Divergence theorem:

$$\int_V \nabla \cdot \vec{A}\, dV = \oint_S \hat{n} \cdot \vec{A}\, dS, \tag{E.1}$$

$$\int_S \nabla \cdot \vec{A}\, dS = \oint_c \hat{n} \cdot \vec{A}\, dl. \tag{E.2}$$

Stokes theorem:

$$\int_S \hat{n} \cdot \nabla \times \vec{A}\, dS = \oint_c \hat{l} \cdot \vec{A}\, dl. \tag{E.3}$$

Vector Stokes theorem:

$$\int_V \nabla \times \vec{A}\, dV = \oint_S \hat{n} \times \vec{A}\, dS. \tag{E.4}$$

Scalar Stokes theorem:

$$\int_S \hat{n} \times \nabla \psi\, dS = \oint_c \hat{l}\, \psi\, dl. \tag{E.5}$$

Gradient theorem:

$$\int_V \nabla \psi\, dV = \oint_S \hat{n}\, \psi\, dS. \tag{E.6}$$

Green's first identity:

$$\int_V \left[\phi \nabla^2 \psi + \nabla \phi \cdot \nabla \psi \right] dV = \oint_S \phi \nabla \psi \cdot d\vec{S}, \tag{E.7}$$

$$\int_S \left[\phi \nabla^2 \psi + \nabla \phi \cdot \nabla \psi \right] dS = \oint_c \phi \nabla \psi \cdot d\vec{l}. \tag{E.8}$$

Green's second identity (Green's theorem):

$$\int_V \left[\psi_1 \nabla^2 \psi_2 - \psi_2 \nabla^2 \psi_1 \right] dV$$

$$= \oint_S (\psi_1 \nabla \psi_2 - \psi_2 \nabla \psi_1) \cdot d\vec{S}$$

$$= \oint_S \left(\psi_1 \frac{\partial \psi_2}{\partial n} - \psi_2 \frac{\partial \psi_1}{\partial n} \right) dS, \tag{E.9}$$

$$\int_S \left[\psi_1 \nabla^2 \psi_2 - \psi_2 \nabla^2 \psi_1 \right] dS$$

$$= \oint_c (\psi_1 \nabla \psi_2 - \psi_2 \nabla \psi_1) \cdot \hat{n}\, dl$$

$$= \oint_c \left(\psi_1 \frac{\partial \psi_2}{\partial n} - \psi_2 \frac{\partial \psi_1}{\partial n} \right) dl. \tag{E.10}$$

Vector analogue of Green's first identity:

$$\int_V \left[(\nabla \times \vec{A}) \cdot (\nabla \times \vec{B}) - \vec{A} \cdot \nabla \times \nabla \times \vec{B} \right] dV$$

$$= \oint_S \hat{n} \cdot (\vec{A} \times \nabla \times \vec{B})\, dS. \tag{E.11}$$

Vector analogue of Green's second identity:

$$\int_V \left[\nabla \times \nabla \times \vec{A} \cdot \vec{B} - \vec{A} \cdot \nabla \times \nabla \times \vec{B} \right] dV$$

$$= \oint_S \left[\vec{A} \times (\nabla \times \vec{B}) + (\nabla \times \vec{A}) \times \vec{B} \right] \cdot d\vec{S}. \tag{E.12}$$

E.2 VECTOR IDENTITIES

$$\vec{A} \cdot \vec{B} = \vec{B} \cdot \vec{A}, \tag{E.13}$$

$$\vec{A} \times \vec{B} = -\vec{B} \times \vec{A}, \tag{E.14}$$

$$\vec{A} \cdot (\vec{B} \times \vec{C}) = \vec{B} \cdot (\vec{C} \times \vec{A}) = \vec{C} \cdot (\vec{A} \times \vec{B}), \tag{E.15}$$

$$\vec{A} \times (\vec{B} \times \vec{C}) = \vec{B}(\vec{A} \cdot \vec{C}) - \vec{C}(\vec{A} \cdot \vec{B}), \tag{E.16}$$

$$\nabla \cdot (\phi \vec{A}) = \vec{A} \cdot \nabla \phi + \phi \nabla \cdot \vec{A}, \tag{E.17}$$

$$\nabla \cdot (\vec{A} \times \vec{B}) = \vec{B} \cdot \nabla \times \vec{A} - \vec{A} \cdot \nabla \times \vec{B}, \tag{E.18}$$

$$\nabla (\phi \psi) = \phi \nabla \psi + \psi \nabla \phi, \tag{E.19}$$

$$\nabla \times (\phi \vec{A}) = \nabla \phi \times \vec{A} + \phi \nabla \times \vec{A}, \tag{E.20}$$

$$\nabla \times (\vec{A} \times \vec{B}) = (\vec{B} \cdot \nabla)\vec{A} - (\vec{A} \cdot \nabla)\vec{B} + \vec{A} \nabla \cdot \vec{B} - \vec{B} \nabla \cdot \vec{A}, \tag{E.21}$$

$$\nabla(\vec{A} \cdot \vec{B}) = (\vec{A} \cdot \nabla) \vec{B} + (\vec{B} \cdot \nabla) \vec{A}$$
$$+ \vec{A} \times \nabla \times \vec{B} + \vec{B} \times \nabla \times \vec{A}, \qquad \text{(E.22)}$$

$$\nabla \times \nabla \times \vec{A} = \nabla \nabla \cdot \vec{A} - \nabla^2 \vec{A}, \qquad \text{(E.23)}$$

$$\nabla \cdot \nabla \phi = \nabla^2 \phi, \qquad \text{(E.24)}$$

$$\nabla \times \nabla \phi = \vec{0}, \qquad \text{(E.25)}$$

$$\nabla \cdot \nabla \times \vec{A} = 0. \qquad \text{(E.26)}$$

E.3 FORMULAS INCLUDING POSITION VECTOR AND/OR CONSTANT VECTOR

$$\nabla(R\hat{R} \cdot \hat{a}) = \nabla(\vec{R} \cdot \hat{a}) = \hat{a}, \qquad \text{(E.27)}$$

$$\nabla \cdot (\hat{a} R) = \hat{a} \cdot \hat{R} = \hat{a} \cdot \nabla R, \qquad \text{(E.28)}$$

$$\nabla \times (\hat{a} R) = \hat{R} \times \hat{a}, \qquad \text{(E.29)}$$

$$(\hat{a} \cdot \nabla) R = \hat{a} \cdot \hat{R}, \qquad \text{(E.30)}$$

$$\nabla \times (\hat{a} \times R\hat{R}) = \nabla \times (\hat{a} \times \vec{R}) = 2\hat{a}, \qquad \text{(E.31)}$$

$$\nabla \cdot (\hat{a} \times R\hat{R}) = \nabla \cdot (\hat{a} \times \vec{R}) = 0, \qquad \text{(E.32)}$$

$$\nabla^2(\hat{a} G) = \hat{a} \nabla^2 G = -\hat{a} k^2 G. \qquad \text{(E.33)}$$

E.4 RECURSIVE FORMULAS INCLUDING POSITION VECTOR AND/OR SCALAR 3-DIMENSIONAL FREE-SPACE GREEN'S FUNCTION

$$\nabla(R^n) = n\vec{R} R^{(n-2)}, \qquad \text{(E.34)}$$

$$\nabla^2(R^n) = n(n + 1) R^{(n-2)}, \qquad \text{(E.35)}$$

$$\nabla\left(\frac{G}{R^n}\right) = -(n + 1 + jkR)\frac{G\vec{R}}{R^{(n+2)}}, \qquad \text{(E.36)}$$

$$\nabla \cdot \left(\frac{\vec{R}}{R^n}\right) = \frac{3 - n}{R^n}. \qquad \text{(E.37)}$$

Finite Element Matrices

F.1 SHAPE FUNCTIONS OVER TRIANGLE

Shape functions expanded over a triangular element can be expressed using a polynomial

$$N_i(x, y, z) = a_i + b_i x + c_i y, \quad i = 1, 2, 3, \quad (F.1)$$

where a_i, b_i and c_i are the coefficients to be found [1,2]. Since the shape function is of local support, it is convenient to derive it using the so-called *simplex* or local coordinates. The local shape function has a unit value at a particular simplex node while it vanishes at all other nodes.

The location of a point from the global coordinate system can be easily expressed via local coordinates using the following relation [1]:

$$\zeta_i = \frac{\sigma(S_i)}{\sigma(S)}, \quad (F.2)$$

where σ denotes the size of the simplex S from N-dimensional space (length of a line, surface of a triangle, volume of a tetrahedra), defined using

$$\sigma(S) = \frac{1}{N!} \begin{vmatrix} 1 & x_1^{(1)} & x_1^{(2)} & \cdots & x_1^{(N)} \\ 1 & x_2^{(1)} & x_2^{(2)} & \cdots & x_2^{(N)} \\ \vdots & \vdots & \vdots & \ddots & \vdots \\ 1 & x_{N+1}^{(1)} & x_{N+1}^{(2)} & \cdots & x_{N+1}^{(N)} \end{vmatrix}, \quad (F.3)$$

where elements of the determinant represent the vertex coordinates of the simplex.

The local coordinate from (F.2) actually represents the ratio between the subsimplex surface (which forms a point within the N-dimension simplex and other N points from the original simplex) and the surface of the original simplex.

In case of a triangle, where $N = 2$, the local coordinates for the three vertices are

$$\zeta_1 = \frac{\begin{vmatrix} 1 & x & y \\ 1 & x_2 & y_2 \\ 1 & x_3 & y_3 \end{vmatrix}}{\begin{vmatrix} 1 & x_1 & y_1 \\ 1 & x_2 & y_2 \\ 1 & x_3 & y_3 \end{vmatrix}}, \quad \zeta_2 = \frac{\begin{vmatrix} 1 & x & y \\ 1 & x_1 & y_1 \\ 1 & x_3 & y_3 \end{vmatrix}}{\begin{vmatrix} 1 & x_1 & y_1 \\ 1 & x_2 & y_2 \\ 1 & x_3 & y_3 \end{vmatrix}},$$

$$\zeta_3 = \frac{\begin{vmatrix} 1 & x & y \\ 1 & x_1 & y_1 \\ 1 & x_2 & y_2 \end{vmatrix}}{\begin{vmatrix} 1 & x_1 & y_1 \\ 1 & x_2 & y_2 \\ 1 & x_3 & y_3 \end{vmatrix}}, \quad (F.4)$$

where the determinant from the denominator is $2A$, A being the triangle surface area.

Expanding the numerator of the first local coordinate ζ_1 by minors of its first column [1] leads to:

$$\zeta_1 = \frac{1}{2A} \left(\begin{vmatrix} x_2 & y_2 \\ x_3 & y_3 \end{vmatrix} - \begin{vmatrix} x & y \\ x_3 & y_3 \end{vmatrix} + \begin{vmatrix} x & y \\ x_2 & y_2 \end{vmatrix} \right), \quad (F.5)$$

which, after expanding the above determinants, results in

$$\zeta_1 = \frac{1}{2A} \left[(x_2 y_3 - x_3 y_2) + (y_2 - y_3)x + (x_3 - x_2)y \right], \quad (F.6)$$

representing the sought linear form (F.1).

Following the same procedure for the remaining local coordinates, the expressions for the shape functions N_1, N_2 and N_3 over a triangle are obtained, written in matrix form as

$$\begin{bmatrix} N_1 \\ N_2 \\ N_3 \end{bmatrix} = \frac{1}{2A} \begin{bmatrix} x_2 y_3 - x_3 y_2 & y_2 - y_3 & x_3 - x_2 \\ x_3 y_1 - x_1 y_3 & y_3 - y_1 & x_1 - x_3 \\ x_1 y_2 - x_2 y_1 & y_1 - y_2 & x_2 - x_1 \end{bmatrix} \begin{bmatrix} 1 \\ x \\ y \end{bmatrix}. \quad (F.7)$$

F.2 SHAPE FUNCTIONS OVER TETRAHEDRA

Following the above procedure, the shape functions expanded over a tetrahedral element, representing the simplex in a 3-dimensional space, can be expressed using a polynomial

$$N_i(x, y, z) = a_i + b_i x + c_i y + d_i z, \quad i = 1, 2, 3, 4, \quad (F.8)$$

where the local coordinates within a tetrahedron can be expressed in terms of the vertex coordinates.

Using (F.2), the local coordinates for vertex 1 are obtained as

$$\zeta_1 = \frac{\begin{vmatrix} 1 & x & y & z \\ 1 & x_2 & y_2 & z_2 \\ 1 & x_3 & y_3 & z_3 \\ 1 & x_4 & y_4 & z_4 \end{vmatrix}}{\begin{vmatrix} 1 & x_1 & y_1 & z_1 \\ 1 & x_2 & y_2 & z_2 \\ 1 & x_3 & y_3 & z_3 \\ 1 & x_4 & y_4 & z_4 \end{vmatrix}}, \tag{F.9}$$

where the determinant from the denominator is $6V$, and V is the volume of the tetrahedron.

Expanding the determinant from the numerator of (F.9) by minors of the first column results in

$$\zeta_1 = \frac{1}{6V} \left(\begin{vmatrix} x_2 & y_2 & z_2 \\ x_3 & y_3 & z_3 \\ x_4 & y_4 & z_4 \end{vmatrix} - \begin{vmatrix} x & y & z \\ x_3 & y_3 & z_3 \\ x_4 & y_4 & z_4 \end{vmatrix} \right.$$
$$\left. + \begin{vmatrix} x & y & z \\ x_2 & y_2 & z_2 \\ x_4 & y_4 & z_4 \end{vmatrix} - \begin{vmatrix} x & y & z \\ x_2 & y_2 & z_2 \\ x_3 & y_3 & z_3 \end{vmatrix} \right). \tag{F.10}$$

If determinants of the last three terms from (F.10) are expanded by minors of their respective first rows, x, y and z can be taken out, while the following linear form is obtained:

$$\zeta_1 = \frac{1}{6V} (a_1 + b_1 x + c_1 y + d_1 z), \tag{F.11}$$

where the coefficients a_1, b_1, c_1 and d_1, expressed via coordinates of the four tetrahedron vertices, can be determined using:

$$a_1 = \begin{vmatrix} x_2 & y_2 & z_2 \\ x_3 & y_3 & z_3 \\ x_4 & y_4 & z_4 \end{vmatrix}, \tag{F.12}$$

$$b_1 = - \begin{vmatrix} y_3 & z_3 \\ y_4 & z_4 \end{vmatrix} + \begin{vmatrix} y_2 & z_2 \\ y_4 & z_4 \end{vmatrix} - \begin{vmatrix} y_2 & z_2 \\ y_3 & z_3 \end{vmatrix}, \tag{F.13}$$

$$c_1 = + \begin{vmatrix} x_3 & z_3 \\ x_4 & z_4 \end{vmatrix} - \begin{vmatrix} x_2 & z_2 \\ x_4 & z_4 \end{vmatrix} + \begin{vmatrix} x_2 & z_2 \\ x_3 & z_3 \end{vmatrix}, \tag{F.14}$$

$$d_1 = - \begin{vmatrix} x_3 & y_3 \\ x_4 & y_4 \end{vmatrix} + \begin{vmatrix} x_2 & y_2 \\ x_4 & y_4 \end{vmatrix} - \begin{vmatrix} x_2 & y_2 \\ x_3 & y_3 \end{vmatrix}. \tag{F.15}$$

Repeating the same procedure for the remaining local coordinates, the expressions for the shape functions over a tetrahedron are obtained, written in matrix form

as [2]:

$$\begin{bmatrix} N_1 \\ N_2 \\ N_3 \\ N_4 \end{bmatrix} = \frac{1}{6V} \begin{bmatrix} a_1 & b_1 & c_1 & d_1 \\ a_2 & b_2 & c_2 & d_2 \\ a_3 & b_3 & c_3 & d_3 \\ a_4 & b_4 & c_4 & d_4 \end{bmatrix} \begin{bmatrix} 1 \\ x \\ y \\ z \end{bmatrix}, \tag{F.16}$$

where the coefficients a_i, b_i, c_i and d_i, $i = 2, 3, 4$, from (F.16), can be determined in a similar way using (F.9) and (F.10), followed by the subsequent determinant expansion over their respective first rows.

F.3 SOLUTION OF CHARACTERISTIC INTEGRALS

The finite element matrix $[K]^e$ from (6.90) requires computing the integrals of the following form:

$$D = \int_{\Omega_e} N_m N_n \, d\Omega_e, \tag{F.17}$$

$$S = \int_{\Omega_e} \nabla N_m \cdot \nabla N_n \, d\Omega_e, \tag{F.18}$$

while the finite element source vector $[P]^e$ from (6.91) and the flux vector $[M]^e$ on the boundary of the finite element (6.92), respectively, feature the following integrals:

$$P = \int_{\Omega_e} N_m \, d\Omega_e \tag{F.19}$$

and

$$M = \int_{\partial\Omega_e} N_m \, dS_e. \tag{F.20}$$

Additionally, if Neumann boundary conditions are applied, the following integral appears:

$$M' = \int_{\partial\Omega_e} N_m N_n \, dS_e. \tag{F.21}$$

The following expression using local coordinates ζ_i is used to compute the integrals (F.17), (F.18), (F.19), (F.20), and (F.21) over tetrahedra [1,3]:

$$\int_V \zeta_1^a \zeta_2^b \zeta_3^c \zeta_4^d \, dx \, dy \, dz = \frac{a! \, b! \, c! \, d!}{(a + b + c + d + 3)!} 6V, \tag{F.22}$$

while, in case of integration over a triangle, one uses

$$\int_S \zeta_1^a \zeta_2^b \zeta_3^c \, dx \, dy = \frac{a! \, b! \, c!}{(a+b+c+2)!} 2A. \qquad \text{(F.23)}$$

Using (F.22) and (F.23), computing integrals (F.17)–(F.21) is very straightforward.

The computed integral from (F.17) has the following matrix:

$$D = \frac{6V}{120} \begin{bmatrix} 2 & 1 & 1 & 1 \\ 1 & 2 & 1 & 1 \\ 1 & 1 & 2 & 1 \\ 1 & 1 & 1 & 2 \end{bmatrix}, \qquad \text{(F.24)}$$

while (F.19) and (F.20) result in vectors

$$P = \frac{6V}{24} \begin{bmatrix} 1 \\ 1 \\ 1 \\ 1 \end{bmatrix} \qquad \text{(F.25)}$$

and

$$M = \frac{2A}{6} \begin{bmatrix} 1 \\ 1 \\ 1 \end{bmatrix} \qquad \text{(F.26)}$$

respectively, and the computed characteristic integral (F.21) has the matrix

$$M' = \frac{2A}{24} \begin{bmatrix} 2 & 1 & 1 \\ 1 & 2 & 1 \\ 1 & 2 & 2 \end{bmatrix}. \qquad \text{(F.27)}$$

Calculating integral (F.18) requires cross-multiplication of all components of the shape function gradient, resulting in a square matrix whose elements are given by [1]

$$S_{mn} = \frac{1}{36V^2} \sum_{i=1}^{4} \sum_{j=1}^{4} \left(b_i b_j + c_i c_j + d_i d_j \right), \qquad \text{(F.28)}$$

or in the expanded form,

$$S = \frac{1}{36V^2} \begin{bmatrix} b_1 b_1 + c_1 c_1 + d_1 d_1 & \dots & b_1 b_4 + c_1 c_4 + d_1 d_4 \\ b_2 b_1 + c_2 c_1 + d_2 d_1 & \dots & b_2 b_4 + c_2 c_4 + d_2 d_4 \\ b_3 b_1 + c_3 c_1 + d_3 d_1 & \dots & b_3 b_4 + c_3 c_4 + d_3 d_4 \\ b_4 b_1 + c_4 c_1 + d_4 d_1 & \dots & b_4 b_4 + c_4 c_4 + d_4 d_4 \end{bmatrix}. \qquad \text{(F.29)}$$

REFERENCES

1. P.P. Silvester, R.L. Ferrari, Finite Elements for Electrical Engineers, third edition, Cambridge University Press, 1996, 516 pp.
2. O.C. Zienkiewicz, R.L. Taylor, The Finite Element Method, Volume 1: The Basis, fifth edition, Butterworth-Heinemann, 2000, 689 pp.
3. M.A. Eisenberg, L.E. Malvern, On finite element integration in natural coordinates, International Journal for Numerical Methods in Engineering 7 (4) (1973) 574–575.

Index

Printed in the United States
By Bookmasters